Domains and Major Transitions of Social Evolution

Domains and Major Transitions of Social Evolution

Jacobus J. Boomsma
University of Copenhagen, Denmark

OXFORD
UNIVERSITY PRESS

Great Clarendon Street, Oxford, OX2 6DP,
United Kingdom

Oxford University Press is a department of the University of Oxford.
It furthers the University's objective of excellence in research, scholarship,
and education by publishing worldwide. Oxford is a registered trade mark of
Oxford University Press in the UK and in certain other countries

Published in the United States of America by Oxford University Press
198 Madison Avenue, New York, NY 10016, United States of America

British Library Cataloguing in Publication Data

Data available

Library of Congress Control Number: 2022933828

ISBN 978–0–19–874617–1 (hbk)
ISBN 978–0–19–874618–8 (pbk)

DOI: 10.1093/oso/9780198746171.001.0001

Printed and bound by CPI Group (UK) Ltd, Croydon, CR0 4YY

For Ada, my better half since we were students

Preface

Social evolution has multiple meanings, many of which are studied in the social sciences. Meaning can be sociological, addressing our behavior in relation to the institutions that characterize nation states and public bodies. It can be political, emphasizing the dynamics of private and public interests and their attendance to, or neglect of, governance, legislation, diplomacy, or military confrontation. And it can be economical, highlighting the production, exchange, and consumption of resources that determine personal and collective prosperity. Social evolution also features prominently in biology, where agents are animals, plants, fungi, algae, and microorganisms, rather than our own species. Here, social interactions are about mate choice, family life, and cooperation with third parties—a subset of the richer human spectrum that was also shaped by the accumulation of culture. This book is primarily about the nonhuman biology domains of social evolution where natural selection is the only force that modifies social adaptations because differential reproductive success is all important.

Given that all social evolution is ultimately about cooperation and conflict, no form of cooperation can be understood unless its possible corrupting forces are made transparent. Prime examples of social corruption that occur throughout living systems are kin-biased nepotism and overexploitation of public goods, which violate forms of merit-based fairness. That notion may strike us as a depressing natural legacy of our deep ancestry, but should also instill a sense of urgency to understand why various forms of social conflict are ubiquitous and why they can rarely be eliminated. Neo-Darwinian evolutionary biology has made significant progress in understanding, from first principles, that natural selection maintains cooperation when exploitative tendencies can be controlled at manageable costs. But can such insights explain why the different levels of organizational complexity arose? Why life has organized itself as simple cells, complex cells, multicellular organisms, caste-differentiated colonies, and, finally, as a single species that lives in symbiosis with its cumulative culture?

First-principle Darwinian explanations must be phrased in terms of differential reproductive success of allelic gene variants for the simple reason that only gene copies can be preserved over evolutionary time. The consequences of this gene's eye abstraction of adaptive information, first developed by William D. Hamilton and George C. Williams in the 1960s, have been tremendous. Their insights revolutionized organismal biology, turning it into a hard science that can make testable predictions about complex social and life-history traits, shaped by natural selection to improve their adaptive design. In such processes, self-preserving genes operate like team players that can each excel, but where none would achieve anything if their larger whole would fail to win. Phrased less materialistically, the allelic variants at gene loci function like the letters in a Shakespeare sonnet that was carefully put together into exquisite meaning and rhyme, but could only be securely preserved for posterity because it had been written up in these rather few symbolic characters.

Because adaptations are always shaped by natural selection, all organisms express phenotypic agency, no matter whether they are cells, multicellular bodies, or caste-differentiated colonies. As Robert Axelrod and William Hamilton put it in their 1981 *Science* paper, "[a]n organism does not need a brain to employ a strategy" because "[a] gene . . . looks beyond its mortal bearer to interests of the potentially immortal set of its replicas existing in other related individuals." This principle forms the foundation of inclusive fitness theory, the general insight that gene copies realized directly via offspring and indirectly via the reproduction of relatives add up to generate social reproductive efficiency capital. Hamilton's theory that natural selection will always reward biological agents that maximize inclusive fitness, the sum of directly and indirectly generated gene copies in future generations, is therefore easy to grasp in economic cost/benefit terms. But can this economic reasoning be extended to explain life's stepwise progress in organizational complexity?

Questions about the hierarchy of life, of how parts and wholes interact, have occupied biologists for centuries. A monograph by John Maynard Smith and Eörs Szathmáry (1995), titled *The Major Transitions in Evolution*, reviewed the kinds of discontinuous change that happened when cells became complex cells, and complex cells formed multicellular bodies, but the ensuing research tradition did not develop a general predictive framework for why such major transitions could happen. This book attempts to fill that vacancy by hypothesizing that the origins of all nonhuman evolutionary transitions that deserve the adjective *major* can be explained with inclusive fitness theory, provided ancestral states secured higher-level organismality through exclusively closed partnerships for life. The overarching idea is that only pairwise lifetime commitment at the foundation of individual organisms could ever breed obligate reproductive division of labor, as we find in domesticated mitochondria, in the somatic cells of our bodies, and in the sterile workers of a termite colony.

The core of my argument will be that major transitions were ratchet clicks of adaptive social evolution that can be explained by partitioning Hamilton's rule into a necessary condition of maximal relatedness and a sufficient benefit/cost condition. I will argue that these conditions, in that particular order, specified the narrow trajectories of change that forged major evolutionary transitions to irreversibly enhanced organizational complexity. Development of that conjecture required a review of currently available theory and evidence. A deeper historical perspective was also necessary because the principles of social evolution were surprisingly well understood by early 20th-century neo-Darwinians such as William Morton Wheeler and young Julian Huxley. For a variety of reasons, their insights faded into oblivion before they could be combined with the gene's eye view of evolution introduced by Hamilton and Williams, and be tested in Niko Tinbergen's manner, which demands that adaptations are also evaluated from angles of proximate causation, ontogenetic development, and phylogenetic history.

The century-long perspective of this book also invited a more general review of the achievements of the neo-Darwinian synthesis. Although often considered to be synonymous to the so-called modern synthesis, the neo-Darwinian synthesis is more encompassing because of its explicit focus on naturally selected adaptation in addition to all other angles that make evolutionary biology interesting. My analysis of the major transitions starts with Julian Huxley's (1912) book, *The Individual in the Animal Kingdom*, a concise but relatively complete treatment of life's gradual and stepwise changes of individuality, just reissued by MIT Press. Huxley was the first to use closure as a key principle for explaining major transitions in organizational complexity, a concept that I redefined in terms of genetic information. This made the connections to inclusive fitness theory straightforward, but also meant a reappraisal of evolutionary progress and a break with recent major transition research that lacks specific focus on adaptative design.

Identifying the major transitions in evolution as having required new levels of genetically closed information implies that there are only four domains of social evolution, each uniquely specified by nested layers of organizational complexity that enabled domain-specific adaptive syndromes: prokaryote cells, eukaryote cells, permanently multicellular bodies, and caste-differentiated colonies. Lower-level entities such as viruses cannot be independently shaped by natural selection and neither can collectives beyond the core family, such as populations, communities, ecosystems, or biospheres. The crucial characteristic of informationally closed life is that each individual has an autonomously sufficient genetic endowment, sampled from a population gene pool which is not a unit of adaptation itself. Outside these four fundamental levels of life's organizational complexity, units can have ecological, chemical, physical, geological, and historical dependencies but they cannot be autonomously designed through natural selection. Any interpretations suggesting self-sufficient adaptive design at these levels must therefore be dismissed.

The first three chapters of this book are largely historical. They reflect on the fundamental difference between first-principle Darwinian *understanding* of life's organization on the one hand, and *interpretation* of patterns in extant and historical biodiversity on the other. Confusion between these complementary opposites has haunted organismal biology for more than a century. It likely reflects that organized biology started out when Van Leeuwenhoek invented the first professional tools and Linnaeus began to organize biodiversity data, while the discipline only became a first-principle science a century later, when Darwin and Wallace framed the theories of natural selection and descent with modification – these traditions should remain complementary and not be uncritically merged. In Chapters 4–7, I then proceed reviewing, extending, and validating the gene's eye theory of social adaptation. I will maintain that the current body of neo-Darwinian theory is both special and fundamental, because its predictions are inductively logical and vulnerable to formal refutation—a hallmark of any good scientific theory as proposed by Karl Popper.

In Chapter 8, I will argue that fundamental causation in biology is always about genetic information, not about environmental variation, for the simple reason that—outside the human mind—environments do not exist until they are discovered by genes as opportunities to replicate. This insight, made explicit by Helena Cronin, logically leads towards renewed interest in how adaptive information virtually accumulates over time, and how it translates into actual control information that enables phenotypes to maximize inclusive fitness. It is in abstractions of this kind, devoid of anthropomorphic bias, that inclusive fitness theory achieves its most generally incisive form, a development that started in the 1970s when Hamilton's original theory was rephrased in the terms of George Price's statistical covariance equation with only selection and transmission variables. It is no surprise, therefore, that inclusive fitness theory now seems underway to be reconciled with Claude Shannon's general statistical theory of information and that consilience between Darwinism and thermodynamics may be possible in the distant future.

After having removed anthropomorphic bias as much as possible, I will briefly evaluate the human condition in the last section of Chapter 8. Conceptualizing the origins of the other major transitions in terms of closed genetic information makes clear why the human advance to unprecedented organizational complexity cannot have the same explanation. Our ancestral societies were already complex when *Homo sapiens* arose and started to coevolve with cumulative culture, a development that never annihilated ancestral individual agency as happened, for example, when the ants evolved specialized germline queens and somatic workers from simple maternal care. As our legal inheritance statutes attest, our customs have been shaped by Hamiltonian kin selection, but no human being is an altruistic servant by birth and our manners of cooperation and conflict are always context-dependent and uniquely co-defined by cultural norms. However, it appears worthwhile to tentatively apply the closure concept to our brains' symbiosis with the algorithmic carriers of cultural information to shed light on our self-domestication and self-deception.

Many feel that we live in a century of biology because massive sequencing and huge computational advances make most of the news. The progress enabled by these developments is enormous but, to retain scientific coherence and secure its status as a hard theory-driven science, biology must remain faithful to its encompassing neo-Darwinian framework developed in the 20th century. Neither data nor computation can by themselves generate synthetic understanding or suggest matching new research priorities. It is crucial that biology abandons the postmodern version of the *Baconian myth*, the idea that the field represents little more than data, analyzed data, and meta-data, accompanied by ad hoc hypotheses. It was the conceptual advance by Darwin and Wallace, not the descriptive tradition established by Linnaeus, that committed later generations of biologists to being part of a theory-driven science—ignoring Weismann's dictum of the *all-sufficiency of natural selection* reduces biology to a mere appendix of geology, chemistry, and physics.

This book is an attempt to extend the neo-Darwinian theoretical biology tradition further into the 21st century, so it can help answer and direct the multitude of empirical questions that biologists will come to ask in the years ahead. But no scientific framework is set in stone. The first-principle Hamiltonian explanation of adaptive social trait design across the domains of biological organization deserves to be challenged continuously so it can be generalized, refined, or refuted. During this process, Karl Popper's principle that no scientific theory will ever provide definite answers will prevail. However, having a well-established provisional theory will be highly beneficial when a truly unexpected finding, such as the possible discovery of lifeforms that originated on other planets, would rattle the foundations of biology and offer a completely novel opportunity for testing the generality of neo-Darwinian social evolution theory.

Jacobus J. (Koos) Boomsma
Copenhagen,
February 2022

Acknowledgments

I want to thank many people and institutions who made vital contributions during the seven years this book has been underway.

The manuscript could only be completed because of a Newton Abraham visiting Professorship at Lincoln College and the Zoology Department of the University of Oxford (2016), a fellowship by the Wissenschaftskolleg zu Berlin (WIKO) (2018/2019), and a Semper Ardens fellowship by the Carlsberg Foundation (2020/2021). Ashleigh Griffin, David Queller, Joan Strassmann, and Stuart West were Oxford hosts and/or WIKO focus group companions who enlightened my winding trajectories towards the final versions of the chapters. The WIKO librarians chased up many older books and papers, also during a brief revisit in June 2021.

In a more overarching sense, my thinking about social evolution has been nurtured by the many excellent graduate students and postdocs that joined the Copenhagen Centre for Social Evolution after the turn of the millennium. They pursued research programs throughout the domains of social evolution that increasingly confronted my naturalist mind with the molecular and omics territories of the modern biology landscape. Many dispersed to niches far afield and some became colleagues down the corridor where they secured a continuing atmosphere of shared scientific curiosity and stimulating complementarity.

During the writing process, Rick Gawne offered a stream of insightful ideas and references on the history and philosophy of early 20th century organismal biology, and Guojie Zhang made me understand the potential of genomic approaches for testing hypotheses of adaptive design. Over the years, many others shared their thoughts on topics relevant for this book: Duur Aanen, Ehab Abouheif, Rachelle Adams, Sandra Andersen, Sophie Armitage, Serge Aron, Boris Baer, Madeleine Beekman, Peter Biedermann, Trine Bilde, Andrew Bourke, Charlie Cornwallis, Sylvia Cremer, Henrik de Fine Licht, Patrizia D'Ettorre, Paul Eggleton, Jørgen Eilenberg, Tom Fenchel, Hermógenes Fernández-Marín, Berti Fisher, Kevin Foster, Raghavendra Gadagkar, Jürgen Gadau, Andy Gardner, Alan Grafen, Ian Hardy, Jürgen Heinze, Heikki Helanterä, Allen Herre, Jens Høeg, Bert Hölldobler, David Hughes, William Hughes, Laurent Keller, Toby Kiers, Judith Korb, Daniel Kronauer, Joanito Liberti, David Nash, Randy Nesse, Ben Oldroyd, Jes Søe Pedersen, the late Christian Peeters, Michael Poulsen, Thomas Pradeu, Bitao Qiu, Francis Ratnieks, Birgitte Regenberg, Panos Sapountzis, Lukas Schrader, Morten Schiøtt, Ted Schultz, Jonathan Shik, Shannon Smith, Víctor Sojo, Steve Stearns, Seirian Sumner, Lotta Sundström, Ulrik

Søchting, Robert Trivers, Bill Wcislo, Doris Williams, and Doug Yu.

Rick Gawne and Steve Stearns read the entire draft manuscript and offered incisive comments, constructive criticism, and editorial advice that improved the book's coherence. Duur Aanen, Alan Grafen, Ashleigh Griffin, Thomas Pradeu, Dave Queller, Joan Strassmann, and Stuart West read a variable number of draft chapters, weeding out errors and inconsistencies that enabled better final versions—any remaining oversights or inaccuracies are solely my own responsibility.

Geoff Avern, Boris Baer, Sandra Baldauf, Hans Henrik Bruun, Casey Dunn, Rick Gawne, Bill Hughes, Patrick Keeling, Klaus Lechner, Margaret McFall-Ngai, David Nash, Michael Poulsen, and Mariya Zhukova kindly allowed me to include organism images or advised on figure design. The families of William Hamilton (via the British Library Board), David Lack (confirmed by the estate of Helen Muspratt), Niko Tinbergen, and George Williams, as well as Princeton University and the AAAS, gave permission to reproduce portrait photographs.

At Oxford University Press, Ian Sherman made me finally embark on this project and kept believing I would complete the book even though target deadlines always turned out to be elusive. As the book neared completion, Charlie Bath advised on the final polishing of the manuscript and John Smallman guided me through the production phase.

Finally, none of this would have happened without the continuing support of my wife Ada Kramer—my most profound thanks go to her.

Koos Boomsma,
Copenhagen,
February 2022

Contents

List of Boxes

List of Figures

List of Tables

CHAPTER 1

A punctuated history of understanding social adaptation

Summary

Towards the mid 20th century, the so-called modern synthesis established evolutionary biology as a materialistic Darwinian and Mendelian science in terminology acceptable to systematists, geneticists, and paleontologists. This consensus is sometimes referred to as a neo-Darwinian synthesis, but that is misleading. A formal theory of adaptation and a phylogenetically controlled comparative method for analyzing adaptive states were lacking and developmental biology was left out. A neo-Darwinian synthesis combining all these approaches was not accomplished until the end of the 20th century and was ultimately based on the insight that natural selection for maximizing inclusive fitness explains adaptive design—the only fundamental process that sets biology apart from physics and chemistry. Conceptually, the interaction between randomly generated genetic information, natural selection, and historical contingency necessitates that biological phenomena attain complementary proximate and ultimate explanations, a form of scientific reciprocity that is irrelevant for understanding the inanimate world. The neo-Darwinian synthesis was initiated by Alfred Russell Wallace and August Weismann in the late 19th century and revitalized in the 1960s by three major developments: (1) Niko Tinbergen's framing of the four complementary types of questions that define organismal biology as a hard science while identifying the study of adaptation as an indispensable component; (2) William D. Hamilton developing inclusive fitness theory, which allowed social adaptations to be captured in a single gene-copy currency no matter whether individuals reproduce themselves or help others to do so; and (3) George C. Williams conceptualizing the gene's eye view of adaptation as a stringent hypothesis-driven and non-reductionist approach for understanding life's organization at the cell, individual, and family level. This chapter summarizes these historical events and explains the logical principles that I will adhere to in the social evolution analyses of this book.

> We must take the theory of natural selection in its simplest and most austere form, the differential survival of alternative alleles, and use it in an uncompromising fashion whenever a problem of adaptation arises. When such usage results in a simple and plausible explanation, the theory will thereby have demonstrated its strength.
>
> **Williams, *Adaptation and Natural Selection* (1966, p. 270)**

The cosmos is universally characterized by spontaneous decay. Atoms lose their radioactivity at set rates and molecules tend to fragment into lower energy states, a process that the second law of thermodynamics captures as the inevitable increase of randomness over time. At first glance, life seems exceptional because it recurrently builds organic structures of increasing order during development and lineage diversification, but this inconsistency, known as Schrödinger's (1944) paradox, disappears when we note

Domains and Major Transitions of Social Evolution. Jacobus J. Boomsma, Oxford University Press. © Jacobus J. Boomsma (2022).
DOI: 10.1093/oso/9780198746171.003.0001

that organisms are closed compartments that actively maintain internal organization, a concept first developed scientifically by Huxley (1912). Structural building blocks and energy that ultimately derives from solar radiation are captured while heat is dissipated into the surroundings (Atkins, 2003). Death inevitably returns bodies to the path of decay, but life as a whole persists because organisms reproduce at rates that exceed their mortality. As Schrödinger noted and Pross (2012) recently updated, life is physics and chemistry, but it differs in the process of evolution by natural selection that has shaped the complexity of its organization. This difference between the animate and inanimate world is so fundamental that the physical sciences and the life sciences are separate scholarly domains. They rest on the same logical principles of theory development and testing of conjectures, but the dynamics of biological complexity cannot be mathematically deduced from first axiomatic principles (Corning and Kline, 1998a). Rather than being a deterministic set of consequences, naturally selected adaptive design accumulates through a statistical process of relative improvement (Fisher, 1930; Grafen, 2019). Williams (1985) aptly captured this difference when he wrote that biological explanations "should invoke no factors other than the laws of physical science, natural selection, and the contingencies of history."

As far as we know, Aristotle was the first to write coherently about biology (Mayr, 1982; Williams, 2010; Leroi, 2014; Russell, 1916), but extensive knowledge of natural history must have been part of our tribal hunter-gatherer legacy for hundreds of thousands of years. However, the systematic scientific classification of biological diversity only began with Linnaeus in the mid 1700s, and the correct functional interpretation of biological change did not emerge until Darwin

and Wallace formulated the theory of adaptation through natural selection a century later. It is impossible to overstate the significance of their achievement. Darwinism made mythical and romantic interpretations of life's adaptive design and diversification obsolete so that biology could commit to the hypothetico-deductive method of science as a matter of principle. Biologists started grouping observations to foment explanatory hypotheses and subjected their conjectures to relentless empirical testing, a transition that physics and chemistry had made a century or two earlier (Weinberg, 2015). If pre-Darwinian biology had any functional paradigms in the sense of Kuhn (1962), they were overthrown by Darwin's and Wallace's set of materialistic concepts, which became self-correcting rather than remaining hostage to predetermined systems of static belief. Darwin's theory has survived 16 decades of empirical onslaught seeking to disprove its claims in the sense of Popper (1975), and it has consistently been expanded by the accumulation of refined first-principle understanding. As Monod (1975) wrote: "[A] good theory [...] will always be much wider and much richer than even the inventor of the idea may know at his time. The theory may be judged precisely on this type of development, when more and more falls into its lap, even though it was not predictable that so much would come of it."

To appreciate progress in scientific understanding it is important to realize that "knowing" is something else than "understanding" (Laplane et al., 2019). As Williams (1966a, p. 20) stated: "Many things were scientifically known of human anatomy and the motions of the planets before they were scientifically explained." Niels Bohr's biographer concluded that information is necessary but not sufficient for insight (Pais, 1997) and Ernst Mayr (1982, pp. 23–24) noted that most progress in biology has been achieved by the introduction of new concepts or the improvement of existing ones. As an example, he mentioned that Darwin obtained 3:1 dominant:recessive ratios in his plant breeding work but was unable to interpret them because he lacked Mendel's particulate inheritance concept (Howard, 2009). Explaining always involves abstraction, and the most powerful scientific abstractions explain many seemingly unrelated phenomena in common-sense logic, at least for the prepared and genuinely interested mind. Once theoretical concepts in biology have achieved that status, they are often supported by mathematical models, but such formalism does not necessarily play a central role in empirical testing. Darwin used strong inferential logic (Platt, 1964) to interpret patterns that no one else had recognized before and then recurrently and rhetorically asked his readers whether they were explained better by special creation or by natural selection (Mayr, 1982). His rigorous approach was rooted in a thorough understanding of comparative natural history, which appeals to us more than 150 years later because of the parsimonious conceptual narrative, despite its lack of mathematical models (Glass and Hall, 2008; Otto and Rosales, 2020). As Williams (1975, p. 7) noted, "for answering questions on function in biology, comparative evidence is more reliable than mathematical reasoning."

That fundamentally new concepts are primarily obtained by intuition rather than by a gradual accumulation of facts continues to deserve emphasis. In Mayr's (1982, p. 30) words, "[t]he creative mind is able, as Schopenhauer has stated it, 'to think something that nobody has thought yet, while looking at something that everybody sees.'" Imagination is thus the most important prerequisite of scientific progress. The hypothetico-deductive method is, in essence, the modern scientific method of discovery,

although the establishment of a tentative hypothesis is invariably preceded by observations and the posing of questions. Mayr echoes Karl Popper's (1962) refutation of the "Baconian Myth," the idea that truth will automatically emerge from data—even though Francis Bacon never seems to have defended that notion (Ghiselin, 1974). As Provine (1988, p. 61) put it succinctly, "[p]henomena that appear so complex as to baffle the imagination [...] have been shown by science to depend essentially upon surprisingly few variables." Conceptual innovations of this kind have become rarer in our present era of data- and methods-driven biology (see also Nurse, 2021), partly because the core of Darwinism has become less well known. In this chapter, I review two developments that were crucial for a Darwinian understanding of social evolution and which, to my knowledge, have not been considered jointly before. The first section addresses the five decades of almost total neglect of social adaptations from the second decade of the 20th century to the early 1960s. In the sections to follow, I then evaluate how the study of animal behavior re-emerged in the middle of the 20th century and why it took until the 1960s before the "gene's eye view" of evolution that I will use throughout this book emerged. In so doing, I hope to make clear that the evolutionary biology of social interactions is as logically rigorous and empirically confirmed as any other part of science.

The eclipse of early understanding of adaptive social evolution

Although Darwin had no concept of genes, there is little in the first edition of *On the Origin of Species by Means of Natural Selection* (1859) that evolutionary biologists would not endorse as correctly argued today—an amazing power of narrative that has been emphasized by many (e.g., Dennett, 1995; Reznick, 2011). Darwin's fundamental theoretical advance is a good illustration of novel insights in biology being possible while remaining ignorant about inheritance mechanisms, molecular aspects of ontogenetic development, and mathematics. All that—and much more—came to "fall into the theory's lap" as time went by, in the typical punctuated manner that characterizes all major progress in science (see full quote by Monod, 1975 above). However, the trajectory by which Darwinism developed was unusually erratic. When what has later been referred to as the *modern synthesis* integrated systematics, Mendelian genetics, and paleontology (Huxley, 1942; Mayr, 1982; Provine, 1988; Smocovitis, 1994), many complementary insights were not incorporated (Box 1.1). For example, the analyses of social evolution in Darwin's *Origin*

that Weismann, Wheeler, and young Julian Huxley had ably elaborated (see Chapter 3) were not addressed, nor was developmental biology (embryology until the 1950s). Other fields did not exist before 1960, such as cladistic phylogenetics and statistical analysis of comparative data, and only became included slowly in the modern synthesis. Incoherent integration of sub-disciplines has contributed to opinions that biology is a rather soft science (Mayr, 1982), a prejudice that persists today. Not appreciating that the discipline is united by an encompassing, universal, and first-principle theory of adaptation by natural selection is a particularly sour point when biology's ability to explain the origin and elaboration of life's organizational order represents its key conceptual difference from physics and chemistry.

Box 1.1 The modern synthesis of evolutionary biology: a retrospective

Evolutionary textbooks often have introductions or later sections on what came to be known as the *modern synthesis*, presenting a narrative or image gallery of the contributing architects (e.g., Ridley, 2004; Stearns and Hoekstra, 2005; Barton et al., 2007; Futuyma, 2013; Bergstrom and Dugatkin, 2016). In these overviews we usually meet the geneticists Ronald A. Fisher, J.B.S. Haldane, and Sewall Wright, the systematist Ernst Mayr, the paleontologist George Gaylord Simpson, and the botanist G. Ledyard Stebbins. Other geneticists such as Theodosius Dobzhansky and the Nobel laureates Hermann J. Muller and Thomas H. Morgan are occasionally mentioned as well. The merits of the modern synthesis have been exhaustively reviewed by Mayr (1982) and Provine (1971, 1986). These authors agreed that it is difficult to pinpoint the actual achievements of the modern synthesis, but mention that over several decades in the early and mid 20th century, biologists came to agree on particulate genetics being the basis of evolutionary change,

on there being enough evidence to no longer doubt that micro- and macro-evolution are complementary processes of change, and on fully dismissing Lamarckian ideas driven by metaphysical élan vital, landslide genetics, or suddenly appearing hopeful monsters. Both also concluded that the synthesis was incomplete, but decades later the study of behavior generally remains underrepresented in evolutionary biology texts, in contrast to behavioral ecology introductions that are usually explicit when they mention that they are a subfield of evolutionary biology (e.g., Davies et al., 2012).

Others have continued to use the term neo-Darwinian synthesis, which seems reasonable because Darwin started the pursuit of understanding evolution as a product of natural selection. If the modern synthesis emphasized Mendelism as a necessary element to be "modern," it seems only fair to bring Darwin back as the universal anchor of the discipline. Yet, there is a strange ambiguity

---

Box 1.1 *Continued*

about the terms "modern" and "neo-Darwinian." Some tend to intuitively feel they are synonyms, while others like the former adjective better because it allows them to refer to Mayr's and Provine's evaluations of what that synthesis actually was. However, if developmental biology was missing, what perspectives did it bring when that field was included and how far did these perspectives go back in time? If the history of population genetics ultimately refers back to Fisher, Haldane, and Wright in the 1920s and 1930s, why does developmental biology coverage systematically neglect pioneering comparative embryologists such as Conrad Waddington, Paul Weiss, Joseph Needham, and Joseph Woodger? Their organicist school was critical of genetic reductionism, but many of them were very clear thinkers about biology as a conceptual discipline (Nicholson and Gawne, 2014, 2015). An even stranger neglect was that of Darwin's concept of adaptation through natural selection—textbooks covering the modern synthesis at best haphazardly cover Crafoord Prize winners in evolutionary biology like William D. Hamilton, George C. Williams, and John Maynard Smith, let alone Nobel laureates Niko Tinbergen, Konrad Lorenz, and Karl von Frisch, who showed that behavior evolves to be adaptive.

The modern synthesis has more recently been criticized for underappreciating epigenetics and plasticity (e.g., Pigliucci and Finkelman, 2014; Laland et al., 2015) which is puzzling because modern epigenetics is often just molecular biology without overarching

hypotheses (Peterson, 2016; Henikoff, 2018). This underlines that the modern synthesis was never a conceptual framework, but mostly a pragmatic conglomerate of relevant approaches. Although some of these critics are aware of the conceptually more balanced neo-Darwinian approach of Tinbergen (1963), they seem dissatisfied with that framework as well (e.g. Bateson and Laland, 2013) in the belief that biology is only about proximate mechanisms and that encompassing understanding can be achieved without considering natural selection. A deeper reason behind this bias may well be that also philosophers often emphasize the material aspects of genes while ignoring the information that they carry. As Williams (1992, p. 12) wrote in an explicit attempt to separate the material and informational aspects of genetics, "Brandon (1988, 1990) and Sober (1984) concede the usefulness of the replicator–interactor distinction but consistently regard replicators as material objects and miss the codex concept" (see also Maynard Smith, 2000). While the neo-Darwinian synthesis could continue to develop by taking the codex concept seriously, the modern synthesis failed to modernize (Gawne et al., 2018), leaving many confused and believing that population genetics is unjustifiably reductionist while Darwinian concepts of adaptation are hand-waving story telling. This book challenges that postmodern attitude by arguing that biology disqualifies itself as a hard natural science if it fails to pursue an integrated conceptual framework that is neo-Darwinian in outline.

Even the integration of Darwinism with Mendelism was slow. It took many decades until particulate genetics was unambiguously shown to match, and provide the proximate mechanism for, adaptive evolutionary change. These intellectual conquests have been summarized in textbooks of evolutionary biology, and for further details I refer to Henry Bennett's 1999 variorum edition of Fisher

(1930), Egbert Leigh's 1990 edition of Haldane (1932), the 2001 edition (Cold Spring Harbor Laboratory Press) of Sturtevant (1965), and to reviews by Mayr (1982) and Provine (1971, 1986). Important other monographs of the time were Theodosius Dobzhansky's (1937) *Genetics and the Origin of Species*, Ernst Mayr's (1942) *Systematics and the Origin of Species*, George Gaylord Simpson's (1944) *Tempo and*

Mode in Evolution, and G. Ledyard Stebbins' (1950) *Variation and Evolution in Plants*. Wright's mathematical theory is often singled out as emphasizing non-adaptive evolution in small subpopulations subject to random genetic drift (Wright, 1931), but Provine (1986) and Skipper (2009) showed that disagreements between Fisher and Wright were about the most prevalent forces of selection in natural populations—in local demes versus panmictic populations, rather than about mathematical substance (Wright, 1949). Kohn (2004) argued that Fisher and Wright were competing for the "grand theorist" position of the modern synthesis (see also Leigh, 1999) while Provine (1986) gave reasons to believe that also confrontations between Wright and Mayr were largely driven by priority claims for "rightful" historical positions as early architects of Huxley's (1942) modern synthesis, similar to the Mayr–Haldane dispute about "beanbag genetics" (Rao and Nanjundiah, 2011). In a deeper sense, it was Fisher's genetic codex narrative of selection on additive genetic variation that clashed with Wright's focus on material genotypic interactions (Crow, 2008; Otto and Rosales, 2020).

While the contributions by Fisher, Haldane, and Wright were crucial in providing the mathematical underpinning of Darwinism, what consensus Huxley's modern synthesis really achieved is not clear (Provine, 1971; Smocovitis, 1996), nor is it obvious that it remains relevant today (Box 1.1). In many respects, the synthesis started with Alfred Russell Wallace and August Weismann towards the end of the 19th century. It included substantial differences of opinion right from the beginning, not so much on the universal significance of evolutionary change, but on the relative importance of natural selection acting on variation among individuals. There was, in fact, an "eclipse" of Darwinism in general around the time that Mendel's laws were rediscovered in 1900 (Bowler, 2005), so the 1909 Cambridge celebration of Darwin's 100th birthday (50 years after *On the Origin of Species* was published) marked in many ways the rise of a new generation who would resolve the most pressing issues (Richmond, 2006). In Chapter 2, I will return to the pervasive discrepancy between the mostly British biologists working in the neo-Darwinian adaptation-focused tradition and the mostly North American biologists who claimed the main credit for the modern synthesis while emphasizing processes other than individual adaptation. Many of these differences have deeper roots in how T.H. Morgan deterministically focused *Drosophila* genetics on allelic transmission, which ignored physiological development (Allen, 1979; Amundson, 2014; Peterson, 2016). Before taking a closer look at what the modern synthesis was, it is important to establish that insights from social evolution—the main focus of this book—were missing because of neglect, not because of evidence showing that behavior was unimportant (see also Boomsma and Gawne, 2018).

It is remarkable that Huxley's (1942) modern synthesis remained almost completely silent about behavior and that his emphasis on genetic mechanisms coincides with a very loose treatment of biological adaptation. Adaptation was considered to be omnipresent and non-Lamarckian, but no leading biologist of the time appeared to realize that competitive and cooperative social traits demand rigorous first-principle explanations (Cronin, 1991). This is particularly surprising because young Julian Huxley—embryologist by academic training and an able ornithologist—had published detailed descriptions of reproductive behavior in several bird species (e.g., Huxley, 1916) that inspired Niko Tinbergen's and David Lack's later analyses of the survival value of animal behavior (Birkhead et al., 2014). Huxley's earlier career had also produced lucid reviews of the major transitions to multicellularity (Huxley, 1912) and of the origins of organismal coloniality in ants and other social insects (Huxley, 1912, 1930). Around the same time, he proved himself to be the world's most broadly read polymath of biology as key author of *The Science of Life* (Wells et al., 1931), an assignment that led him to relinquish his position as Chair of Zoology at King's College London; he never returned to academia. His early contributions left few traces in the modern literature, in part because Huxley did not summarize their merits in his later books and papers (e.g., Gayon, 2000). As it happened, David Lack (1949) actually criticized Huxley (1942) for a speciation review that appears to dismiss adaptation. Huxley's shift, in his 1942 monograph, from organismal biology, in which he was an expert, to disseminating reductionist population genetics, a field to which he had made no research contributions, seems puzzling at first glance and will be analyzed further in Chapter 3.

August Weismann's (1893) understanding of social evolution, expressed in his essay "The all-sufficiency of natural selection: a reply to Herbert Spencer," has been similarly neglected. Weismann is universally acknowledged for being the first to recognize the distinction between germline and soma, but he received scant credit for his insights into colonial social insects and multicellularity (e.g., Weismann, 1889, 1904). Mayr (1982) praised Weismann for correctly rejecting any form of Lamarckian inheritance, for discovering crossing over between recombining chromosomes (indicating that sex generates almost infinite genetic variation among individuals), and for careful rational analyses of every evolutionary problem that he encountered—he forced his opponents to produce hard evidence when they disagreed with his early but comprehensive theory of genetics based on the fundamental distinction between what

would later be called genotype and phenotype. Mayr quoted Correns, one of the biologists who reclaimed Mendel's papers, for the statement that "the rediscovery of the Mendelian rules in 1900 was no great intellectual achievement after Weismann had paved the way." It was George Romanes, Darwin's last pupil and collaborator, who coined the term neo-Darwinism to characterize Weismann's stringent emphasis on adaptive evolution by individual natural selection at the end of the 19th century (Mayr, 1982; see Bell, 1997, for a kaleidoscopic modern overview). We are thus justified in asking why neo-Darwinism disappeared from the modern synthesis after Wells et al. (1931) had reviewed it so well. How could it be that Julian Huxley, "the only senior British zoologist in the inter-war years who thought ecology and behaviour important" (Marren, quoted in Cain, 2010), had abandoned most of that emphasis when he published his *Modern Synthesis* book (Huxley, 1942)?

A recurrent theme throughout the first history chapters of this book will be that the modern synthesis never included elements of the neo-Darwinian approach because that part of the synthesis had been interrupted for decades almost everywhere. This may also explain that the modern "gene's eye view" of adaptive evolution received little appreciation from the last surviving architect of the modern synthesis (Mayr, 1982, 1983a, 2001) and from American historians interpreting that synthesis (Provine, 1971, 1986; Smocovitis, 1996). Up to the present day, encyclopedic summaries of the modern synthesis usually suffice with spelling out the contributions of population genetics and the additions by Mayr, Stebbins, and Simpson (Box 1.1). Some have recently claimed that there is a need for extension into other mechanistic directions (e.g., Pigliucci, 2007; Laland et al., 2014; see Wray et al., 2014; Futuyma, 2017 for responses).

However, such discussions fail to consider that Darwin's *Origin* was fundamentally about the origin of adaptations (Charlesworth et al., 2017; Coyne, 2009), a topic that neither Huxley's modern synthesis monograph nor its supposedly desirable extensions help us to understand better as long as emphasis remains focused exclusively on molecular mechanisms (Box 1.1). In the next section, I will show that the 1959 Darwin Centennial Celebration symposia at the University of Chicago contained several contributions that refocused on adaptation by natural selection, suggesting that Huxley's modern synthesis was already outdated in 1959 when the innovations that would establish a broader neo-Darwinian synthesis were about to begin. These new directions were to emphasize understanding of social traits in the broadest possible sense, of phenotypic life histories, and of the ontogenetic development of adaptive phenotypes.

Another recurrent theme will be that the ca. five decades-long interruption of the neo-Darwinian synthesis, from 1914 onwards, not only meant that the early contributions of key figures such as Huxley (Dronamrayu, 1993) and Wheeler (Evans and Evans, 1970) were lost, but also that transparent terminology for social adaptations disappeared. Boomsma and Gawne (2018) have documented several striking examples of biologists in the late 19th and first half of the 20th century who understood the prevailing patterns of social evolution across the levels of organizational complexity in clearer terms than the modern terminology of sociobiology allows. Here and in the next two chapters I unearth this intellectual fossil record because it helps to explain why it took so long for the connections between Darwinian adaptation and Mendelian genetics to include the fundamental social interactions that pervade all levels of biological organization.

The 1959 Darwin Centennial Celebration: mainstream thought and emerging neo-Darwinian innovation

A century after Darwin's *Origin*, a series of symposia gathered many of the world's influential evolutionists in Chicago and produced three edited volumes of synthesis and outlook papers (Tax, 1960a, 1960b; Tax and Callender, 1960). The opening chapter of the first volume (Huxley, 1960) offers a wide-ranging account of what is being celebrated, but a strange paradox is also apparent. Huxley explicitly states that the centennial is a milestone of neo-Darwinism, but emphasizes the reconciliation of Darwin's verbal concepts of natural selection with genetics, ecology, and paleontology with little mention of adaptation, the topic that Wallace, Weismann, and his younger self had been writing about profusely (Huxley, 1912, 1923a, 1923b, 1930; Wells et al., 1931). Retrospective reading suggests that the centennial contributions with lasting neo-Darwinian significance came from authors who represented ideas that were absent from both Huxley's (1942) monograph and his 1959 opening address. Five of these are important for this book, for they contained remarkably modern visions on how natural selection shapes adaptations, anticipating developments that were to come. These chapters illustrate that the neo-Darwinian perspective was in fact waiting to be redeveloped shortly afterwards when the gene's eye view of evolution by Hamilton (1963, 1964a, 1964b) and Williams (1966a) saw the light. It is interesting to note that these five chapters had authors who were either European or had been profoundly influenced by European evolutionary thinking. They all implicitly assumed that adaptations are based on genetically encoded information that must involve Mendelian genetics. However, they also appreciated that our limited knowledge of the physical properties of these genes should not constrain theorizing about adaptation—what Williams (1992) would later refer to as the approach based on the codex of genetics (Box 1.1) (see also Maynard Smith, 2000).

The first of these novel views is a chapter by Niko Tinbergen (1960) entitled "Behavior, systematics, and natural selection," emphasizing that many behaviors have explicit survival value. They would thus be subject to individual natural selection in the same way as morphological traits, so behavioral adaptations for increasing reproductive success are expected to be commonplace. He also elaborated that we need to define how the traits under study affect survival before we can understand how selection affects both these focal traits and any correlated traits that may indirectly determine the phenotypic outcome. As he went along, Tinbergen insisted that biologists need to trace phylogenies if they are to understand why behaviors change over time and in what direction, and he argued that trained naturalists tend to see many more signs of adaptive design, and of their phenotypic compromises when they study species in the field, than do laboratory or museum biologists. It is here, I believe, that the first post-World War II reconnection with the naturalist views of Darwin, Wallace, and Weismann is being made for a global audience, by an author who in doing so is paying homage to a well-established Oxford tradition (Ford, 1938). Tinbergen's address just preceded his inaugural lecture at the University of Leiden, which was entitled "Nature Is Stronger Than Nurture" and subtitled "In Praise of Fieldwork." Shortly afterwards, Tinbergen moved to Oxford, where he joined like-minded colleagues who were equally interested in the adaptive significance of animal behavior under natural conditions and in explaining these behaviors as products of individual natural selection (Kruuk, 2003; Burkhardt, 2005).

The second contribution that deserves mention is Ernst Mayr's (1960) chapter entitled "The emergence of evolutionary novelties." Like Tinbergen, Mayr referred directly to Darwin. He emphasized that both saltationism and Lamarckianism had to be refuted before progress in understanding evolutionary novelty became possible. Mayr defined an evolutionary novelty by its adaptive function and argued that the origin of a novel trait may be haphazard (e.g., be a pleiotropic byproduct of another genotypic change). However, he also maintained that we should always expect novel traits to owe their persistence to either an intensification of adaptive function or to a change of function, but in such a way that none of the intermediate stages would ever have been non-adaptive. Mayr concluded that evolutionary novelty only rarely becomes an adaptive radiation because key innovations need to achieve general "perfection potential" to allow adaptive diversification. He echoed Tinbergen's and Darwin's notions that adaptive shifts are often initiated by changes in behavior that are later followed by structural phenotypic changes, that is, that function precedes form. He also agreed with Tinbergen that "the phenotype is a compromise between conflicting selection pressures and that every specialization is bought at a price," arguing that this may explain that ecological non-specialization requires the maintenance of large brains, whereas specialization is often seen in birds with small brains. These insights are remarkable but they do not appear to have become important in Mayr's later work, perhaps because he misunderstood (Mayr, 1983b) the role of genotypes versus genes (Maynard Smith, 1988; Dawkins, 2013, p. 269) and did not strictly separate individual and species-level selection. The term "perfection potential" is in fact unhelpful as adaptations are never perfect and always relative (Pittendrigh, 1958, Cain, 1964). Ernst

Mayr and Niko Tinbergen had known each other since 1946 (Kruuk, 2003; Burkhardt, 2005); their 1960 chapters show that Tinbergen was an explicit neo-Darwinian while Mayr was not.

Invited by Dobzhansky (Peterson, 2016), the third remarkable chapter was presented by the pioneer of developmental biology Conrad H. Waddington (1960) under the title "Evolutionary adaptation" (*sic*). It starts out summarizing Darwin's well-known arguments for adaptive design, which Waddington explained as the unambiguous product of natural selection, and as being incompatible with natural theology predicting perfect adaptation (Paley, 1809; Kirby and Spence, 1818; Gardner, 2009). He then proceeded to argue that gene-by-environment interactions are crucial for understanding how the phenotypes of breeding individuals arise during development. He further noted that organisms are very able to adaptively react to alternative living conditions with matching modes of behavior, suggesting he endorsed the idea of behavioral strategies being subject to individual selection. He also presented his famous epigenetic landscape, which would feature prominently in the synthesis of developmental biology that he provided a few years later (Waddington, 1966). His evolutionary approach to development replaced the more limited discipline of comparative embryology and remained profoundly organismal, unlike many later purely molecular approaches to the study of ontogenetic development. That Waddington adhered to selection at the level of the individual also emerged earlier when he wrote, "given that more offspring are conceived than are necessary to preserve the numbers of the population, and that there are inherited variations between individuals, natural selection *must* occur" (Waddington, 1958, p. 5). His approach to "genetic assimilation" became controversial because it retained suggestions of belief in *élan vital* (Williams, 1966a; Mayr, 1974; Haig, 2007),

but that should not detract from the value of his other insights (Slack, 2002; Baedke, 2013).

The fourth chapter worth highlighting was by Alexander J. Nicholson (1960), chief entomologist of the Commonwealth Scientific and Industrial Research Organisation (CSIRO) in Australia, and was titled "The role of population dynamics in natural selection." It challenged beliefs in adaptation for the good of the species, based on a series of population cage experiments with the Australian sheep blowfly *Lucillia cuprina*— probably the first well-replicated attempt at experimental evolution. He argued that there is "no justification for the belief that possible benefit to the species ever influences the direction of selection." Nicholson launched a well-argued attack on the prevailing idea at that time that species are matched to their environments, emphasizing that the struggle for existence is active rather than passive and that individual success is "analogous to the 'fitness' of winning athletes rather than to the 'fit' of a key to a lock." He further analyzed how naturally selected adaptation interfaces with development, writing that "[a]lthough selection is limited to preserving and developing advantageous phenotypic properties, by doing so it also preserves and tends to create physiological harmony between the underlying developmental processes. When a gene change occurs, not only must it initiate changes in developmental processes appropriate to produce an advantageous character in the phenotype, but, to confer selective advantage, these changes must also be compatible with all the other developmental processes which underlie the essential properties of the phenotype." The latter echoes Woodger (1945) and Waddington (1958) who discussed the same issue from a developmental perspective. Williams (1966a) cited Nicholson only once, but his strong first-principle endorsement of competitive and environmental selection of individuals must have inspired him when he

THE 1959 DARWIN CENTENNIAL CELEBRATION

started to write *Adaptation and Natural Selection* in the early 1960s.

The fifth noteworthy chapter is by Hermann J. Muller (1960), Nobel laureate (1946) for his 1920s work on X-ray-induced mutations in *Drosophila*, entitled "The guidance of human evolution." He demonstrated a keen conceptual understanding, *avant la lettre*, of the interplay between kin selection and cultural selection when he wrote that "the cultural development of hunting and its derivative activity, warfare, served as the foundation for the intensified selection of predispositions to combativeness, xenophobia, and related impulses, which made intergroup antagonism an active complement of intragroup cohesion," even though "one can no longer disentangle the influence of these two interwoven sets of factors in changing both human culture and the human genotype." Muller connected this notion to what would soon become established as inclusive fitness theory by Hamilton (1963, 1964a, 1964b). "Certainly, as Darwin pointed out," he wrote, "the family, long before the arrival of man, must have afforded the primary unit for intergroup selection whereby the genetic basis of altruistic proclivities became developed." This general endorsement of the operation of Darwinian family selection is then expanded to the human domain by noting that "as the groups came to include a number of families, they were still small enough and numerous enough to allow effective selection for the traits that predisposed their members for the wider co-operation and altruism here in order. Even if the individual sacrificed himself for the small group, he tended to foster the multiplication, through the others of that group, of genes like those that had predisposed him to this behavior. With the formation of towns and large civil units in general, this kind of influence on selection tended to disappear." Much of what we consider as later insights into human evolution (see also Chapter 8) is

already captured here, but Muller has received no credit for these insights.

Against this background, the gene's eye view of evolution that originated in the 1960s did not come out of the blue. The five leading biologists highlighted here saw clearly what was unaccounted for in the modern synthesis, most notably the connection between genes and phenotypes, social and life-history evolution, behavior, and the study of development. They were extremely clear thinkers even though one can always find flaws in hindsight, and they were all well-read outside their core disciplines. Ten years earlier, Hermann Muller (1949) had already provided a transparent agenda for developmental biology, writing that "the organism cannot be considered as infinitely plastic and certainly not as being equally plastic in all directions, since the directions which the effects of mutations can take are, of course, conditioned by the entire developmental and physiological system resulting from the action of all the other genes already present." Waddington (1958) has almost identical phrasing about limits on evolvability (see also Uller et al., 2018), but he understood "adaptation as a Brit" and "Americans did not like to see development distract from genetics" (Peterson, 2016, pp. 196, 201). As Provine (1988) noted, the modern synthesis ended up being a scholarly constriction that liberated evolutionary biology from what Dobzhansky (1958) called "a tremendous mass of intellectual rubble," or in Mayr's (1982, p. 570) words, it "empathically confirmed the overwhelming importance of natural selection, of gradualism, of the dual nature of evolution (adaptation and diversification), of the population structure of species, of the evolutionary role of species, and of hard inheritance." However, as Williams (1966a) was soon to demonstrate, sweeping away the rubble of uncritical thinking about the default level at with natural selection forges adaptations remained to be done.

Although the gene's eye view of evolution was tightly argued from its very inception (see next section), the intellectual substrate for reviving neo-Darwinism remained deprived of nutrients for quite a while. None of the senior architects of the modern synthesis attending the Chicago symposia had a coherent idea about the way in which natural selection produced adaptations, except for George Gaylord Simpson who, as paleontologist, was probably not considered to be an "expert", in spite of a series of monographs (Simpson, 1944, 1949, 1953) that Pittendrigh (1958), the first to frame the gene's eye view of adaptive organization (see next section), highlighted as exemplary. It appears that the organizers of the centennial meeting had been keen to avoid controversy, which implied that neither Fisher nor Haldane were invited (Smocovitis, 1999). The idea that younger scholars could be logical Darwinians without being systematists, paleontologists, or geneticists was far from obvious, which may have contributed to reducing the British and continental European delegations. Many of those biologists were only mid-career scholars at that time, including Colin Pittendrigh, a US-based British PhD student of Dobzhansky (Menaker, 1996), and Peter Medawar, whose innovative work on the evolution of aging (Medawar, 1952) was soon to become part of new approaches to understand life-history evolution (Williams, 1957). Also John Maynard Smith did not attend even though he published his own concise Pelican Book *The Theory of Evolution* just before the centennial meeting (Maynard Smith, 1958). Trained as a population geneticist by J.B.S. Haldane, it is striking how much emphasis Maynard Smith gives to adaptation and natural selection in wild populations, underlining that British evolutionary biologists differed in important ways from mainstream thought in the US at that time (Peterson, 2016, pp. 196–201). I will return to the conceptual divide across the Atlantic in Chapter 2.

Teleonomy, proximate versus ultimate causation, and the problem of panselectionism

The reintroduction of neo-Darwinism was initiated by William D. Hamilton and George C. Williams, who combined the pivotal elements of gene's eye understanding in a formal theory of adaptation based on the informational principles of particulate inheritance. In the final chapter of his book Williams (1966a), recommended that theory to be developed further in teleonomic language to avoid confusion with conscious (teleological) or inanimate (teleomatic) processes, referring to a seminal book chapter by Pittendrigh (1958) (Box 1.2). Eight years later, Ernst Mayr (1974) published an essay, "Teleological and teleonomic, a new analysis," outlining the Aristotelian roots of teleological thinking and the positive implications for philosophy of biology of replacing that term by teleonomy. However, he did not refer to Williams, which illustrates the divide between the modern synthesis architects and the new gene's eye view of adaptive evolution. Around the same time, Hamilton's inclusive fitness theory allowed all social interactions to be expressed and analyzed in a single gene-copy currency, an innovation that Williams (1966a) reviewed for the first time in a broad context. It seems likely that Tinbergen's (1963) explicit distinction between proximate and ultimate causation, rather than Mayr's (1961) less conducive paper, provided a frame of reference for Williams and Hamilton. However, it was Huxley (1916) who made this distinction for the first time, which led the proximate–ultimate terminology to pervade thinking among Oxford zoologists ever since (Baker, 1938; Lack, 1949; Mayr, 1982). Tinbergen (1963) and Mayr (1961) thus came to independently elaborate an already known and highly fruitful way of addressing complementary causation, Mayr with a set of rather

Box 1.2 Teleonomy and the formal theory of adaptation

In the closing chapter of *Adaptation and Natural Selection*, Williams (1966a) focused on stringent terminology to ensure that adaptive phenotypic traits appear "goal-directed" in the sense of having functions that ultimately enhance reproductive success, but in scientific language free of anthropomorphic or vitalist connotations. He found that terminology in Pittendrigh (1958) and applied it to explain what adaptations are and what they are not. Williams envisaged a theory that would explain the general functional principles of adaptation from a single set of parsimonious concepts, a set of first principles. As he wrote: "Natural selection arises from a reproductive competition among the individuals, and ultimately among the genes, in a Mendelian population. A gene is selected on one basis only, its average effectiveness in producing individuals able to maximize the gene's representation in future generations. The actual events in this process are endlessly complex, and the resulting adaptations exceedingly diverse, but the essential features are everywhere the same." This phrasing represented a firm break with any organismal biology so far in North America. None of the contributors to the Darwin Centennial Celebration meeting just seven years earlier had been so explicit about the ultimate unit of natural selection being a unicellular or multicellular individual's genes. On the contrary, most of the 1959 centennial contributions were implicitly and sometimes explicitly (e.g., Emerson, 1960) assuming that adaptations evolved for the good of populations if not entire species (see also Chapter 2).

Williams' book became influential among phenotypically focused (often younger generation) evolutionary biologists (e.g., Stearns, 1976; Alexander and Borgia, 1978), but his approach was never accepted by the senior biologists who organized the Darwin Centennial Celebration. As Ghiselin (1974, p. 7) notes, Mayr only appreciated selection as acting on complex integrated genotypes, a view he shared with leading US scholars at the time (e.g., Gould and Lewontin, 1979). This split between scholarly generations partly subdued the transparent logic of Williams' gene's eye view of adaptation, in spite of Michael Ruse's (1988) explicit judgment that "[a]fter that book, you could never think of group selection with ease again." Many never understood the fundamental non-teleological soundness of the gene's eye teleonomic approach, and perhaps merely remembered Mayr's (1974) quotation of Haldane's quip that "[t]eleology is like a mistress to a biologist: he cannot live without her but he's unwilling to be seen with her in public." However, teleonomy became a metaphorical matriarch in the hands of Pittendrigh and Williams because it implemented a scientific abstraction of the earlier organicist insight that biology is about dual teleological and materialistic causation (Woodger, 1929). Skepticism continued, however, also among population geneticists who would often rather maintain that selection is unimportant than admit that biology cannot be a serious science without a first-principle theory of adaptation—John Maynard Smith (e.g., 1978a) and James Crow (e.g., 1991) were noteworthy exceptions at that time.

It is worthwhile to briefly dwell on why Pittendrigh (1918–1996) was such a pivotal trail-blazing figure. He was born and educated in Britain, completed a PhD with Dobzhansky in 1947, settled down in Princeton and later Stanford, founded the field of circadian biology, and coined the term teleonomy as a necessary and respectable alternative to teleology (Menaker, 1996). His 1958 book chapter, "Adaptation, natural selection, and behavior," is a delight of forcefully coherent logic on how to understand the neo-Darwinian principle of adaptation through natural selection. Pittendrigh started defining adaptation as the total organization of a living system, where organization is always relative to an end, like an army is organized relative to the end of defeating an enemy, and where the converse of organization is randomness. He further noted that "the essential nonrandomness of adaptation is due entirely to the organism's (not the environment's) capacity to accumulate and retain information both phylogenetically and ontogenetically" (see also Chapter 8). He emphasized that the 1930s books of Haldane and Fisher replaced 19th-century notions of struggle and survival of the fittest by the more encompassing variable of differential reproductive success, but without most biologists noticing this crucial distinction. He

Box 1.2 *Continued*

explicitly acknowledged multiple causation in biology and recognized the importance of phylogenetic comparison, saying that "an exclusively causal explanation of life is possible [. . .] only if organisms are abstracted from their concrete history." This clearly paved the way, not only for Williams' (1966a) monograph, but also for Tinbergen's four-quadrat formalization of asking questions in biology (Figure 1.1).

Pittendrigh's information interpretation of adaptation is clearly Fisherian in suggesting links with the second law of thermodynamics, saying that "[a]s novelty is thrust into the inherited message by mutation and recombination, there is a slow net gain by the population gene pool of information with respect to reproductive efficiency." In Pittendrigh's words, this "implies that adaptation is a relative matter in many senses; that it cannot be discussed in absolute terms; and specifically that adaptation cannot be recognized with a criterion of necessity—for—viability." The point that only differential reproductive success matters, not mere viability, was made by Nicholson (1960) as well. Pittendrigh also elegantly acknowledged developmental and historical constraints saying that "[a]daptation is thus relative to past genetic alternatives realized by the population, as well as to the particular environment the population inhabits." He anticipated the later metaphor of replicators and vehicles (Dawkins, 1976), recognizing that "in a very real sense the developed organism is no more than a vehicle for its genotype." And he also noted that "[t]he concepts of adaptation and natural selection are so interwoven that it is impossible to misunderstand one without doing violence to the other," to finally conclude that "[t]he study of adaptation is not an optional preoccupation with fascinating fragments of natural history; it is the core of biological study," because "the living system is all adaptation insofar as it is organized." I will have more to say on the informational aspects of adaptation in Chapter 8.

Pittendrigh may have been the first to assume, albeit implicitly, that adaptive phenotypic traits are predictable from optimality logic (see Chapter 2). As (Williams, 1992, p. 61) would later write, "calling a character *optimized* is always more realistic than calling it *optimal*," but this had not distracted

him from using optimality logic in his first book Williams (1966a). He was explicit about gene pools of species and their breeding populations merely passively accumulating adaptive information from the genetic endowments of breeding individuals as they are shaped by natural selection to maximize reproductive success. In his words, the species "is not an adapted unit and there are no mechanisms that function for the survival of the species. The only adaptations that clearly exist express themselves in genetically defined individuals and have only one ultimate goal, the maximal perpetuation of the genes responsible for the visible adaptive mechanisms, a goal equated to Hamilton's [. . .] 'inclusive fitness.'" (Williams, 1966a, p. 252). Adherents of this general (codex) gene's eye view were inspired by the idea that prediction of adaptive phenotypes was now possible, while those using material genotype selection arguments often failed to see the importance of this novel perspective. Dual ultimate (teleonomic) and proximate causation are complementary by definition, as both Tinbergen (1963) and Mayr (1974) appreciated (Box 1.3). It is futile, therefore, to a priori ignore the predictive power of the gene's eye view of adaptation just because the conceptual logic requires simplifying assumptions about genetic mechanisms. Only empirical tests can decide whether such shortcuts are reasonable or not, and if they confirm predictions such assumptions must have been sensible approximations.

Taking a gene's eye teleonomic position forces a biologist to be critical towards the level of organization where natural selection has caused adaptive functionality. This is of key relevance when evaluating whether groups of cells or multicellular individuals can ever have organizational complexity that represents specific adaptations that exclusively evolved for the higher level. In Williams' (1966a, p. 257) own phrasing: "Each type of group must be examined separately to determine whether its characteristics make functional sense. Such an examination of a family group of birds or bees in a hive unmistakably favors the recognition of a functional organization. An examination of other groups, such as a swarm of moths around a lamp or a mass of mussels on a piling, force[s] no such conclusion. It is certainly possible that some groups have a

---

Box 1.2 *Continued*

---

functional organization that is too subtle to be appreciated by the conceptual and technical equipment of the observer. Considerations of parsimony, however, demand that we not recognize a functional organization unless we have definite evidence for it." Williams' (1966a) book is full of stringent teleonomic statements like this, so it is no surprise that many at the time found his approach to be refreshing and inspiring. His insights are as relevant today as they were half a century ago, but have faded from collective memory when biology became increasingly methods and data driven. One of the aims of this book is to restore that balance. A complementary essay on the influence of Williams' *Adaptation and Natural Selection* book is provided elsewhere (Boomsma, 2016).

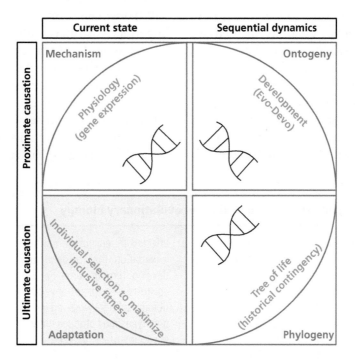

Figure 1.1 Tinbergen's four complementary ways of asking questions in organismal biology. The blue lettering summarizes what Tinbergen's complementary quadrats stand for. The blue adaptation quadrat is special because it is the only one that asks why-questions about ultimate causation and therefore relates directly to what occupied the minds of Darwin, Wallace, and Weismann in the 19th century. This was before biology had an agreed-upon concept of particulate Mendelian genetics and before phylogenies were interpreted as historically contingent bifurcations instead of orthogenetic chains of being (Haeckel, 1879). Double helix symbols illustrate that molecular and computational biology have made biological knowledge increasingly complex, but that this has not affected the blue quadrat because the neo-Darwinian theory of adaptation has become increasingly unified and simpler to understand, particularly when focusing on the codex aspects of adaptive information accumulation (Box 1.2, Box 1.3; see also Chapters 3 and 8). Reprinted with minor adjustment from *Current Biology*, Volume 26, J.J. Boomsma, Fifty years of illumination about the natural levels of adaptation, R1250-R1255, 2016, with permission from Elsevier.

general comments and Tinbergen with specific focus on the practical study of behavior.

Details of how the distinction between proximate and ultimate causation was instrumental in paving the way for the gene's eye view of adaptive evolution are summarized in Box 1.3, emphasizing that it was Tinbergen's version that offered the crucial novel insights for framing the overall agenda of the neo-Darwinian synthesis. Tinbergen not only identified the four complementary ways of asking questions in biology; he also made explicit that both proximate and ultimate causation have two different levels that refer to fundamentally different time scales (Figure 1.1). One of the four Tinbergen quadrats focused on the proximate physiological mechanisms of behavior; another accommodated Waddington's developmental biology. Behind both of these we would now acknowledge the importance of differential gene expression in gene regulatory networks, either instantaneously or across ontogenetic time, but the outlook remains organismal and is not reductionist in a molecular biology sense. Similarly, he identified two ultimate quadrats, one for current adaptive state and one for phylogeny to understand the historical connections between present and ancestral adaptive states, the latter to be reconstructed by mapping traits on phylogenetic trees. Tinbergen's framework is as fundamental today as it was in 1963, and deserves to be well known because it captures the best way for organismal biology to be recognized as a fundamental, theory-driven natural science (Boomsma, 2016). Ryan (2009), Barrett et al. (2013), Nesse (2013), and Strassmann (2014) offer complementary evaluations of the importance of Tinbergen's four-tier approach. Neurobiologists (Pfaff et al., 2019), artificial intelligence researchers (Rahwan et al., 2019), and game theoreticians (McNamara and Leimar, 2020) have recently used Tinbergen's quadrats to frame general research agendas, confirming the lasting scientific merit of Tinbergen's thinking for biology as a whole.

Box 1.3 Framing and resolving questions in evolutionary biology

Rational thought requires broad frames of reference (Kahneman, 2012). The modern evolutionary synthesis of the mid 20th century was pragmatic and partial (Box 1.1) and thus in need of a broader frame of reference—one that captures all of biology and with a proper balance between ultimate and proximate causation approaches. Tinbergen (1963) was the first to recognize that achieving such a frame required that organismal biologists address four complementary types of questions (Figure 1.1): How does it work? How does it develop? How did it evolve? Why does it make adaptive sense? None of these questions is superior to any of the others because all four levels of inquiry are needed for a comprehensive understanding of variation in morphological, physiological, behavioral, and even molecular traits. In modern translation, the quadrats of Figure 1.1 are all affected by environmental and genome-wide genetic variation that co-determine phenotypes as they develop through time. In the mechanism quadrat, epistasis and pleiotropy affect the rate of evolutionary change while epigenetic factors ensure that genotypes may produce different phenotypes through environmentally induced plasticity; in the development quadrat, the expression of genes interacting with a nested set of genetic, social, and purely external environments shapes how gene regulatory networks mediate physiological processes during ontogeny; in the phylogeny quadrat, historical isolation and subsequent changes in environment determine which lineages radiate or go extinct; and in the adaptation quadrat, some phenotypes make more adaptive sense than others so the frequency of their genes increases.

Box 1.3 *Continued*

Also today, the research program and scholarly interests of an organismal biologist can be captured by a set of x, y coordinates, each with a mean and a variance, and be mapped in two-dimensional Tinbergen research space as laid out in Figure 1.1. A series of vectors then specify the connectedness of her research niche within and across the four quadrats, including the centripetal and centrifugal forces that she might use to optimize scholarly objectives. The centripetal forces will seek interdisciplinary crosstalk with one, two, or three other Tinbergen quadrats, while the centrifugal forces will emphasize exclusivity and specialization towards the periphery of a single quadrat. There are no principal criteria to validate the coordinates and vectors of one research program relative to another, but it is of fundamental importance that organismal biologists realize that their research niche will always be part of Tinbergen's map and that it is futile to claim that any position should have priority by default. The simplified double helices in the white quadrats of Figure 1.1 illustrate that, since the 1960s, new molecular and computational technology has enormously enriched the two proximate causation quadrats while also giving us much more reliable trees of life in the phylogeny quadrat. Concurrently, the foundational concepts underpinning the theory of adaptation (the blue quadrat) have become synthesized and unified into a simpler and more comprehensive framework of first-principle understanding, as I will detail in Chapters 4, 5, and 8, consistent with what Popper (1962) envisaged as typical for a frugal scientific theory.

In all this, it is important to realize that general scientific concepts are fundamentally different from mathematical models of particular situations or scenarios (Levins, 1968; Frank, 2013a, 2013b; Otto and Rosales, 2020). Models will typically contribute to a precise but partial understanding of adaptive dynamics in a forward, predictive direction. That is something else than understanding the evolutionary stability of complex adaptive states that resulted from selection in the past, which is feasible only with comparative data analysis driven by accurate predictive concepts. Thus,

every step in making models more specific than the foundational theory on which they rest makes them less generally applicable (Levins, 1968; Frank, 2013a, 2013b). The most powerful concepts are therefore those that capture a huge amount of reality with relatively few predictors, both in physics and chemistry and in organismal biology. As attributed to Einstein around 1950, only when we "[m]ake everything as simple as it can be, but no simpler," are we likely to "cover the greatest possible number of empirical facts by logical deductions from the smallest possible number of hypotheses or axioms." In Ghiselin's (1974) words, "[c]alculation is precise, experience is accurate." The implications are that abstract overarching concepts in biology can always be presented as simple plots or equations with just a few variables, and that more complex modeling can explain additional fractions of hitherto unexplained variance only within those overall settings. Finally, as Popper (1962, p. 36) stipulated, theories need to be operational, that is, their conjectures need to be framed in terms that can actually be measured, which implies that Darwinian research questions need to be about phenotypic traits that can, in fact, respond to selection as I explain in more detail in Box 1.4. This separates evolutionary biology from ecology, which has many informative variables that are not organismal traits (e.g., density, species richness, and ecosystem service). Confusion has often arisen when ecological attributes became reified as if they were characteristics of phenotypes. This happened, for example, with the term eusociality—as defined by E.O. Wilson it is not a trait that can respond to selection (Boomsma and Gawne, 2018). As Williams (1966a) took great pains to emphasize, neo-Darwinian theory can only be general when it remains precise in its functional definitions. His stringency of argumentation implies that, to uncover deeper Darwinian understanding, one needs to pursue explanations equivalent to a greatest common divider in calculus, a reductive factor that has the same explanatory significance for many different and seemingly more complex phenomena, consistent with another Einstein dictum that says

---

Box 1.3 *Continued*

"[i]f you can't explain it simply, you don't understand it well enough." The opposite of a greatest common divider (GCD) is a least common multiple (LCM), which captures a diversity of phenomena in a higher number that somehow relates to all of them but without offering root-cause understanding. This is often the required and preferred approach in ecology.

For example, the Shannon information index (Sherwin et al., 2017) can be used as a composite measure of species diversity but not as an organismal trait that responds to selection. Confusion between reductive GCD and inclusive LCM understanding has often plagued organismal biology (see also Chapters 6 and 8).

I will use Tinbergen's logic throughout this book in a long argument for the legitimacy of a formal theory of social adaptation based on both inductive and hypothetico-deductive logic (Box 1.4), supported by, but not primarily dependent on, mathematical formalism. Looking back more than half a century, it is clear that the contributions of Williams and Hamilton were the first to give explicit gene's eye content to Tinbergen's adaptation quadrat. As I noted in the previous section, some of these developments were waiting in the wings during the 1959 centennial meeting, although precursors cannot always be reconstructed via direct citations. For example, Williams' strong emphasis on the function of adaptations is represented in several chapters of the Darwin Centennial Celebration volumes, but he cites other work by Tinbergen and Mayr. Only later (Williams, 1985) did it emerge how important Tinbergen's (1963) paper was. The preface to the second edition of *Adaptation and Natural Selection* (Williams,

1996) further shows that Williams was strongly motivated to rebut Alfred E. Emerson's (1960) explicit good-for-the-species approach in his Darwin Centennial Celebration chapter, which was why he gave his book the subtitle *A Critique of Some Current Evolutionary Thought*. Hamilton's inspiration was similarly complex as he relates himself (Hamilton, 1996). Haldane and Fisher are sometimes mentioned as having foreshadowed inclusive fitness theory, and I just added Hermann Muller as possibly the most explicit precursor, complementing another such case highlighted by Dugatkin (2006, p. 165). However, there can be no doubt that it was Hamilton who spent years of his graduate studies to develop the first formal and general population genetics model to prove that any conditionally expressed gene for altruism will spread if relatives contribute more of its copies to future generations than own offspring could achieve, independent of the frequency of such a gene. An extensive justification of this conclusion is given by Segerstrale (2013).

Box 1.4 The stringency and testability of neo-Darwinian predictions

The teleonomic theory of adaptation turned evolutionary organismal biology into a hypothesis-testing and hypothesis-generating field addressing generally

important phenomena of adaptive design. However, the approach can only work for predictions about functional traits that are precisely defined and free

Box 1.4 *Continued*

of anthropomorphic bias, as pointed out by Williams (1966a) and more recently reiterated by West et al. (2007a, 2011). Transfers of terminology are justified only "when there is a real functional analogy between what man's reasoning (and trial and error) can produce and what natural selection can produce" (Williams, 1966a, p. 261). Williams referred to Tinbergen's hierarchical classification of instincts in a statement consistent with his four complementary quadrats (Figure 1.1; Box 1.3), writing that "no complete explanation of a biological phenomenon can be achieved without an inquiry into its evolutionary development, and a teleonomic analysis would not proceed far without the use of historical data." In the preface of the second edition of *Adaptation and Natural Selection*, Williams (1996) emphasized that adaptation theory works both ways: "It logically predicts that there are certain sorts of properties that organisms must have, and others, such as adaptations for the 'benefit to the species' [. . .] that they could not possibly have." Ten years later, sociobiology (Wilson, 1975) arose with an anthropomorphic agenda and a number of poorly defined "human construct" definitions that violated Williams' principle. Notwithstanding its "new synthesis" label, sociobiology stripped from its Hamiltonian elements therefore contributed little to the neo-Darwinian theory of adaptation, as a later review made clear, albeit unintentionally (Alcock, 2001).

The theory of adaptation needs to be constructively dogmatic if the goal is to develop and test predictions that are maximally amenable to empirical refutation (Popper, 1975). While this principle is generally accepted in physics and chemistry, it has been a hard nut to crack in biology, where forms of pluralism thrive, often as sitting-on-the-fence attitudes disguised as open-mindedness (Bondi, 1975). Williams hit the nail on the head with the opening quote of this chapter and by concluding that adaptation theory should be its own worst critic, undoubtedly in response

to the fuzzy panselectionist spirit among many biologists around him. For example, Sewell Wright (1953) has a few lines on social insect colonies indicating that he understood the significance of biparental colony founding and full-sibling relatedness for the evolution of castes, but his interpretation is one for the good of the species. Williams' concept of adaptation radically opposed such lack of stringency when he wrote that "[e]volutionary adaptation is a special and onerous concept that should not be used unnecessarily, and an effect should not be called a function unless it is clearly produced by design and not by chance. When recognized, adaptation should be attributed to no higher a level of organization than is demanded by the evidence." Although direct references are missing, one has the impression that Tinbergen adopted the hierarchical system of hypotheses of the English philosopher of biology Joseph H. Woodger (1929) for his understanding of instincts, and that also Williams strictly adhered to the unambiguous terminology that Woodger felt necessary and sorely missing in biology (Nicholson and Gawne, 2014).

As with Hamilton's inclusive fitness theory, Williams' arguments are firmly rooted in Fisher's fundamental theorem of natural selection, which stipulates that a naturally selected increase in mean fitness of a reproducing agent at any time is equal to the genetic variance in fitness at that time, and that any such increase proceeds through changes in gene frequencies (Edwards, 1994; Crow, 2002). Fisher (1930) considered this theorem to be his most important insight because it appeared to offer an analogy with the second law of thermodynamics. The theorem assumes that fitness is a complex polygenic trait determined by many loci each with small phenotypic effects, so that only the additive genetic variance of fitness is important. Assuming change is continuous, the additive genetic variance is equivalent to the Malthusian parameter, defined as an individual fitness

Box 1.4 *Continued*

exponent in the absence of limiting factors. Fisher's fundamental theorem has been criticized as being only approximate, but Crewe et al. (2018) appear to have recently resolved these ambiguities. Indeed, the theorem is not mathematically precise when dominance, epistasis, and deviations from random mating are included, but it is always the additive genetic variance that determines responses to selection, which is what counts in the daily lives of field and experimental biologists who need accurate general predictions. As James Crow (2002), one of the population geneticists supporting the gene's eye view of adaptive evolution, expressed it: "Is there any other quantity that captures so much evolutionary meaning in such a simple way?" (see Box 2.3 for further details).

Hamilton's rule (see Chapter 2), which will play a key role throughout this book, is in fact a general extension of Fisher's fundamental theorem, as George Price (1972) was the first to realize and recent reviews have confirmed and elaborated (Frank, 2013a, 2013b; Grafen, 2014a, 2014b; Gardner, 2015a, 2015b; Queller, 2017). An important characteristic of all these abstract theoretical approaches is that they combine optimality principles with inclusive fitness maximization subject to any local constraints that may apply, as I will discuss in more detail in Chapters 2 and 4. Given these connections, it is hard to see how Williams' statement about reproductive competition and gene-level selection cited in the first paragraph of Box 1.2 would fail to be sufficiently correct to explain a large amount of interesting biology. However, too few attempts have as yet been made to actually proceed with the explicit further development of teleonomic adaptation theory, with the papers cited above as important exceptions (see also Chapters 4, 5 and 8). Physics has come a long way in replacing human-construct definitions such as ether and celestial bodies by hierarchical concepts of atoms comprising fundamental particles, and galaxies consisting of solar systems with planets. Likewise, chemistry replaced alchemy with first-principle theory where periodic table atoms are nested within inorganic molecules which make up the organic molecules that provide the physical building blocks of life. In a similar way, formal adaptation theory should consider particulate genes as the fundamental units of adaptive information and natural selection as the all-pervasive non-reductionist mechanism for maintaining organized life (Box 1.2).

Throughout this book I will use a combination of inductive, deductive, abductive, and analogical arguments to make the case for a single causal explanation for the major evolutionary transitions in organizational complexity. As explained by Juthe (2005), induction is reasoning from the peculiar to the general, something that most naturalists do following Darwin's example. In contrast, deduction starts out from general principles to derive a series of universal and testable predictions that can challenge established concepts as imagined by Popper (1975). In recent decades, it has become increasingly clear that Hamilton's rule, the mathematical formalization of inclusive fitness theory via the Price equation, and their philosophical embedding in Alan Grafen's formal Darwinism (see Box 2.3) are sufficiently axiomatic to be used as foundations of hypothetical-deductive arguments. In modern language, we can summarize the neo-Darwinian axiom as natural selection being expected to optimize adaptive phenotypic design because it rewards gene replicators that induce their somatic vehicles (or interactors when using Hull's (1988a) terminology) to maximize inclusive fitness under average environmental conditions. I will also use abduction as a hybrid form of reasoning going from particular facts via the general principles to other particular facts, challenging the reader to refute my inferences with more plausible and parsimonious arguments that explain as much of social organization with as few key variables. Finally, I have sometimes used direct analogy arguments, well knowing that this way of reasoning requires particularly close scrutiny. Such logic extrapolates from one (set of) particular(s) to

Box 1.4 *Continued*

another, either within the same domain of social evolution or *mutatis mutandis* across such domains, to lay out striking patterns that may have been neglected or overlooked. As Juthe (2005) explains, analogies are either correct or incorrect with some likelihood but they cannot be partially correct. This notion appears to have disappeared in our current era of data- and methods-driven biology, where rhetorical use of shallow metaphor has become rampant particularly in review papers, editorials, and titles of research contributions that are less thoroughly policed during peer review than the actual content of primary research papers. In some cases, authors and editors may even actively promote fancy-looking but unwarranted analogies to increase citations or sales (Nicholson, 2013; see Haig, 2012; Henikoff, 2018, for

similar critical notes arrived at from different perspectives). Juthe asserts that conclusive analogies need to be based on 1:1 cross-correspondence of what in this book would be sets of core functions shaped by natural selection rather than effects that can be dismantled to be mere accidental byproducts (Williams, 1966a). At the end of this book, I will summarize five overarching testable predictions (see Box 8.3) to remain faithful to Williams' dictum at the heading of this chapter and to Popper's (1962, 1975) demand of scientific transparency. By then, I hope to have convinced the reader that the neo-Darwinian synthesis, adhered to and developed further in this book, is the only coherent theory framework to secure that biology remains a hard natural science in the overarching conceptual sense.

The early work by Hamilton and Williams was so novel that none of them owes a direct depth of gratitude to Tinbergen. Both worked in isolation and were initially not aware of each other's insights, and both were frustrated about the lack of interest, even outright hostility, that they encountered when they tried to share their ideas. Hamilton scrambled to obtain a PhD position because hardly any senior academic could be persuaded by the interest or appropriateness of asking how altruistic traits evolve (Hamilton, 1996; Segerstrale, 2013). Williams was ten years older than Hamilton and had a faculty position, but painted a bleak picture of the convenience Darwinism in the 1960s (Williams, 1992, pp. 46–47) when he wrote that "the main use of the theory of natural selection in the early and middle years of this century was as an excuse to believe in evolution. Of course there was some use of the theory to explain adaptation, often in relation to climatic differences over a geographic range. [...] Thereupon any

axiomic use of the theory was abandoned, and it was merely assumed that natural selection always promoted what was in some way good." In his obituary of Stephen J. Gould, Stearns (2002) wrote that "[t]he complacency and rigidity of evolutionary biology in the 1960s were real. The consistency of evolutionary phenomena with population genetics was incorrectly extended to a general belief that population genetics was sufficient to account for evolution. [...] [I]t created an atmosphere in which important evolutionary phenomena not directly tied to genetic mechanisms were often defined away or ignored, to the great frustration of those interested in them." This underlines that only the broadening of evolutionary biology during the later decades of the 20th century allowed the completion of a neo-Darwinian synthesis, as I will detail in Chapter 2.

However much Tinbergen remained an ethologist throughout his life, he came to greatly appreciate the conceptual innovation embodied in the gene's eye view. His biography (Kruuk, 2003, p. 332) mentions that he wrote to Hamilton in 1980 that "I have often 'kicked' myself for not having cottoned on at the time to the importance of the work you, Trivers, Maynard Smith, did when you did it, and that it took Richard Dawkins' book [*The Selfish Gene*, 1976] to make me see the importance of the direction you three symbolize for me." I am not aware of any direct contact between Tinbergen and Williams, but the Human Sciences Library in Oxford has a copy of Williams' 1966 book with a posted note saying "This book was purchased from a gift made by Professor N. Tinbergen for Human Sciences." In North America, Williams appears to have been one of the very few to expose group selection arguments when Wynne Edwards (1962) finally made them so explicit that they could no longer be ignored. However, Gould (1983) merely complained about a "hardening of the modern synthesis" towards the late 1940s because Sewall Wright had also shifted to emphasizing selection and the importance of adaptation, albeit at the group-selected level (Provine, 1986) that Williams refuted. Neither Gould nor Mayr (1982) showed any appreciation for selection needing to be at the individual or gene level by default (Mayr, 1983a, 1983b), and Mayr's (1974) teleonomy paper focused on rebutting physics envy and defending the goal-directedness of species-specific genetic programs, rather than on individual selection being required to forge organismal adaptations[1]. In sum, uncritical pan-selectionist attitudes persisted, and Gould (1983) blamed mostly his British colleagues for "hardening the synthesis."

Williams' urge, in the final chapter of *Adaptation and Natural Selection*, that evolutionary biology needed a formal teleonomic theory of adaptation (Box 1.2), was more recently repeated by Pross (2012), who argued that a teleonomic approach is also important for understanding the very origin of life, because it defines the difference between the inanimate teleomatic world of physics and chemistry and the organizational realm of biology driven by natural selection. Pross shares the view that all organisms in the widest sense of the term are expected to have teleonomic "agency" so that natural selection will tend to make their strategies effective within the caveats of phylogenetic contingency and proximate constraint. In the remainder of this book, I will explore the extent to which Williams' vision has borne out, staying loyal to his other principle that teleonomic explanations should always be maximally parsimonious. The subjection of scientific ideas to Occam's razor (Sober, 2015) seems the only objective criterion for generating priority in the busy calendars of researchers, because both confirmation and refutation of parsimonious theory are intellectually rewarding. In several ways, this book builds on Andrew Bourke's (2011) *Principles of Social Evolution*, but the deeper historical anchors of my approach allow a reframing of inclusive fitness theory that is more specifically applicable to life's major discontinuous transitions in organizational complexity. More details of my approach are provided in Box 1.4, where I emphasize the need for strict

---

[1] It is striking that Mayr (1982) never relates to the gene's eye view of adaptive evolution as developed by Williams and Hamilton. In a *Science Sendings* message of September 11, 2010, Neal Smith (STRI, Panama) refers to a personal conversation with Ernst Mayr that strongly suggested that, by the early 1980s, he had not read WILLIAMS, G. C. 1966. *Adaptation and Natural Selection*, Princeton, N.J., Princeton University Press.

Their views were very different because MAYR, E. 1961. Cause and effect in biology - kinds of causes, predictability, and teleology are viewed by a practicing biologist. *Science*, 134, 1501–1506. had defined functional causation as belonging to the proximate domain, diametrically opposite to Williams' definition.

definitions of adaptation in terms of gene's eye neo-Darwinian theory, and discuss the lack of such stringency that has occasionally confused Darwinian concepts in recent decades. Also, the avoidance of anthropomorphic elements in definitions and concepts remains a key objective when developing and interpreting social evolution theory (West et al., 2007a, 2011; see also Chapter 8).

Williams (1985) noted that: "[t]he only experiments that can provide evidence on evolutionary history are those set up by nature in the remote past," and explained that monitoring such natural experiments for evidence of teleonomic predictions requires the comparative method. While Darwin had only very crude hypothetical phylogenies to mentally map adaptations for their homology or analogy, our present molecular trees of life have become very precise and allow comparative analyses with unprecedented statistical power. Felsenstein (1985, 1988) was the first to realize that the statistical weight of comparative data points was dependent on the extent of co-ancestry with branches that provide other data points, which led to the development of multiple approaches for unbiased analysis based on statistically independent contrasts (Grafen, 1989; Harvey and Pagel, 1991). Since then, the field of comparative analyses has grown consistently, supported by a variety of sophisticated software packages (e.g., Huey et al., 2019). As Cronin (1991) summarized, the first independent-contrast analyses resolved an important controversy over the causation of allometry in antler size of deer, which arose from Huxley's (1931) paper that had failed to obtain an adaptive explanation. The new analyses showed that Huxley's original proximate explanation had an ultimate sexual selection complement because the overall allometry relationship appeared to consist of three regression lines with large males inseminating more females in each of

the three categories (Clutton-Brock and Harvey, 1979; Clutton-Brock et al., 1980; Clutton-Brock, 1982; Harvey and Clutton-Brock, 1983). Both antler size and male body size thus turned out to be independent adaptations for increasing individual male reproductive success (see also Harvey and Mace, 1982), illustrating the power of modern comparative analysis.

As many evolutionary biologists have done earlier, I will follow the phenotypic gambit (Grafen, 1984) and isolate the dynamics of particulate genetics in black boxes (Holm, 2019), assuming that genetic details of proximate causation will rarely be crucial for understanding the broader principles of biological organization. When I occasionally peek into these black boxes in the later empirical Chapters 6 and 7 it will be to look for consistency evidence; that is all one can realistically wish for because the complexities of proximate causation are endless. It is, however, a worthy ambition to develop a single theory for the ultimate causation of social adaptations following in the tracks that Tinbergen, Hamilton, and Williams carved out in the 1960s. A traveler on what Hamilton (1996) termed the narrow roads of gene land needs to equip herself with a head not unlike that of the Roman mythical god Janus, but with asymmetrical faces because the odds of understanding adaptation in retrospective analysis are much better than any prospective predictions can be. There is nothing puzzling in this assertion—it is analogous to reconstructing the causal history of personal wealth or bankruptcy, which is also much easier than predicting such trajectories some decades ahead. My approach will have a number of parallels with the economic and political sciences, because the neo-Darwinian synthesis that unfolded during and after the 1960s was ultimately based on the insight that there is almost no form of cooperation that does not have some inbuilt potential conflicts.

Any endeavor to understand the evolutionary stability of cooperation therefore has to consider the corrupting forces that threaten it from within, or fail to do so at major transitions in social evolution.

What I hope to achieve is an updated neo-Darwinian synthesis that allows evolutionary biology to better fulfill its scientific mission of providing the conceptual foundation for all of biology. Stearns and Medzhitov (2016) characterized this very well when they wrote that "[e]volutionary biology, like physics and chemistry, is a basic science that underpins all of biology. It is more than a specialty like genetics, biochemistry, development, or physiology. Evolution has that status because it helps to explain both the origins and the current state of everything we find in organisms, including the subjects of such special fields." In this view, understanding social adaptations from first principles is a crucial component of neo-Darwinian theory. The synthesis that this book will pursue focuses on inclusive fitness theory. I will review the theory's wide-ranging significance for explaining condition-dependent cooperation in societies of cells or multicellular individuals and will extend that gene's eye understanding to capture the origins of the major transitions in life's organizational complexity that produced the various forms of somatic altruism. I will argue that obligate cooperation always arose as irreversible ratchet clicks from simple ancestry and not by gradual emergence in societies practicing condition-dependent cooperation. This conjecture will expose a considerable amount of contemporary thinking about major transitions as unduly anthropomorphic, for only the culturally driven human major transition in organizational complexity originated from an already complex non-cultural form of society. But before embarking on the contours of an inclusive fitness theory for the origins of the major transitions, the next two chapters will review what the 20th-century neo-Darwinian synthesis actually achieved and how problematic it has been that relevant analyses of organizational complexity from the beginning of the 20th century have been ignored.

CHAPTER 2

The gene's eye view that forged a neo-Darwinian synthesis

Summary

The gene's eye view of adaptive evolution was conceptualized in the 1960s by W.D. Hamilton and G.C. Williams, inspired primarily by R.A. Fisher's theoretical genetics and David Lack's and Niko Tinbergen's empirical research. In the decades that followed, the new inclusive fitness view of individuals as optimizing agents facing trade-offs, opportunity costs, and reproductive conflicts was further developed and disseminated mainly by Robert Trivers, Richard Dawkins, John Maynard Smith, and Helena Cronin. Because neo-Darwinism originally arose in the UK and the modern synthesis was forged in the US, differences in emphasis on adaptive versus non-adaptive evolution persisted across the Atlantic throughout the 20th century. The gene's eye theory of adaptation offered ground-breaking and often counterintuitive understanding of the evolution of somatic senescence, parent–offspring conflict, clutch-size optimization, and the costs and benefits of animal aggression—shedding general light on how cooperation and conflict are shaped by natural selection. I explain the fundamental logic of Hamilton's rule as it emerged from inclusive fitness theory, using the most general notation possible, and I briefly review the principle of co-replication as a key concept for understanding how potentially conflicting social and life-history traits are maintained by natural selection. I also highlight the two major assumptions behind the gene's eye view of adaptation: optimality principles as drivers of naturally selected adaptive design and the simplification of genetics to breeding values for complex phenotypic traits. I then summarize theoretical advances and empirical research efforts that were instrumental for completing a neo-Darwinian synthesis of organismal biology towards the end of the 20th century.

> My ideas about kin selection were at last written down and submitted to a journal. I was pretty sure they were right—that is, I was sure they were correctly argued. If they were it was clear that no amount of evidence from nature would make them wrong; or, if it did, then at least for my comfort, Darwin's and Fisher's versions of evolution, perhaps along with all of Mendel's genetics, would have to come crashing down as well.
>
> **Hamilton, *My Intended Burial and Why* (2000, pp. 116–117)**

The neo-Darwinian approach based on individual natural selection was clearly European, with Wallace and Weismann as proponents (Churchill, 2015). It was therefore natural that Darwin's 100th birthday in 1909, also the 50th anniversary of the *Origin*, was celebrated in Cambridge, UK (Richmond, 2006). By that time, evolution in the sense of descent with modification had become scientifically uncontroversial, but natural selection as the universal mechanism for adaptation had not (Ghiselin, 1974; Padian, 2009). Oxford professors Edward Poulton and Ray Lankester (and Raphael Weldon until his untimely death in 1906) were among the few who consistently defended adaptation-focused Darwinism. They shaped a matching zoological research environment, of which Julian Huxley would soon become an exponent (De Beer, 1938; Provine, 1986; Morrell, 1997; Kohn, 2004; Burkhardt, 2005; Churchill, 2015). In Cambridge, however, William Bateson adhered to saltational genetics and remained averse to Darwinism, which inhibited a Cantabrigian connection between genetics and Darwinism until 1940, when R.A. Fisher was appointed to the Cambridge chair of genetics. As Richmond (2006) relates, Lankester told the 1909 jubilee gathering at Cambridge that there was no doubt that "Darwin would have been deeply interested in Mendel's results, but [that] these, although throwing light on the mechanisms concerned in hereditary transmission, were not in any way opposed to Darwin's great theoretical structure." This illustrates that many Oxford professors considered Mendelism to be a key complement of Darwinism, and not a challenge. Rather than only being concerned with discrete allelic variation at specific loci, the Oxford school was equally inspired by the biometrical aspects of particulate genetics as they were developed by Fisher (1918, 1930) to become known as quantitative genetics, a formalization of the codex approach focusing on adaptive genetic information (see Chapter 1).

Between the 1909 Cambridge Celebration and the 1959 Darwin Centennial Celebration in Chicago, adaptation-focused zoology in Oxford was consolidated via a series of chairs, including Edwin S. Goodrich, Edmund B. Ford, David Lack, and the junior positions of Arthur Cain (later in Manchester and Liverpool) and Julian Huxley, who mentored the early careers of David Lack at Oxford and of Niko Tinbergen while he was still in Leiden. Oxford zoology established a wide spectrum of research programs on birds, insects, and other invertebrates that were impregnated by conceptual thinking about adaptive traits and their explanation as products of natural selection under field conditions. Poulton studied mimicry which inspired Fisher, who later became a close collaborator of E.B. Ford, ultimately leaving a bigger scholarly footprint in Oxford than in Cambridge (De Beer, 1938; Kimler, 1986; Morrell, 1997; Kohn, 2004; Burkhardt, 2005; Anderson, 2013). By comparison to Oxford and some other UK universities, there was little interest in individual natural selection in the wild among biologists in the US until G.C. Williams' (1966a) *Adaptation and Natural Selection* appeared. As the Darwin centennial volumes (see Chapter 1) show, systematics, paleontology, and particularly population genetics dominated evolutionary biology in North America, established by Thomas H. Morgan, Hermann J. Muller, Theodosius Dobzhansky, Alfred Sturtevant, and others (Sturtevant, 1965). The prime importance of agriculture and a less traditional academic culture predisposed US academia to lead animal and plant genetics (Lerner, 1950; Allen, 1979; Pauly, 1984; Carlson, 2004) and cement a modern synthesis based on experimental population genetics (Huxley, 1942), while the UK extended ecological Darwinism into ecological genetics to study polymorphisms in field populations (Maynard Smith, 1958; Reznick and Travis, 1996).

That Darwin's theory originated in England was also no coincidence because it was inspired by natural theology (Paley, 1809; Kirby and Spence, 1818) and a distinctly British version of Enlightenment (Himmelfarb, 2004). Natural theology had gone extinct in 19th-century continental Europe (Mayr, 1983b) to be replaced by non-empirical rationalism in France going back to Descartes (Grosholz, 1991) and romanticism in Germany, which tended to deny that objective truth existed (Berlin, 1959; Watson, 2010). Why Darwin's legacy was maintained so consistently in Oxford, rather than in Cambridge, may be contingent, but historical analysis has confirmed that Poulton and Lankester were directly influenced by the first neo-Darwinian, Alfred Russell Wallace (Provine, 1983). In particular, as Kimler (1983) asserted, Poulton and Huxley kept coherent thinking about adaptation alive until Ford established his school of ecological genetics with R.A. Fisher as key collaborator. A direct connection with field ecology and biogeography was always present in the Oxford approaches to evolutionary biology, a combination that was lacking elsewhere (Ford, 1938, 1971). All five contributors to the 1959 Darwin Centennial Celebration meeting in Chicago highlighted in Chapter 1 appear to have shared or been familiar with this bottom-up natural history approach where individuals vary and compete for survival and reproductive success (Ghiselin, 1974). This included Hermann Muller, Huxley's assistant during his brief tenure at Rice University, who had published on these topics well before the Chicago meeting (e.g. Muller, 1949). However, top-down "systems" ecology, established at the same University of Chicago, took a very different perspective and would become a dominant force in the US. This would overshadow the legacy of Darwinism for decades, while often turning the logic of natural selection into a byproduct of ecology or genetics—the conversion that Williams referred to as providing a mere excuse to believe in evolution (see Chapter 1).

Conceptual divide across the Atlantic

As of ca. 1930, a "new ecology" came to dominate developments in the USA, driven by a desire to understand population regulation as a form of homeostasis (Kimler, 1986). The Chicago school of ecology played a leading role in this movement (Banks, 1985) until their concept of group-selected ecological communities was decisively challenged by Alexander J. Nicholson's (see Chapter 1) and David Lack's strict views of individual selection and the importance of density dependence (Anderson, 2013). Nicholson upset several prominent readers of his Darwin centennial contribution when he gave evidence for naturally selected traits not being good for the species—Haldane, Huxley, and Wright apparently offered comments suggesting that the individual selection logic by Darwin, Wallace, and Weismann was still controversial (Kimler, 1986). As with Gould's (1983, 1988) skeptical attitude to natural selection, the Chicago school's anthropomorphic belief in nature always working for the greater homeostatic good was politically or religiously inspired (Mitman, 1988; Dugatkin, 2006; Harman, 2010). The influence of the Chicago school petered out in the 1960s, but left signatures of ambivalence about the level at which adaptations evolve in *The Theory of Island Biogeography* (MacArthur and Wilson, 1967). The concept of r and K selection, which combines two factors, the Malthusian parameter (r) and carrying capacity (K), that are characteristics of populations rather than individuals, lingered for decades in spite of a distinct group selection bias. This emerged, for example, when Pianka (1970) endorsed MacArthur and Wilson's (1967) phrase that "as an ecologic vacuum is filled, selection will shift a population from the r- toward the K-endpoint." By this time, Hamilton and Williams had built their theories of inclusive fitness and senescence by following Fisher who considered the Malthusian parameter to be a measure of individual fitness, while K always remains an ecological population-level measure. It was no surprise therefore that the original r/K approach needed to be rephrased in terms of individual survivorship and growth to allow testable predictions (Sibly and Calow, 1985).

Diverging research traditions meant that emerging fields of evolutionary and behavioral ecology focused their research priorities differently. Evolutionary ecologists in the US were rarely explicit about the levels at which natural selection operates, while behavioral ecology as it was first developed in the UK focused primarily on adaptation by individual selection. As Kimler (1986) summarized, American ecology textbooks continued to marginalize processes related to individual fitness (see also Mitman, 1988) and came to develop a composite population biology centered on mathematical models in population ecology and population genetics (Wilson and Bossert, 1971; Collins, 1986; Levallois, 2018). Here, evolutionary biology was a mere appendix meant to keep pattern-focused ecology compatible with Sewall Wright's group selection in deme-structured populations (Kimler, 1986; Odenbaugh, 2013), and with general "levels" thinking in community ecology that merely acknowledged that heredity exists (e.g., Odum, 1971). Ambiguity with regard to the default level of selection also characterized E.O. Wilson's *Sociobiology* (1975), in contrast to the explicit gene's eye individual logic in Dawkins' (1976) *Selfish Gene*, which appeared almost simultaneously.

Given its scholarly history since the 19th century, it was almost unavoidable that Oxford zoology came to drive the European revival of neo-Darwinism, and it was the field of ethology that would play a major role in this development. The first (1952, Germany) and second (1953, Oxford, UK) International Ethological Conferences had later (1973) Nobel laureates Konrad Lorenz, Niko Tinbergen, and Karl von Frisch in prominent roles, but it was particularly Tinbergen (Figure 2.1) who would continue to promote ethology as a neo-Darwinian hypothesis-testing field throughout his career (Kruuk, 2003; Kohn, 2004; Burkhardt, 2005).

David Lack (Figure 2.1) made two important, albeit indirect, contributions to the foundation of the gene's eye view of evolution.

During World War II, he revised his monograph on Darwin's finches to concentrate on adaptive explanations (Gould, 1983; Kohn, 2004; Anderson, 2013; Provine, 1983). He made this shift after George Varley, the Hope Professor of Entomology at Oxford, had introduced him to the work of Alexander J. Nicholson, whose career-long research (see Chapter 1) had focused on density-dependent population regulation and life-history adaptation in blowflies (Anderson, 2013). In turn, Lack's work inspired G.C. Williams (1966b) (Figure 2.1) to elaborate on some of Lack's reproductive life history data for British birds—Williams mentions Lack as an inspiration in the preface of the second edition of his 1966 book (Williams, 1996). Second, David Lack was instrumental in creating a position for Niko Tinbergen to establish ethology in the Oxford Department of Zoology (Burkhardt, 2005). It was his group of Oxford ethologists that would primarily revitalize the neo-Darwinian synthesis, because American ethology had become restricted to studying the comparative psychology of laboratory animals after World War I (Pauly, 1984; Burkhardt, 2005). The gene's eye perspectives were developed to full maturity at University College London (Hamilton) (Figure 2.1) and the University of Sussex (Maynard Smith) (Kohn, 2004). As already noted in Chapter 1, John Maynard Smith's (1958) *The Theory of Evolution* book firmly supported the emphasis on individual selection and adaptation, and the second (1966) edition added incisive updates on the increasing importance of molecular biology, on how molecular approaches appeared to vindicate Weismann's views, on why the study of altruism and other social behaviors was important, and on why the maintenance of sex was poorly understood.

From Oxford and Sussex, the gene's eye perspective spread across the UK, North America, and the European continent during the 1970s and 1980s via edited volumes that defined behavioral ecology (Krebs and Davies,

Figure 2.1 The pioneers of ecological Darwinism in the early 1960s: from left to right, Niko Tinbergen (1907–1988), David Lack (1910–1973), George C. Williams (1926–2010), and William D. Hamilton (1936–2000), photographed in the period in which they wrote their seminal contributions, when based at the University of Oxford, UK (Tinbergen and Lack), Stony Brook University, New York (Williams), and University College London, UK (Hamilton). Permissions: Niko Tinbergen 1964, photographer unknown, the Tinbergen Family Archive (attempts have been made to contact all possible copyrightholders); David Lack ca. 1964, the Lack family by kind permission of the estate of Helen Muspratt; George Williams 1962, the Williams family; William Hamilton 1967, the Hamilton family with © of the British Library Board. The images of Tinbergen, Williams and Hamilton are also reprinted from *Current Biology*, Volume 26, J.J. Boomsma, Fifty years of illumination about the natural levels of adaptation, R1250-R1255, 2016, with permission from Elsevier.

1978, 1984). Also the first books by Richard Dawkins—*The Selfish Gene* (1976) and *The Extended Phenotype* (1982)—were instrumental in this process. Early dissemination in the Americas depended mainly on Richard Alexander, Eric Charnov, and Robert Trivers in North America, and former Tinbergen student Martin Moynihan as the first director of the Smithsonian Tropical Research Institute in Panama. Alexander's inspiration likely began with Konrad Lorenz's (1966) monograph *On Aggression* (Alexander and Tinkle, 1968), and Charnov and Krebs started to collaborate when they were both at the University of British Columbia (Charnov and Krebs, 1974), just before the establishment of the marginal value theorem as one of the foundational pillars of behavioral ecology (Charnov, 1976). Robert Trivers learned to think about individual natural selection from his mentor William H. Drury, a friend of David Lack who gave him a Xerox copy of the Appendix of Lack's then "in press" (1966)

book taking down Wynne-Edwards' (1962) group selection theory. This was well before Trivers met Hamilton and Williams in person (Trivers, 2015; R.L. Trivers, personal communication, June 26, 2019). John Maynard Smith, William D. Hamilton, and George Williams spent semesters at the Zoology Museum of Ann Arbor in the second half of the 1970s, and Hamilton became a tenured professor there for six years (1978–1984). Ann Arbor graduate students trained in that period would disperse to tenured professorships across the Americas in a, for that time, remarkably even balance between female and male researchers (Summers and Crespi, 2013). Thinking about the origins of adaptations also progressed more generally (e.g., Futuyma, 2013, and previous editions), but the Hamiltonian approaches to social evolution were decisive for establishing the gene's eye view of evolution.

It was in this context that Edward O. Wilson (1971) published *The Insect Societies*, a monograph mainly about the social biology of ants, bees, wasps, and termites with a final chapter promising a more encompassing approach that came with *Sociobiology* (Wilson, 1975). The book's objective was to promote "the systematic study of the biological basis of all social behavior in all organisms, including humans," claiming that all social systems could be explained by the same evolutionary "gene logic" (Levallois, 2018). This led to a third monograph, *On Human Nature* (Wilson, 1978), completing a social gradient narrative from insects to men. *Sociobiology* was heavily criticized for being a compilation of *just-so* stories (e.g., Gould and Lewontin, 1979; Segerstråle, 2000), for evading scholarly accountability (Lyne and Howe, 1990), and for trying to supersede ethology (Baerends et al., 1976; Burkhardt, 2005, pp. 462–463)—a clear provocation just after the 1973 Nobel Prize to Lorenz, Tinbergen, and Von Frisch for establishing that field internationally. In contrast, *The Selfish Gene* (Dawkins, 1976) was primarily censured for the adaptation-centered gene's eye focus (e.g., Gould and Lewontin, 1979) that it shared with Williams (1966a), but that critique never challenged the deeper genetic codex abstraction behind the gene's eye perspective (Daly, 1991; Queller, 1995). In retrospect, it seems difficult to understand how the *Sociobiology* narrative came to overshadow Jerram Brown's (1975) *The Evolution of Behavior*, a comparable book that avoided sweeping statements about the human condition. A few years later, Brown (1980) offered a critical analysis of sociobiology's terminology and definition issues that would likely have benefitted subsequent developments had that book chapter become more widely known.

It is a common misunderstanding that sociobiology and the gene's eye view of social evolution are two sides of the same coin (see e.g., Gibson, 2013). Using my Box 1.3 comparison, inclusive fitness is a deep scientific theory with an explicit greatest-common-divider focus, while sociobiology emphasized shallow analogies of the least common multiple type. As I will show below and in subsequent chapters, inclusive fitness theory developed gradually to explain a huge breath of social phenomena from first principles, while sociobiology's legacy remained a mere all-inclusive narrative of social role variation (Boomsma and Gawne, 2018). While his North American predecessor recognized the fundamental discontinuities between societies and (super)organismal colonies throughout his career (Wheeler, 1926, 1934) (see Chapter 3), Wilson incorrectly emphasized smooth social gradients, writing that "a single strong thread does indeed run from the conduct of termite colonies and turkey brotherhoods to the social behavior of man" (Wilson, 1975, p. 129). Consistent with a tradition of mostly implicit Darwinism, sociobiology demoted inclusive fitness theory right from the start, implying it was an appendix of imprecise group selection logic, a development that largely disabled the theory from delivering new testable predictions unless readers continued to focus on the original inclusive fitness and sexual selection contributions (e.g. Alcock, 2001). When the two incompatible approaches finally clashed explicitly (Nowak et al., 2010; Abbot et al., 2011), the more than 150 authors who wrote in defense of inclusive fitness theory were an even split between Europeans on one hand and Americans plus Australasians on the other, implying that the conceptual cross-Atlantic divide was finally fading away. A recent analysis (Kay et al., 2020) confirmed that mathematically explicit models such as Nowak et al. (2010) have never been able to show that altruism can evolve without relatedness in spite of claims to the contrary that keep emerging from time to time (see also Bourke, 2021).

Enter evolutionary conflicts: in life-histories, mating systems, and social evolution

In a landmark monograph, *The Ant and the Peacock* (1991), Helena Cronin explained that confusion about the organizational levels at which adaptations evolve often originates from not appreciating that adaptive traits never come for free. Recognition of opportunity costs, that is, the forgoing of benefits from possible alternative behaviors, was a major conceptual advance during the first decades after World War II. This simple notion made the quantitative evaluation of alternative behaviors in economic cost–benefit terms a natural pursuit for pioneers such as Tinbergen and Williams (see Chapter 1), and it likewise pervaded Hamilton's groundbreaking social evolution (Hamilton, 1963, 1964a, 1964b) and senescence (Hamilton, 1966) papers. For example, Williams' teleonomic analysis of adaptive function (see Box 1.2) asked why a putative adaptive trait continues to be maintained rather than being allowed to degenerate over evolutionary time when selection is relaxed (see also Lahti et al., 2009). Framing questions like this automatically necessitated inclusion of a phylogenetic perspective to "distinguish between the forces that initiated the development of an adaptation and the secondary degenerations that the adaptation, once developed, permitted" (Williams, 1966a, p. 266). This logic is obvious when realizing that the origin of flight was crucial for the adaptive radiation of the modern birds, but that island birds could lose that function secondarily and some marine birds could modify wings into paddles. The adaptiveness of these later changes requires an opportunity cost explanation based on recovering investments that no longer give returns. Interestingly, natural selection can occasionally tinker even with vestigial organs so they become engines of innovation in unexpected directions (e.g., Rajakumar et al., 2018), not unlike the scenario that Mayr (1960) imagined (see Chapter 1).

The evolutionary theory of aging aptly illustrates the need to consider opportunity costs when trying to make sense of adaptive functions and byproduct effects. The first advance by Peter Medawar postulated that aging is unavoidable because mutations expressed late in life are "invisible" for selection when reproduction is all but completed for other reasons (Medawar, 1952). Williams (1957) extended that insight by noting that relaxed selection would even accelerate aging when negatively pleiotropic genes that express deleterious phenotypes late in life positively affect early reproductive success. This conjecture was synthesized in a proximate perspective as the disposable soma theory (Kirkwood, 1977); in ultimate social evolution terms, infinite somatic repair will fail to be selected because genes that make soma "volunteer" to die at some point will prevail (Figure 2.2a). When that point is reached depends on Hamiltonian costs and benefits of somatic repair, but the implication is that somatic altruism towards a germline weakens over an organism's lifetime. Williams' gene's eye conjecture has been a very useful heuristic, but Hamilton's (1966) derivations captured the predictions more precisely (Charlesworth, 2000; Moorad et al., 2019). Interpreting senescence as a manifestation of altruism helps to see the analogy with Hamilton's (1963, 1964a, 1964b, 1972) approach to explain reproductive altruism among individuals rather than somatic cells (Box 2.1). The insight that "no life schedule, even under the most benign ecology imaginable could escape [the] forces of senescence [. . .] in the farthest reaches of almost any bizarre universe" (Hamilton, 1996, p. 90) stands tall among the gene's eye view achievements. An intriguing implication, originally suggested by Medawar, is that human reproductive aging was possibly modified by altruistic grandmothering (see Cant and Croft, 2019 for a summary account).

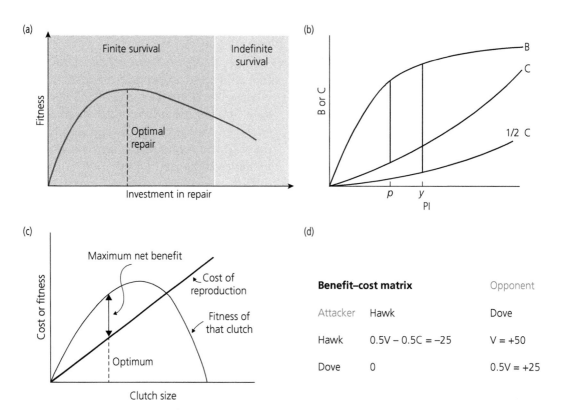

Figure 2.2 Four classic examples of social interaction benefits and their opportunity costs: the trade-off logic of somatic senescence, conflicting returns on parental investment, optimal clutch size, and the costs and benefits of aggression. (a) The principle that somatic deterioration with age is unavoidable because the extent to which selection will maintain a somatic support system will never be sufficient for indefinite survival. Reasons are that deleterious genes expressed late in life will be largely invisible for negative natural selection (Medawar, 1952) and that some of these deleterious genes will be actively maintained by positive selection when they have negatively pleiotropic effects that enhance reproductive success early in adulthood (Williams, 1957). Animal life spans may therefore vary greatly across species depending on extrinsic mortality risks early and later in life (Charnov and Schaffer, 1973), but without ever allowing indefinite survival. The image illustrates a synthesis of these complementary factors known as the disposable soma concept (Kirkwood, 1985). Redrawn from Stearns (1992). (b) Trivers' (1974) principle of parent–offspring conflict over optimal parental resource investment (PI) conjecturing that both parties have the same benefit curve (B) of decreasing returns to parental investment, but different curves of opportunity costs (C). The latter depend on each party's relatedness to the next sibling or clutch of siblings because each offspring (young; y) is 100% related to itself but only 50% to its subsequent full-sibling, whereas the investing parent (p) is equally related to all its current and future offspring. The offspring cost curve therefore has only half the slope of the parent cost curve (would be one-quarter if later offspring were to be half siblings), so that parents are selected to terminate investment at PI = p (the value that maximizes parental reproductive success), whereas offspring are selected to resist weaning until PI = y (the value that maximizes offspring reproductive success). Parent–offspring conflict is therefore expected for p < PI < y, whereas parties are predicted to agree on continued parental investment as long as PI < p and on its termination when PI > y. Redrawn from Trivers (1974). (c) Lack's (1954a) principle conjecturing that "the clutch size of each species of bird has been adapted by natural selection to correspond with the largest number of young for which the parents can, on average, provide enough food," which predicts that parental fitness is a dome-shaped function of clutch size. However, the optimal clutch size (dashed line) is often not found at the top of the curve (where the first derivative equals zero) but at a lower value where the first derivative equals c_R, the average cost of reproduction slope by which parental mortality between the present and the next clutch increases with current clutch size (Williams, 1966b). Reprinted from *Ecology: From Individuals to Ecosystems*, 4th edition, C.R. Townsend, J.L. Harper, M. Begon (2005), with permission from Wiley, after an original insight by Charnov and Krebs (1974). (d) Maynard Smith and Price's (1973) principle of individual selection restraining the benefits of escalated aggressive contest because the risk of costly injury quickly becomes

(Continued)

Understanding why aging is unavoidable was just one of a number of early milestones realized by applying the gene's eye view of evolution. Another example (Figure 2.2b) concerns parent–offspring conflict. Inspired by inclusive fitness theory, Robert Trivers published a series of papers, while still a graduate student, with trail-blazing impact. The first (Trivers, 1971) showed that cooperation can evolve without relatedness provided there are direct mutual benefits and partners have ways to know who they are loyal to, which paved the way for important later extensions by Axelrod and Hamilton (1981). The second argued that parental asymmetries in offspring care are the very foundation of sexual selection (Trivers, 1972), a contention that was later argued to need qualification (Kokko and Jennions, 2008), although these reservations were not upheld (Fromhage and Jennions, 2016). The third showed that mothers can be selected to adjust the sex ratio of their offspring depending on the variance in reproductive success of daughters versus sons (Trivers and Willard, 1973). This argument predicts condition-dependent sex ratio adjustments, particularly in species not constrained by rigid (e.g., X/Y or X/Z) sex chromosome systems, a contention that stimulated much new research (see Hardy, 2002 and Wild and West, 2007, for recent reviews). His fourth paper (Trivers, 1974) showed that parents and offspring are often in conflict over the amount of parental investment to be given or received. It made a hitherto unappreciated consequence of inclusive fitness theory explicit—that individuals within families have been selected to express disharmonious traits during specific time windows of offspring provisioning because offspring are more related to themselves than to their siblings (Figure 2.2b). These novel insights became almost immediately influential, particularly in field research of primates (e.g. Hrdy, 1980).

In *The Selfish Gene*, Richard Dawkins (1976) compellingly argued that the reproductive efforts of animals and plants are almost always shaped by potential conflicts, capitalizing on all four of the Trivers papers available at that time. In *The Extended Phenotype* (1982), he also discussed Trivers' fifth paper (Trivers and Hare, 1976), which showed that workers and queens of ants, bees, and wasps have fundamental conflicts over sex allocation and the parentage of males because haplodiploidy induces asymmetries in relatedness between nursing workers and the female and male reproductives that they raise. This fifth paper was the pinnacle of Trivers' entire series because it combined: (1) Fisher's insight that equilibrium sex ratios must be expressed in terms of parental investment rather than as offspring numbers, (2) Hamilton's insight that gene-level reproductive success can be both

---

Figure 2.2 (*Continued*)
prohibitive when the frequency of aggression increases in the population. These early applications of evolutionary game theory that used imaginary but illustrative pay-off matrices (as given here for the hawk–dove game) established the concept of the evolutionary stable strategy (ESS) to determine optimal phenotypes that would be impossible to outcompete by mutant strategies pursuing a slightly different optimum. The present simple example assumes that a victorious contest winner (V) gains 50 fitness units, while a loser gets 0, except when she pays the cost of being wounded (C) and gets −100. ESS logic appeared to be a powerful tool for generating testable predictions that were often met with accurate support, particularly in applications where frequency dependent interactions were the rule, as for example in sex ratio evolution (West, 2009). Reprinted from *An Introduction to Behavioral Ecology*, 4th edition, N.B. Davies, J.R. Krebs, S.A. West (2012), with permission from Wiley. Numerical data reprinted from *Evolution and the Theory of Games*, J. Maynard Smith (1982a), with permission from Cambridge University Press.

Box 2.1 The fundamentals of inclusive fitness theory and Hamilton's rule

The four principal types of social interaction were summarized in a two-by-two table by West et al. (2007a), emphasizing direct fitness benefits (+) or losses (−).

	Effect on recipient	
	+	−
Effect on actor		
+	Mutual benefit	Selfishness
−	Altruism	Spite

This shows that individual actions that produce mutual benefit (+/+) differ from selfish actions (+/−), which are the default because natural selection normally favors traits that benefit personal survival and reproduction in competition for limited resources (Darwin, 1859; Nicholson, 1960; Lack, 1966). Understanding the upper row in the table is therefore straightforward, but the bottom row was long enigmatic. Altruism (−/+) and its nasty mirror-image (−/−) spite (Lehmann et al., 2006) were not understood from first biological principles until Hamilton (1964a) formulated inclusive fitness theory, and captured the evolution of conditional altruism in what came to be known as Hamilton's rule (Charnov, 1977). In its simplest form, the rule stipulates that costly behaviors that reduce personal (actor) interests (e.g., survival) can only evolve when that cost (c) is less than a compensating benefit (b) gained by increasing the survival of others to whom the altruist is related (r). Hamilton (1964a) thus obtained the inequality:

$$br > c \qquad (2.1)$$

that must be fulfilled for altruistic phenotypes to be maintained by natural selection. His rule implies that benefits and costs need to be understood in gene-copy currency and differential reproductive success, confirming that "survival of the fittest" is not a useful term (see also Box 1.2).

Relatedness can often be inferred from pedigree information but is, in a more general sense, a statistical property that reflects limited offspring dispersal or general population viscosity (West et al., 2007a; Foster, 2009; Gardner et al., 2011). In terms of allelic distribution, relatedness is, by definition, zero between any two individuals sampled at random from the breeding population, so Hamilton's rule (equation 2.1) never applies to mutual benefit cooperation between non-relatives for altruism cannot evolve when Hamilton's rule reduces to $0 > c$. However, we always need Hamilton's rule to explain the second row of the table. Spite can evolve only when recipients are predictably less related to the actor than a random individual and when the benefit of harming such (relatively) negatively related individuals exceeds the cost of harming self or self's relatives (West and Gardner, 2010). Hamiltonian explanations therefore consider both altruism and spite to be self-preserving behaviors for the genes encoding these traits. In a gene's eye perspective, mutually beneficial interactions (+/+) must also be encoded by self-preserving genes, consistent with cooperation for direct benefit being just reciprocal exploitation (Herre et al., 1999). Hamilton's rule, and the inclusive fitness theory behind it, thus leave no room for any type of naturally selected social behavior to be fundamentally benign in an anthropomorphic moral sense.

To summarize, the table captures the two distinct kinds of elementary acts within a cooperative relationship, mutual benefits and altruism, under the same (+) column heading, illuminating that they are fundamentally different from the interaction categories in the second (−) column where there is never a recipient that gains fitness. However, as we will see in Chapter 5, the conceptual difference between altruism and mutual benefit may disappear when we move from interactions within the same species to actors and recipients belonging to different species, because the same conditions that promote intraspecific altruism can—under some additional assumptions—facilitate interspecific cooperation provided intraspecific altruism is maintained as well.

Box 2.1 *Continued*

It is important to acknowledge that terminology in the past has not always been consistent (for a recent review see Foster, 2009). For example, Trivers (1971) coined the term reciprocal altruism for interactions of mutual benefit (+/+) without long-term costs to any of the partners, which is inconsistent with altruism in the sense of Hamilton (1964a) and West et al. (2007a, 2007b). Further ambiguity may arise because the Hamiltonian altruism definition is about "others only" benefits (Gardner et al., 2011), while approaches that start from a group selection perspective consider average benefits to all group members, including "self" as an actor. This has led some authors to use the terms "strong" and "weak" altruism to characterize these scenarios (Foster, 2009). I will use the strong Hamiltonian definition of "others only" altruism, because most of my arguments will be about family selection, where it is key to keep subjects and objects of altruism apart.

The significance of inclusive fitness theory is that its definition of total reproductive success in joint gene-copy currency applies across all domains of social evolution from microbes to men (Foster, 2009; West and Gardner, 2010; Bourke, 2011a). Inclusive fitness (IF) can be written as the sum of *direct fitness* via own offspring and *indirect fitness* via gene copies passed on to offspring of others to whom the focal organisms is related by any positive relatedness $0 < r_x \leq 1$:

$$\text{IF} = r_o n_{own\ offspring} + r_x n_{offspring\ others} \quad (2.2)$$

where r_o is the typical relatedness to outbred offspring and n is the number of offspring raised relative to the population average. The second (indirect fitness) term is zero when a focal individual interacts with random individuals from the breeding population to whom she is not related ($r_x = 0$ by definition) or when she is unable to ever contribute to the reproductive success of others ($n_{offspring\ others} = 0$). However, even when both terms are positive, natural selection will often not establish traits that compel actors to help others reproduce, because Hamilton's inequality (equation 2.1) is not fulfilled. Then it does not pay to be a conditional altruist, and so inclusive fitness reduces to the first

term, that is, classical direct fitness. If, on the other hand, individuals or cells live in families or other kin groups of common descent, traits that mediate altruism can be positively selected via what Maynard Smith (1964) coined kin selection, that is, selection acting on the second (indirect) term in equation 2.2. In all this it is important to distinguish between the process of (kin) selection on the one hand, and inclusive fitness as the quantity expected to be maximized by natural selection (i.e., inclusive fitness as a maximand; Box 2.3) on the other (West et al., 2007a; Grafen, 2014a).

We have now seen how Hamilton's inequality (equation 2.1) relates to the inclusive fitness equation 2.2. If natural selection generally works towards maximizing IF and the first term of equation 2.2, capturing direct, non-social reproduction, is the Darwinian default, the second term will only become relevant if it yields more gene-level reproductive success per unit of investment than the first term, which is true when equation 2.1 is fulfilled. This shows that Hamilton's rule is fundamentally about the expression of conditional, context-dependent altruism because gains of indirect reproductive success can easily be frequency dependent. That is, they can decrease when more actors express altruism, so at some point it becomes best for the remaining actors not to be altruists. Another point worth making explicit is that there are always two relatedness terms in equation 2.2. This has often gone unnoticed because one of these r-terms = 1 when the cost of altruism is paid in terms of personal survival because one is related to oneself by 100%, so we obtain equation 2.1. Early tests of kin selection theory tended to address such situations, for example, when evaluating the kinship contexts that would select for individuals to give alarm call warnings against approaching predators while facing an increased probability of being killed (e.g., Sherman, 1977). However, the theory is ultimately about alternative routes of passing on gene copies to future generations, so it is often useful to make explicit the two relatednesses that scale this trade-off.

Box 2.1 *Continued*

When we write Hamilton's rule in its fully specified form, we obtain an expression showing that relatedness is always relative, a ratio between the gene-transmission efficiencies via non-offspring r_x and offspring r_o:

$$br_x > r_o c \qquad (2.3)$$

As we will see in Chapter 4, this notation is particularly useful when we compare the evolution of colonial superorganismality with the evolution of multicellular organismality, for relatedness to cell copies is 1 (100%) whereas relatedness to outbred offspring is 0.5 (50%). Alternative pathways of indirect reproductive success via full siblings (average $r = 0.5$) rather than offspring ($r = 0.5$ as well) thus makes the relatedness terms cancel, similar to when we compare clonal cell copies that disperse or adhere as somatic tissue.

In both cases, equation 2.3 converges on the same $b > c$ expression. Marshall (2015) offered a mathematically more detailed review of the verbal account provided here. He noted that relatedness is formally a ratio between two heritabilities and showed that this connection is consistent with Grafen's (1985a) geometric view of relatedness. He also confirmed that, while Hamilton's rule specifies the threshold for the expression of altruism, inclusive fitness also represents an overarching concept for understanding adaptive design. Finally, it is worth remembering that, although the theory is phrased in terms of self-preserving genes, such single genes rarely express phenotypic agency by themselves. The complex phenotypic traits that mediate social behaviors normally depend on the co-expression of hundreds of genes in tightly integrated gene regulatory networks.

direct and indirect but always needs to be weighted by relatedness between investors and recipients, and (3) his own insight that parents and offspring are often in conflict over the optimal provisioning level of siblings (Trivers, 1974). As Trivers (2015) later confirmed, his conclusion was that queens tend to win the conflict over male production, whereas workers tend to win the conflict over sex allocation. Such first-principle Darwinian predictions are always all-else-equal insights. For example, they become modified when the implicit assumption of iteroparity is violated, because workers of annual bumblebees and vespine wasps may adaptively kill their mother queen to reproduce themselves before their colony dies at a predictable time (Bourke, 1994; Loope, 2015). Overall, this fifth paper became one of the most convincing showcases for the validity of inclusive fitness theory and the gene's eye view of social adaptation (Bourke and Franks, 1995; Ratnieks et al., 2006; Meunier et al., 2008; West, 2009; Bourke, 2011a).

The study of clutch size variation, which built on a first-principle Darwinian insight

by David Lack (1948, 1954a, 1954b), became another rich test bed for the gene's eye view of adaptation in the face of reproductive conflicts. It made explicit that provisioning parents adjust the number of chicks in a clutch to maximize their total reproductive success in terms of fledglings. This logic became one of the key arguments against the idea that animals restrain their breeding effort for the benefit of the local population (Wynne-Edwards, 1962). Lack's optimal clutch size was later shown to be often affected by parental survival costs, predicting somewhat smaller optimal clutches. That insight was due to Williams (1966b) who formally partitioned Fisher's concept of reproductive value into current and future components of reproductive success that will always trade-off in iteroparous organisms (reviewed in Stearns, 1992; Grafen, 2020) (Figure 2.2c). As it emerged, unavoidable somatic senescence and optimal reproductive allocation are two sides of the same natural selection coin that are easy to grasp with gene's eye logic, while an exclusive focus on proximate mechanisms cannot clarify this connection (Partridge and

Barton, 1993). Clutch size optimization theory stimulated substantial new research on how and when bird parents are selected to adjust reproductive effort (see Birkhead et al., 2014; chapter 5), and to abandon weaker offspring or leave it to the siblings to practice self-thinning (Mock and Parker, 1998; Mock, 2004). While the percentage of clutch size variation that can be explained by Lack's and Williams' principles varies across species, long-term field and laboratory studies showed that animals can have very precise cognitive adaptations for realizing optimal clutch sizes under environmental uncertainty (e.g., Daan et al., 1990; Godfrey et al., 1991). Similar first-principle theory was developed for reproductive investment and sex allocation in plants (Charnov, 1984; De Jong and Klinkhamer, 2005).

A fourth major achievement was the application of game theory to make sense of the adaptedness of behavioral strategies. This approach calculated and interpreted life-history strategies that could not be improved upon by imaginary "mutants" who (within bounds) pursued a different strategy. It produced the concept of evolutionary stable strategies (ESS), yielding predictable equilibria between alternative phenotypes that can be assumed to have different genetic underpinning. John Maynard Smith and George Price developed the outlines of this important tool for the theory of adaptation (Maynard Smith and Price, 1973) (see also Maynard Smith, 1974; Maynard Smith and Parker, 1976), building on pioneering work in economics by John von Neumann, Oskar Morgenstern, and John Nash (Maynard Smith, 1976a, 1978b). These advances, recently reviewed and updated by McNamara and Leimar (2020), translated the dynamics of natural selection into testable predictions about resource acquisition (e.g., optimal foraging), resource allocation (e.g., sex allocation), and specific condition-dependent social behaviors that could be confirmed or

refuted by laboratory experiments and field observations. Such matches between theory and reality would lack the mathematical precision that allows astronomers to predict and reconstruct the trajectories of planets and comets, but they resemble the way in which economists develop and test models about resource acquisition and spending in human populations with even more complex social structures (e.g., Ostrom, 1990; Ostrom et al., 1999). ESS models of aggression (e.g., Maynard Smith, 1974) were particularly effective in completing the demise of naïve group selection arguments because they showed that animals could in principle restrain aggression for their own, rather than their population's, benefit as claimed by Wynne-Edwards (1962) (Figure 2.2d).

In retrospect, the achievements in the 1960s and 1970s made crucial contributions to the reinstallment of neo-Darwinism. The new gene's eye view of evolution phrased questions of adaptation in fundamentally novel ways that neither the architects of the modern synthesis nor verbal theoreticians like August Weismann and young Julian Huxley could have done (see Chapters 1 and 3). Inferred optimal life histories or social behaviors could now be accepted or rejected in a general statistical manner because predictions were accurate (i.e., unbiased), so one did not necessarily need more precise mathematical theory for special cases (see Charnov, 1976). This underlined that natural selection is a process governed by odds ratios. Selection for maximizing inclusive fitness is a necessary but not a sufficient condition for making a success of life, just as human economic prudence is necessary but not sufficient for avoiding bankruptcy. Also human financial prosperity has ultimate explanations of goal directedness, is subjected to historical contingency, and is affected by proximate constraints and the timing of such limiting factors during an individual's life span, very similar to Tinbergen's scheme (see Figure

1.1; Box 1.3). It is remarkable how successful inclusive fitness theory has been in making quantitative predictions of both straightforward and counterintuitive social traits (Abbot et al., 2011; Bourke, 2011b). This success has been particularly striking when predictions were retrospective (interpreting extant adaptive states as products of inferred selection forces in the past) rather than prospective (predicting the future dynamics of maximizing inclusive fitness). Also, in that sense, the successes of life history and inclusive fitness theory are not much different from what the economic sciences can understand backwards and predict forwards about the success of households, companies, or national states.

Towards completion of a neo-Darwinian synthesis

By the early 1980s, the gene's eye view had become part of mainstream organismal biology. The first edited volume *Behavioural Ecology: An Evolutionary Approach* by John Krebs and Nick Davies (1978) had established a field visibly inspired by Tinbergen, Hamilton, and Williams. The subsequent textbook *Introduction to Behavioural Ecology* (Krebs and Davies, 1981) and the second edited volume (1984), with 50/50 contributions by British and North American authors, accelerated hypothesis-driven studies of sexual selection and kin selection (see Sibly and Smith, 1985 for a survey of population dynamical implications). Eric Charnov's (1982) *Sex Allocation* monograph significantly extended Ronald Fisher's (1930) fundamental insights and reviewed Trivers' contributions from the 1970s (Trivers and Willard, 1973; Trivers and Hare, 1976) (see West, 2009, for a later review). Other research extended early work by Maynard Smith (1958) and Williams (1966a) on the selection forces that justify the twofold cost of meiotic sex (Williams, 1975; Maynard Smith, 1976b, 1976c, 1978c, 1984). For

roughly the next decade, considerable research focused on two contending explanations for the maintenance of sex, Darwin's tangled bank emphasizing habitat heterogeneity in space (Bell, 1982) and Van Valen's (1973) Red Queen focusing on variation in living conditions across generations (Stearns, 1987). The issue was perhaps never fully resolved but W.D. Hamilton contributed compelling evidence in favor of the Red Queen, driven by rapidly evolving parasites (Hamilton, 1980; Hamilton and Zuk, 1982; Hamilton et al., 1990), consistent with sex having direct individual fitness benefits, as also Williams had maintained (Howard and Lively, 1994; see Decaestecker et al., 2007 for a recent affirmative field study). A Dahlem conference (Markl, 1980) offered a useful summary of what had been achieved in these years and of questions remaining to be addressed.

New textbooks (Trivers, 1985; Cockburn, 1991) and the third Krebs and Davies (1991) volume deepened the impact of behavioral ecology, while studies of life-history evolution focused on trade-offs between current and future reproduction, offspring size and number, and age and size at sexual maturity, as summarized by Clutton-Brock (1988), Stearns (1992), and Reznick and Travis (1996). Other research elaborated marginal value approaches (Charnov, 1976) for optimal decision-making and evolutionary stable equilibrium (ESS) solutions for resource allocation dilemmas (Figure 2.2). After aging was argued to often depend on the ratio of externally imposed juvenile versus adult mortality (Charnov and Schaffer, 1973), experimental work confirmed that there was heritable variation for life span (Rose, 1991; Partridge and Barton, 1993).

Tests of inclusive fitness theory initially focused on vertebrates, with pioneering studies on kin recognition (Sherman, 1977; Holmes and Sherman, 1982; Fletcher and Michener, 1987), and on humans (Alexander, 1979;

Tooby and Cosmides, 1990). However, this line of thinking became somewhat compromised when it appeared that selection would not necessarily endow non-human animals with abilities to discriminate between degrees of kin, and that much of what looked like kin recognition might be due to assortment for other reasons (Grafen, 1990a) (see Maynard Smith, 1976a for a similar earlier notion). The most incisive early work on field model systems of invertebrates was the confirmation of optimal sex allocation in fig wasps differing in degree of local mate competition (Herre, 1985), consistent with Hamilton's (1967a) predictions. These matches between gene's eye predictions and real-life biology brought renewed interests in studying proximate mechanisms of patch quality perception and kin recognition, in keeping with Tinbergen's approach of asking four complementarity types of questions (see Figure 1.1; Box. 1.3).

Reproductive conflict hypotheses also drove studies of mate choice and sexual selection in new directions (Cronin, 1991). Geoff Parker showed that sexual selection often happens after insemination when ejaculates compete for fertilizing eggs (Parker, 1970; Parker et al., 1972), and William Eberhard (1985) argued that promiscuous females influence fertilization success of sperm from different males by cryptic female choice (Thornhill, 1983). Manipulation of female behavior via components of male ejaculates was another major discovery that conceptualized mating systems as biological arms races (Chapman et al., 1995). While most mammalian studies continued to address the evolution of life-history traits (e.g., Clutton-Brock et al., 1982; Clutton-Brock, 1988), many ornithological research programs refocused on sexual promiscuity after genetic marker studies began to show that extra-pair paternity was common in socially monogamous birds (Westneat, 1987; Burke, 1989; Davies, 1992; Birkhead, 1998; Langefors et al., 1998). Field experiments by Andersson (1982)

confirmed that widow bird females select mates based on secondary sexual characters, and adaptive explanations gained further force in Amotz Zahavi's (1975, 1977) handicap principle, proposing that female choice was driven by males honestly advertising costly traits that compromise survival, rather than by Fisher's (1930) non-adaptive null model of runaway selection. The handicap idea became accepted after Andrew Pomiankowski (1988) and Alan Grafen (1990b) provided mathematical support (see also Iwasa et al., 1991; Pomiankowski et al., 1991), and Hamilton and Zuk (1982) showed that selection favors males whose ornaments signal resistance to parasites. The arms race perspective uncovered constraining mechanisms as well, illustrating the power of pursuing several of Tinbergen's complementary questions simultaneously (Ryan, 2009; Lea and Ryan, 2015).

The gene's eye perspective also entered the realm of cell biology. The conflict between nuclear and cytoplasmic genes was explored (Hurst, 1991), and *Wolbachia* and other manipulative microbes were discovered (Werren et al., 1995). These studies confirmed that symbiotic co-replication often triggers potential genetic conflicts that would have required regulation if evolution were to maintain and elaborate coexistence or active cooperation. Such a stably balanced result had obviously been achieved in the last eukaryote common ancestor (LECA) which co-replicated a nuclear and a mitochondrial genome and later also evolved photosynthesizing lineages by domesticating plastid genomes. These two forms of symbiosis vastly extended the efficiency of unicellular resource acquisition and would later give rise to all macroscopically visible forms of complex life. The Hamiltonian logic of co-replication was worked out in beautiful gene's eye transparency by Cosmides and Tooby (1981) well before most confirmative empirical studies were published (Box 2.2).

Box 2.2 The principle of symbiotic co-replication

Understanding the logic of joint reproduction between independent agents started with a trailblazing paper entitled "Cytoplasmic inheritance and intragenomic conflict" by Cosmides and Tooby (1981). At that time, reproductive conflicts had been identified at the inter-individual level for parental investment (Trivers, 1974), sexual selection (Parker, 1970; Trivers, 1972), and sex ratio (Trivers and Hare, 1976), but this paper took conflict to a more fundamental level that could not be directly observed because it happens inside cells or multicellular bodies. This new approach to gene-level conflict thinking produced a substantial literature, both theoretical and empirical—see, for example, reviews by Hurst (1991), Burt and Trivers (2006), and Werren (2011). It also helped to set the stage for genomic imprinting theory, based on a special form of conflict between autosomal genes whose expression is manipulated by a parent that provided one of the alleles (reviewed by Haig, 2004). Cosmides and Tooby (1981) defined a co-replicon as a faction of replicators with 100% aligned interests operating in an arena of symbiotic reproduction with other such co-replicons. They showed that potential reproductive conflict between symbiotically coexisting co-replicons is always expected, and that the extent of conflict can be predicted by a measure of relatedness between the co-replicon agents in the same selection arena. Co-replicon conflicts may thus create disharmony within cells, multicellular bodies, and caste-differentiated colonies, but such conflicts are always over reproductive resource allocation between committed participants that have already cooperated to maximize joint resource acquisition (see below).

In any form of joint reproduction between coreplicon partners, gene-level interests are only partly aligned. The classic example is that nuclear genes tend to be passed on most efficiently when individuals invest 50/50 in sons and daughters, while mitochondrial or plastid genes are transmitted uniparentally, usually by daughters only. Coexistence of nuclear and organelle genes in cells, nested within the same multicellular body, thus implies that there are conflicts between these genomes about resource allocation towards alternative reproduce goals. However, we need to realize that this is only a potential conflict in

the sense of Ratnieks and Reeve (1992). As Cosmides and Tooby showed, the nuclear genome faction and the organelle genome(s) faction will each be under selection to manipulate the composite phenotype that has the agency to realize future reproductive effort, but whether this will leave signatures of actual conflict remains to be seen. As it happens, so many genes of mitochondrial and plastid origin have normally been lost or transferred to nuclear chromosomes over time that these organelles are often unable to influence phenotypic agency. They have thus become somatically domesticated and their reproductive priorities need to follow those of the nuclear genomes in the host cells (Harold, 2014). Hence, known cases of successful manipulation by mitochondria, that is, of potential conflict having escaped regulation so it became expressed as a lasting actual reproductive conflict, are uncommon (Couvet et al., 1998; Luo et al., 2013). This is because the so-called parliament of genes (Alexander, 1974; Leigh, 1977) is expected to evolve modifier genes to control putative agency of minority organelle factions over evolutionary time (Scott and West, 2019).

Cosmides and Tooby (1981) listed examples of cytoplasmatic genes affecting body size, growth, senescence, competitive ability, and reproductive effort. It is important to note, however, that organelle genes are never under selection to express disharmonious traits outside time windows in which reproductive allocation decisions are made. Nuclear and mitochondrial genes should thus cooperate harmoniously in resource acquisition, that is, mitochondrial genes for root functions and chloroplast genes for photosynthesis should never "disagree" with nuclear genome interests. As we will see in Chapters 4 and 5, this difference between resource acquisition and resource allocation is crucial for understanding the major evolutionary transitions in life's organizational complexity, because the origins of such transitions (e.g., the first eukaryote cell and the handful of transitions both to obligate multicellularity and colonial superorganismality) depended on the efficiency of joint resource acquisition, not on having solved all potential reproductive allocation conflicts. As it appears, maximal harmony between co-replicons

Box 2.2 *Continued*

is achieved only when (1) reproductive arenas containing different co-replicons are 100% closed (Box 2.3), so that optimizing joint resource acquisition is never subject to conflict because co-replicons are "in the same boat" for life; and (2) each co-replicon has an exclusive complementary position within the closed arena, for example, a single nuclear zygote has teamed up with a single mitochondrial and/or plastid clone. I will henceforth refer to this principle as non-redundant (genetic) informational closure (see Chapter 3).

Cosmides and Tooby (1981) identified five types of co-replicons: autosomal genes, maternally transmitted cytoplasmatic genes, paternally transmitted cytoplasmatic genes (if they exist), male sex chromosomes, and female sex chromosomes. They further argued that one should always expect natural selection to improve co-replicon performance by fixing beneficial new mutations and eliminating inferior alleles, but that potential conflict between co-replicons is otherwise inevitable and proportional to the extent to which their reproductive targets differ. The males and females of the haplodiploid social insects could be added as higher-level co-replicons even though males only survive as stored sperm inside queens. Their manipulative agency can thus only be indirect via brood-nursing workers that carry paternal genes and not via haploid brothers who only have maternal genes. The classic worker–queen conflict over sex allocation (Trivers and Hare, 1976) is therefore basically a co-replicon conflict (Queller, 1994; Boomsma, 1996). The co-replication concept can also be used to explain the evolution of anisogamy and uniparental inheritance, because there cannot be harmony in a eukaryote cell when there are two competing mitochondrial or plastid genotypes within the same cell (Frank, 1996a). Cosmides and Tooby reiterated Trivers' (1972) suggestion that the co-replicating sexes behave as two different species that practice mutual exploitation just like partners in other mutualisms that lack obligate commitment for life. However, when a novel microbe obtains a foothold as an intracellular endosymbiont—for example, *Wolbachia* (Werren et al., 1995)—it will join an existing co-replicon, almost invariably the maternally transmitted cytoplasmatic organelle one.

Cosmides and Tooby (1981) concluded that the reproductive success of co-replicons cannot be simultaneously maximized in any form of social reproduction, so that potential reproductive conflict is a fundamental aspect of unicellular and multicellular life. In their closing paragraph, they defined an individual as consisting of a number of co-replicon subsets with sufficient positive fitness correlations among them to allow collective social reproduction to work. Their paper established a gene's eye perspective on cooperation between factions of intracellular non-relatives and suggested that there could be interesting parallels between the genetic accounting required to explain cooperation for mutual benefits on the one hand and altruistic sacrifice (or for that matter spite) on the other, also at higher levels (Box 2.1). The co-replicon paper by Cosmides and Tooby (1981) thus turned out to be an important early milestone in the generalization of inclusive fitness theory, because it captured an enormous amount of organizational complexity in a straightforward narrative that is broadly applicable. A good illustration of its usefulness is a recent study that applied coreplicon logic to explore how the very first enzymatic replicators may have joined forces at the dawn of life before cell membranes existed (Levin et al., 2020).

Towards the end of the 1980s, it had become clear that the optimization assumption behind the gene's eye view was a very reasonable heuristic (Parker and Maynard Smith, 1990) even though it could be used in several ways (Box 2.3). Standard applications at the core of teleonomic adaptive dynamics (see Box 1.2) were usually about minimizing or maximizing a hypothetical fitness function (Figure 2.2), rather than about extreme state description (Sober, 1991). Recently, however, Frank (2013b) offered a useful distinction between comparative dynamics and comparative statics in inclusive fitness theory. In Chapter 4, I will develop a static kin selection conceptualization of the origins of the major transitions in organizational complexity that is based on an extreme state argument.

By the 1990s, the gene's eye study of social evolution had become a successful multifaceted field that had adopted experimental approaches and new phylogenetically based comparative methods for unbiased hypothesis testing (see Chapter 1). The rapid growth of two new learned societies underlined that neo-Darwinism had been reinvigorated with the potential to synergistically coexist and interact with the earlier established fields of population and quantitative genetics: the *International Society for Behavioral Ecology* (ISBE; founded in 1986) and the *European Society for Evolutionary Biology* (ESEB; founded in 1987) (Stearns, 2008). These increasingly global communities used both genetic and phenotypic evolution approaches, similar to their older North American counterparts, the *Society for the Study of Evolution* (SSE) and the *American Society of Naturalists* (ASN). However, a difference in relative emphasis on population genetics and phylogeny reconstruction (dominant in SSE) and adaptive phenotypic evolution (more important in ESEB) continued to exist. These discrepancies almost certainly reflect historical contingencies related to the modern synthesis (Huxley, 1942) having been driven by North American population geneticists while the British traditions of ecological genetics and behavioral ecology remained more influential in Europe. Up to this point in time, the neo-Darwinian state of the art was ably summarized in Helena Cronin's (1991) book *The Ant and the Peacock*. She emphasized how very different the trajectories of evolutionary studies of altruism and mate choice had

been in questions asked, but also made clear that these seemingly disparate fields were complementary parts of the same legacy of 19th-century insights by Darwin, Wallace, and Weismann.

Another significant achievement during the 1990s was the insight that ectosymbiotic cooperation between unrelated agents had to be mutually beneficial to remain evolutionary stable (Herre et al., 1999; Sachs et al., 2004) and not decay into unilateral parasitism (Sachs and Simms, 2006). Cooperation between species was re-conceptualized as mutual exploitation and the extent to which one partner benefitted more than the other could thus be characterized as "virulence" (Frank, 1994, 1996a, 1996b). These new insights ultimately derived from pioneering game theory approaches (Axelrod and Hamilton, 1981) and produced the first glimpses of a social evolution theory that extended towards mutualisms (Frank, 1998; Foster and Wenseleers, 2006). These explorations confirmed the omnipresence of potential conflict in social arrangements while also emphasizing that many such potential conflicts do not translate into much actual conflict because regulating mechanisms have evolved to secure the interests of the majority coreplicon (Box 2.2). The fundamental difference between potential and realized conflict was worked out in considerable conceptual detail for the social insects (Ratnieks and Reeve, 1992) (see Chapter 4) and was applied successfully in evolutionary research on mutualisms (e.g., West et al., 2002; Herre et al., 2008) (see Chapter 5). These studies often showed that one had to experimentally manipulate phylogenetically ancient mutualisms to make the potential conflicts become apparent because the most efficient sanctions against cheating are those that normally reduce cheating tendencies to negligible frequencies. Confusion between the likelihood of potential conflict and the rare expression of actual conflict has

Box 2.3 Optimality, formal Darwinism, the phenotypic gambit, and quantitative genetics

The most thorough discussion of the merits and limitations of optimality approaches that I have come across appeared as a target review (Schoemaker, 1991) in *Behavioral and Brain Sciences* followed by a series of open peer commentaries. The opening review emphasized how the heuristic of optimality has been successfully applied to explain utility maximization in the economic sciences, least effort principles in physics, entropy in chemistry, and adaptive design of forms and functions. This success is remarkable in spite of possible pitfalls relating to the selective pursuit of confirmational evidence and possible confusion between prediction and explanation. Three commentaries by biologists are particularly relevant for this book. First, Crow (1991) showed that Fisher's fundamental theorem of natural selection implicitly assumes that evolving populations follow an optimal path, a contention earlier proven by Svirezhev (1972) who derived "that natural selection operates in such a way that the path followed when a population changes from one state to another is the one that minimizes the total genetic variance over the path. Evolution [thus] gets the biggest fitness bang for the variance buck." He reiterated this argument a decade later (Crow, 2002), albeit with the remark that theory becomes less straightforward for multiple loci despite attempts to make ends meet (Ewens, 1992). As I will elaborate further down, the putative connections between formal population genetics and optimization principles were developed further by Alan Grafen (see also Charlesworth, 1990; Bull and Wang, 2010; Gardner and Welch, 2011; Marshall, 2015).

Second, following in Williams' (1966a) footsteps, Daly (1991) argued that natural selection is the very reason that cells and organisms with apparently goal-directed purposes exist, and that "[a]daptationism, the paradigm that views organisms as complex adaptive machines whose parts have adaptive functions subsidiary to the fitness-promoting function as a whole, is today about as basic to biology as the atomic theory is to chemistry." He concluded that Gould and Lewontin's (1979) call for an alternative paradigm has failed to impress practicing biologists both because

adaptationism is successful (in predicting how wild organisms organize their lives) and conceptually well founded, and because its critics have no alternative research program to offer. Although Daly's argument is correct, it seems hardly useful to polemically insist on being an adaptationist, because it distracts from the necessity to pursue the full spectrum of complementary Tinbergen questions (see Figure 1.1; Box 1.3). In Tinbergen's (1965) own words, "[w]hen the relations of an animal with all aspects of its environment are studied, numerous interactions between selection pressures are discovered that have led to compromises, and this fact should encourage us to look for indirect selection effects." However, he also added that dismissing the need to clarify the complex survival value aspects of a behavioral trait by suggesting it may just be due to genetic drift or non-adaptive pleiotropy "amounts to a refusal to investigate." Openness to complementary answers, provided adaptation is always one of the targets, remains a fundamental and obliging characteristic of Tinbergen's legacy.

Third, Sober (1991) argued that minimality (or maximality) concepts may refer to optimality in three different ways: (1) they may be extreme state descriptions, that is, matter-of-fact statements that the value of some parameter is as low or high as it can get in a particular system; (2) they may refer to ongoing processes of minimizing (or maximizing) some quantity; and (3) they may be methodological recommendations for preferring a particular hypothesis over its alternatives when it matches some optimality/minimality/maximality criterion better. For example, Fermat's principle of least action of a light ray travelling from A to B fits category 1, but not categories 2 and 3, and Fisher's fundamental theorem or the inevitable monotonic increase in entropy in a closed physical system fits category 2, but has no bearing on 1 and 3. Finally, the principle of parsimony, also known as Occam's razor, represents the pursuit of optimality category 3, but does not necessarily refer to an extreme state as in category 1 or to a mini/maximization process as in category 2. While phenomena fitting

Box 2.3 *Continued*

categories 1 and 2 are empirically verifiable as being true or untrue, parsimony principles are based on plausibility—although they are a crucial part of the scientific method in general and of the teleonomic approach in particular (see Box 1.2). The conclusion was that principles of design by natural selection, of maxi/minimization of parameters, and of efficiency of evolutionary change (optimalization given a set of extrinsic and intrinsic constraints) are different but interconnected conceptually.

An encompassing effort to prove that organisms should appear as if designed by natural selection for maximizing inclusive fitness has been made by Alan Grafen in his "formal Darwinism project" (Grafen, 2006, 2009), a logical sequel to his phenotypic gambit (Grafen, 1984). This has offered ways to formalize behavioral ecology reasoning into mathematics and towards population genetics (Grafen, 2007, 2014a, 2014b), a field that has traditionally been skeptical to the idea of fitness maximization (Lehmann and Rousset, 2014; Lessard and Ewens, 2019). Grafen (2007, 2014a) identified that an evolutionary optimization program has three elements: (1) *phenotypes* under the control of optimizing organismal agents that express them with specific values; (2) *feasibility (constraint) sets*, that is, the total ranges of values available to phenotypes, relative to all possible values; and (3) *maximands*, the real-phenotype-value functions within feasibility sets that define the success (inclusive fitness gains) of each possible choice as specified by Price's (1972) equation (Hamilton, 1970, 1975). In models based on these principles, individuals manage to solve optimization challenges if their phenotypes achieve the highest feasible maximand value (Grafen, 2014a). This conclusion is consistent with Steve Frank's (1998, pp. 30–32) assessment, referring to Grafen (1991), that "the phenotypic agency approach is the only viable method for study of social evolution," because social instincts or life-history syndromes are conserved while individuals and their populations turn over rapidly.

Important practical benefits of the formal Darwinism approach are that there now is formal mathematical support for fitness, specified as a quantitative individual trait, to be the same for all genetic loci within a co-replicon (Batty et al., 2014), for deep connections with Fisher's fundamental theorem of natural selection, and for the correctness of Darwin's intuition that natural selection is fundamentally an improvement process (Crewe et al., 2018; Grafen, 2019). This implies that an individual can be conceptualized as expressing its agency as a Bayesian optimizer, possessing a known prior provided by natural selection in the past and sampled from the population gene pool in haploid or diploid form. The individual's agency then consists in optimal updating of the likelihood of events as far as she can draw and process information from the environment via learning or by expressing the right plastic phenotype. The mathematics also confirmed that inclusive fitness is maximized when social interactions are additive (and close to maximized when this is a reasonable approximation). Finally, the formal Darwinism project reinstated arithmetic mean fitness as the key maximand (Grafen, 2020), consistent with Fisher (1930), and abandoned arguments in favor of geometric mean fitness under uncertainty in time (Philippi and Seger, 1989). The formal Darwinism project has provided significant support to the heuristic teleonomic approach behind the neo-Darwinian synthesis. Lehmann and Rousset (2020) recently corroborated Grafen's conclusions in a standard population genetics model combined with explicit game theory formalism, while also showing that the weak selection assumption is key (see also Lehtonen, 2020).

Grafen's formal Darwinism program goes a long way justifying Sober's category 2 optimization when studying adaptations conjectured to be shaped by kin selection. The same applies for a more general spectrum of questions about the sufficiency of inclusive fitness incentives for designing adaptive behavior and life histories (Grafen, 2009). This extends to the general teleonomic theory of adaptation built on these assumptions, which simplifies genetics to the additive component of heritable variation and in practice refers all nonadditive effects of inheritance to Tinbergen's complementary proximate mechanism quadrat (see Figure 1.1; Box 1.3). Thus, as far as genetic arguments are used in the adaptation quadrat they are either statistical (i.e., biometrical) and thus based on quantitative genetics (see below), or they

Box 2.3 *Continued*

address the invasion conditions of a single gene mutant, a useful technique to test whether an adaptive state (ESS) derived from optimization principles is in fact reachable (Christiansen, 1991). Partitioning of additive and nonadditive components of overall genetic variation acknowledges that limited population size, geographic population structure, genetic (over)dominance, and epistasis or pleiotropy may prevent the most adaptive states from being achieved. It also accepts that these genetic details will almost never actually become known (Grafen, 1984). It is therefore ultimately an empirical question how successful abstract teleonomic theory is in explaining nontrivial adaptations in natural populations relative to alternative hypotheses that specific, more precise but less general, genetic modeling approaches might generate (Frank, 2013a, 2013b).

Another commonly used argument in favor of Grafen's *phenotypic gambit* approach is that biologically interesting phenotypes almost invariably consist of highly complex traits no matter whether they are expressed conditionally or unconditionally. This implies that they are encoded by many different genes, of which most if not all contribute with very small effects. Both social and life-history traits are therefore best considered as quantitative or metric characters, because our understanding of their genetics depends on measurement of phenotypic effects rather than on counting gene frequencies (Hill and Mackay, 2004), that is, on the information (codex) concept of genetics rather than on explicit material alleles. For example, body size has a significant heritable component while also being dependent on environmental growth conditions, but no specific genes can normally be singled out as responsible for the overall genetic variation in body size in any breeding population. It remains surprising that the status of biometrical quantitative genetics in evolutionary biology is almost universally considered to be lower than formal mathematical modeling based on rather few specific and equally hypothetical material gene loci. As we will see in Chapter 4, conceptual teleonomic approaches based on the Price equation gain considerable causal weight when formal gene effects are substituted by breeding

values for complex traits. These applications focus directly on the adaptive information that genotypes sample from the population gene pool (see Box 1.2). Even before any forms of particulate genetics were known, Darwin (1859) used the quantitative intuitions of animal and plant breeders to show that rapid responses to selection are possible, and more precise genetics rarely overturned that logic.

The most authoritative classic textbook (Falconer, 1960; Falconer and Mackay, 1996) aptly illustrated that quantitative genetics is based on single-gene genetics and becomes practically operational because it partitions the variance of metric phenotypic traits into fractions that are (1) genetic and responding to selection in a predictable manner; (2) genetic and not responding to selection or doing so in an unpredictable fashion; and (3) environmental effects emanating from other genes in the genome (epistasis), specific social environments, or external abiotic and biotic factors, a hierarchy consistent with Williams' (1966a, p. 259) partitioning of environments. As Hill and Mackay (2004) wrote: "The importance of this branch of genetics need hardly be stressed; most of the characters of economic value to plant and animal breeders are metric characters, and most of the changes concerned in micro-evolution are changes of metric characters. It is therefore in this branch that genetics has its most important application to practical problems and also its most direct bearing on evolutionary theory." This includes inclusive fitness theory, which is fundamentally based on variance and covariance arguments via the Price equation (Gardner and Welch, 2011). However, all this applies to the multicellular lineages with permanent somatic cell differentiation (plants, animals, fungi, some algae) and not necessarily to unicellular organisms. At the unicellular levels of organizational complexity, it may be more compelling to connect social phenotypes with variation at specific gene loci (see also Chapter 7).

A key parameter in quantitative genetics is the narrow sense heritability, often expressed as h^2 in what is generally known as the breeders' equation, an elegantly simple mathematical expression:

Box 2.3 *Continued*

$$R = h^2S \qquad (2.4)$$

This relationship stipulates that the response to selection of any complex phenotypic trait (R) is equal to the product of the selection differential S (the difference between the individuals selected to be parents of the next generation and the current population mean), and the narrow sense heritability (the predictable part of heritability based on V_A, the additive fraction of the total genetic variation) of the trait (Falconer, 1960; Falconer and Mackay, 1996). Although the additive genetic variance does not capture all aspects of the genetic underpinning of a trait and may differ in magnitude across environments (e.g., Feldman, 1992; Wade, 1992), it is the important part for understanding the direction of adaptive evolution, as

has been amply demonstrated in animal and plant breeding programs. As Grafen (1984, 2009) noted, there may be situations in which one or a few genes have disproportional effects on a quantitative trait. Such genes or gene clusters are generally referred to as quantitative trait loci (QTLs), and mapping them thus represents a middle ground between single-gene population genetics and quantitative genetics (Falconer and Mackay, 1996). If one's objective is to understand adaptation per se, and the rate of evolutionary change is of secondary importance, the phenotypic gambit and its quantitative genetics underpinning usually produce accurate predictions that are testable, which is not often the case for more precise population genetic models.

sometimes meant that ecologists distrusted gene's eye theory because they could not find evidence for abundant cheating in the field (e.g., Ferriere et al., 2002; Jones et al., 2015). However, that is actually what the theory predicts.

From the 1980s onwards, research on social insects also became a major engine for investigating the usefulness of the gene's eye view of evolution. Particularly the haplodiploid ants, bees, and wasps offered attractive testing grounds for inclusive fitness predictions because asymmetric relatedness of unmated nursing helpers/workers towards younger female and male siblings increases or decreases the potential to pass on genes of these reproductive altruists to future generations. Inspired by the seminal paper of Trivers and Hare (1976) and facilitated by co-dominant allozyme and later microsatellite markers, explicit tests of predicted biases in sex allocation and the control of worker egg-laying provided consistent support for kin selection theory (e.g., Bourke et al., 1988;

Queller et al., 1988; van der Have et al., 1988; Ratnieks and Visscher, 1989; Strassmann et al., 1989; Boomsma and Grafen, 1990), and led to influential monographs (Bourke and Franks, 1995; Crozier and Pamilo, 1996). Studies of socially breeding birds and mammals yielded similarly encouraging evidence for the validity of the gene's eye view (Wrege and Emlen, 1987; Westneat, 1990; Pemberton et al., 1992), developments that were adequately tracked in the later Krebs and Davies edited volumes (1991, 1997) which now also had chapter authors from continental Europe. The first volume of Hamilton's (1996) autobiography *Narrow Roads of Gene Land* summarized the history and state of the art towards the turn of the century, and a monograph by Steve Frank (1998) further generalized inclusive fitness theory. As the opening quote of this chapter illustrates, Hamilton always remained confident it had to be so—a clear example of first-principle theory in biology being at its best when it offers stringent predictions and actively solicits Popperian refutation attempts.

A relatively late development that only gradually became influential was David Haig's genomic imprinting theory. This concept is an extension of inclusive fitness theory to capture parent-of-origin conflicts at the level of single genes already committed to the same diploid body of an animal or vascular plant (Haig and Westoby, 1989; Moore and Haig, 1991; Haig, 1997). As in earlier inclusive fitness theory, Haig's objective was to explain potential gene-level conflicts, but this time between paternal and maternal genes within the same coreplicon rather than between parents, offspring, or siblings. At first sight, the pursuit of self-preserving gene agendas within a single clonal body appeared astounding, but consistent evidence for the existence and agency of such genes has accumulated over the years. As a crude approximation, imprinting conflicts make sense only when reproductive resource provisioning of embryos happens after rather than before fertilization. When maternal resources are committed to "naked zygotes" (as in mammals and angiosperm plants) rather than to completely provisioned eggs (as in birds, reptiles, and amphibians), selection may favor paternal genes to encode phenotypic agency for demanding more resources from the maternal body than would otherwise have been in the mother's interest to provide. Placentas nourishing mammalian fetuses and triploid endosperm provisioning angiosperm seeds were identified as specific locations where imprinting conflicts can be expected, which has now been amply confirmed (Haig, 2004, 2010). It is increasingly becoming clear that imprinted genes are also expressed in the mammalian and human brain, where they may affect the behavior of dependent offspring in directions that favor either paternal or maternal fitness interests (e.g., Badcock and Crespi, 2008; Haig, 2011; Byars et al., 2014).

General imprinting theory was later extended to cover sex-specific differences in dispersal (i.e., different degrees of population viscosity for males and females), which generates continuous rather than discrete variation in relatedness asymmetry and yields intriguing predictions that apply to most multicellular organisms (Gardner and Ubeda, 2017). Other Hamiltonian insights on potential conflicts associated with meiotic drive, horizontally acquired genetic elements, and vertically transmitted organelles or less permanently domesticated bacterial symbionts were also developed further (Moran, 1996; Burt and Trivers, 2006; Werren, 2011; Haig, 2016). These advances, and the empirical findings supporting them, confirmed that gene-level conflicts are a fundamental aspect of life, even within single and stably developing bodies of animals and plants. Imprinting conflicts were also predicted in the haplodiploid social insects with their asymmetric relatedness of workers towards male and female reproductives (Queller, 2003) (but see Kronauer, 2008), although evidence for this claim mostly remains indirect or ambiguous (Kocher et al., 2015; Remnant et al., 2016; Howe et al., 2020; Oldroyd and Yagound, 2021). By the turn of the 21st century, two decades of research had shown that colonies of social insects with permanent, morphologically differentiated castes have many potential conflicts but that only a few of them are visible to the researcher observing their day-to-day social dynamics. The same is true for vascular plants where mitochondrial genes have no interest in male gametes (pollen) but where few examples of evolutionary stable and lasting cytoplasmatic male sterility are known (e.g., Luo et al., 2013) (Box 2.2).

Although it had taken the entire last third of the 20th century, the inclusive fitness-based view of adaptive evolution had fulfilled its promise of forging a neo-Darwinian synthesis, to such an extent that it revived the notion that Darwin's *Origin* had really been about the origin of adaptations rather than about the origin of species (Charlesworth and Charlesworth, 2009; Coyne, 2009, p. 184). Where the modern synthesis of the mid 20th century had been a constriction towards purity of core (Dobzhansky, 1958; Provine, 1988), the neo-Darwinian synthesis had been positively self-reinforcing in Monod's (1975) sense of a good theory always being "much wider and much richer than even the inventor of the idea may know at his time" (see Chapter 1). Meanwhile, the more restricted modern synthesis continued to be successful in its precise mathematical analyses of Mendelian gene frequency dynamics, but its ambiguous stance towards, or outright neglect of, complex adaptations persisted. In Huxley's (1916), Woodger's (1929), Pittendrigh's (1958), Mayr's (1961), Tinbergen's (1963), and Williams' (1966a) terms, these two syntheses therefore represent different complementary spectra of understanding. Tinbergen captured this duality best, not only by keeping the proximate and the ultimate apart, but also by realizing that teleonomic explanations in the adaptation quadrat will always have been affected by contingent interruptions in the phylogeny quadrat, and that causal explanations of genetic mechanism may be ontogenetically constrained in the development quadrat (see Chapter 1). These developments underlined that an encompassing gene's eye theory is essential to do justice to Darwin's vision of natural selection being the only organizing principle towards adaptive design, as presented most convincingly in the first edition of the *Origin* (Darwin, 1859). I will return to the deeper principles of how genes and environments interact to realize adaptive organizational order in Chapter 8.

My summary of what the gene's eye view achieved is obviously incomplete. A broader spectrum of topics could have been covered and I could have roamed further beyond the confines of the UK and the anglophone Americas to highlight more 20th century research in continental Europe, Asia and South America had space constraints not existed. However, it remains quite striking how dominant was the role played by the native English-speaking nations on both sides of the Atlantic in developing a neo-Darwinian synthesis up to the turn of the 21st century. Ending this chapter at the millennial transition is somewhat arbitrary but defensible because it is questionable whether work in the last two decades can be covered as history—much of it is ongoing discourse that will contribute to my review of recent research in Chapters 4–7. The turn of the millennium also increasingly introduced big data and advanced computation into organismal biology, which has meant that a new generation of evolutionary biologists grew up with less training in theory based on first Darwinian principles and more focus on proximate causation and methodology. I hope that this chapter's summary of how the gene's eye view became established and supported will remind them of the logical coherence and continued relevance of neo-Darwinian theory. However, before I turn to 21st-century research and my inclusive fitness conceptualization of the origins of the major transitions in evolution, we need a third history chapter to summarize how the five-decade-long eclipse of neo-Darwinism between Huxley (1912) and Williams (1966a) affected organismal biologists' thinking about progress in evolution. Why is it that this very idea makes many evolutionary biologists uneasy and how does that question relate to how we distinguish between organisms and societies?

CHAPTER 3

A reappraisal of progress in evolution

Summary

Physics has its universal law of regress, the perpetual increase of disorder formalized by entropy, but biology lacks a generally accepted principle of progress even though it is obvious that life's organizational complexity has increased over evolutionary time. The concept of evolutionary progress has been debated from before Darwin's days, misused for political purposes, and remained controversial. As many have noted, organismal lineages do not necessarily gain organizational complexity over time, but Ronald Fisher's fundamental theorem established that progress should be real, albeit always relative, provided we only consider adaptive evolutionary change shaped by natural selection. Progress is easier to grasp when focusing on the major transitions in evolution (MTEs), the historical ratchet clicks towards higher organizational complexity. The MTE concept is generally assumed to have arisen in the last decade of the 20th century, but goes back to William Morton Wheeler and, particularly, Julian Huxley in the first years of that century. I review these neglected insights to show that they were correctly and transparently argued, that they remained implicitly appreciated in their original form until the 1960s, and that they were, and remain, fully compatible with the neo-Darwinian perspective. In comparison, modern discussions of evolutionary progress and actual research on the MTEs have not been very productive. I follow Huxley and Bonner in considering the four canonical levels of nonhuman hierarchical organization and show how they define life's domains of social evolution and the clicks of the MTE ratchet between them. I then conjecture how these MTE origins can be explained by a single set of genetic information conditions, a basis from which I will develop the contours of a kin selection theory for organismality in the next two chapters.

> To the average man it will appear indisputable that a man is higher than a worm or a polyp, an insect higher than a protozoan, even if he cannot exactly define in what resides this highness or lowness of organic types.
>
> **Huxley, *Progress, Biological and Other* (1923, p. 10)**

Philosophical thinking about natural progress became explicit in the European Renaissance (Huxley, 1923b; Russell, 1945; Hull, 1988b), but opinions on how to understand progress came to diverge significantly during the 18th-century Enlightenment (Berlin, 1959; Himmelfarb, 2004). A detailed historical analysis of these developments would go beyond the subject area of this book, but it is relevant to reassert Hull's (1988b) observation that global optimism about human progress culminated in *la belle epoque* (1880–1914), came to a rather abrupt end during World War I, and was revived as a cultural paradigm in the 1960s.

Domains and Major Transitions of Social Evolution. Jacobus J. Boomsma, Oxford University Press. © Jacobus J. Boomsma (2022).
DOI: 10.1093/oso/9780198746171.003.0003

The approximately five-decade-long eclipse of global belief in progress roughly coincided with the intellectual demise of neo-Darwinian understanding of naturally selected biological adaptation as a relative improvement process (see Box 2.3). The 1959 Centennial Celebration of Darwin's *Origin* shunned explicit arguments of how natural selection actually works, as Julian Huxley's (1942) book *Evolution: The Modern Synthesis* had done as well. That monograph disseminated a large amount of new knowledge on particulate genetics, speciation, unspecified adaptation, and progressive evolutionary trends, but Darwin's and Wallace's foundational insight that adaptations are shaped by natural selection at the individual level was neglected (see also Cronin, 1991). In practical matter of fact, Huxley's (1942, 2010) *Modern Synthesis* ended up burying the neo-Darwinism of his youth under a narrative of Mendelism and overall change that could be fitted into a secular framework of intellectual and methodological progress (see also Huxley, 1960a). Compared to his first book (Huxley, 1912), his perception of evolutionary change had shifted almost completely from logical theory to proximate mechanism.

This chapter will argue that Huxley came close to a correct understanding of evolutionary progress in 1912. However, his narrative style in later public essays loosened (Huxley, 1923a, 1923b), his emphasis on indiscriminate progress gradually increased, and his revisiting of the subject in *The Biological Basis of Individuality* (Huxley, 1926) lacked the logical coherence of his first book, *The Individual in the Animal Kingdom* (Huxley, 1912). Across the Atlantic, William Morton Wheeler also stopped disseminating his neo-Darwinian superorganism concept for the caste differentiated social insect colonies, developed around the same time (Wheeler, 1911). Reviewing the reasons behind these shifting priorities is once more beyond the scope of this book, but changes in sociopolitical thinking may well

have played a role, possibly driven by the anthropomorphic notion that nature's potential to teach a war-torn humanity harmony rather than strife was worth exploration (Dugatkin, 2006; Harman, 2010). Some leading biologists had become Marxists which, in Russell's (1945) analysis, tends to blur the distinction between subject and object, while others must have felt that biology should primarily solve immediately pressing practical questions (see also Chapter 2). In that climate, the pursuit of abstract Darwinian ideas without having many tools to test hypotheses may have seemed an unattractive career path for young biologists. In the next section, I will show how Huxley and Wheeler excelled in first-principle understanding before World War I but lost their neo-Darwinian edge when they, respectively, promoted and reluctantly went along with a softening of ethology, the discipline they co-founded as a distinct branch of Darwinism. These developments may explain in part why less precise and more inclusive definitions of adaptation were adopted, a shift not unlike the kind of changes that Darwin made in later editions of the *Origin* to the detriment of logical coherence.

I will maintain that a meaningful concept of evolutionary progress requires using Williams' (1966a) strictly hierarchical argumentation (see Box 1.2 and Box 1.4), based on Pittendrigh's (1958) definition of adaptive organization. Without such a first-principle teleonomic theory of adaptation, biology fails to be an independent natural science entitled to its own philosophical domain, for "natural selection [is] the only theory that a biologist needs in addition to those of the physical scientist" (Williams, 1985, p. 20). Biology should therefore cherish that focus and for deeper reasons than those brought forward by Ernst Mayr (see Chapter 1). Doing that requires teleonomic theory to address both the dynamics and the statics of evolutionary progress. Dynamic theory will always imply that naturally selected progress (or regress) is relative, as in Fisher's (1930) fundamental theorem and in host–parasite arms races (e.g., Hamilton et al., 1990). However, a complementary static theory of stepwise irreversible progress in organizational complexity was never developed because Huxley's (1912) thinking about the interplay between continuous dynamical change in individuality and discontinuous change in hierarchical organization was never properly re-evaluated (see Huxley, 2022). In the remainder

of this book, I will aim to complete Huxley's early career project by arguing that all non-human major transitions in evolution (MTEs) are strictly defined in terms of nested organizational complexity, and as having originated by uniquely novel, higher-level and lifetime closure around non-redundant carriers of genetic information. At this point, this phrasing may appear unorthodox. I therefore explain these terms, and others frequently used in this book, in Table 3.1 and I will further unpack

Table 3.1 Glossary of frequently used terms

Adaptive design A recognizable phenotypic state that can be assumed, subject to formal testing, to have been produced by a Fisherian process of gradual relative improvement by natural selection; the resulting match with the demands of the typical environment is achieved by maximizing the inclusive fitness of individual organisms.

Agency The dynamics of growth and behavior that organisms express to continue maximizing inclusive fitness under prevailing constraints via the perception and processing of environmental signals and cues to guide acquisition and allocation of resources.

Closure A key characteristic of any individual organism, no matter its level of organizational complexity, defined as the integrity throughout life of the genetic information obtained at its foundation; this integrity is maintained by physical closure (cell membranes) or functional closure via lifetime commitment between a pair of founding gametes or parents.

Continuous versus discontinuous change in organizational complexity Continuous change is gradual and possibly reversible while discontinuous change is stepwise and irreversible. The restriction to naturally selected organizational complexity here means that there is no relation to Dollo's (1893) morphological law.

Domestication The enforcement of hierarchical interdependence, either asymmetrically when hosts maintain societies of redundant symbionts, or symmetrically when a partnership is non-redundant and a previously independent level of agency is subjugated into somatic service during an irreversible major transition towards higher organizational complexity.

First-principle explanation An argument based on the principle of adaptation through natural selection—a Darwinian axiom that allows development of predictive and testable theory for explaining the agency and organizational complexity of closed individual organisms and of the open societies they may form; such explanations need to be phrased in inclusive fitness terms.

Gene regulatory network (GRN) The hypothetical, but increasingly measurable, interactions by which hundreds of differentially expressed allelic effects determine organismal development and trait-based agency; GRNs are the genetic basis of homologous traits because they will remain at least partially conserved within lineages; an organizational complexity GRN, mediating a higher level of division of labor via symmetric mutual domestication of (a) previous level(s) of organismal agency, can be conjectured to uniquely define each major transition in evolution (MTE).

Gene's eye view of adaptive evolution The perspective that allelic variants of genes are the only carriers of adaptive information that naturally persist through evolutionary time, so that their population-level gene pools preserve this information when phenotypes are subjected to natural selection; evolutionary theory thus needs to be formulated at the gene level to understand the adaptedness of complex phenotypes determined by the combined effects of many such genes in interaction with many environmental factors.

Horizontal gene transfer (HGT) Genes that organisms acquire by "infection" rather than by (vertical) descent, a phenomenon that has been common over evolutionary time, albeit decreasing in frequency with increasing organizational complexity of organisms; HGT affects population gene frequencies and mediates many adaptive responses, but can be ignored in organismal genetic information arguments based on closed lifetime commitment between cellular organelles, gametes, or monogamous parents.

continued

Table 3.1 *Continued*

Inclusive fitness theory The gene's eye theory, developed by W.D. Hamilton and elaborated by many others, specifying that self-preserving alleles affecting phenotypic social traits may maximize their representation in future generations by reducing their carriers' direct reproductive success while expressing cooperative behavior towards relatives possessing copies of the same alleles; natural selection shapes organismal phenotypes to appear as if designed for maximizing inclusive fitness, the sum of direct and indirect gene-level reproductive success.

Information A characteristic of any population-level gene pool that has accumulated genome-wide allelic variation from individuals that adapted to the population's environment over past generations because of natural selection; haploid cells, diploid zygotes, or founding parents of superorganismal colonies obtain a single, usually recombined, sample of this information about past adaptations and are subject to phenotypic natural selection based on that unique informational endowment.

Kin selection The process by which gene-level natural selection shapes phenotypes to prioritize indirect reproductive success at the expense of direct reproductive success; kin selection requires positive relatedness between interactants and operates when investment in indirect reproduction has a higher payoff in gene-copy currency than investment in direct reproduction, thus acting to increase inclusive fitness.

Levels of selection and levels of adaptation Nested levels of group structure can contribute to naturally selected allele frequency change and thus to phenotypic change, but such changes cannot produce specific higher-level adaptations; it is important to remember Williams' (1966a) principle that adaptations for the exclusive benefit of the higher level cannot evolve unless the lower level of selection has been eliminated. Much conceptual thinking has spuriously assumed that higher-level adaptations can be achieved by conflict reduction in societies, while I contend that the lower level of selection needs to be removed up front by symmetric, non-redundant closure.

Lifetime commitment (LTC) The irreversible merger of founding units carrying complementary genetic information (organelles, gametes, or monogamous parents) into closed pairs that uniquely define a conjugated protist cell, a multicellular organism developing from a single zygote, or a colonial superorganism in a lineage ancestrally defined by strict lifetime monogamy; I conjecture that LTC was a universal necessary condition for non-human MTE origins across these distinct levels of organizational complexity.

Major transition in evolution (MTE) A discontinuous and irreversible change in organizational complexity characterized by a novel higher-level syndrome of adaptations evolved for the exclusive benefit of that higher level; nonhuman MTEs always domesticated formerly independent lower-level agency into complementary germline–soma functions while securing that the resulting individuals continued to maintain their lower levels of organizational complexity, albeit sometimes vestigially; the human MTE is the only known exception—it represents an irreversible and discontinuous change in organizational complexity (relative to variation across great ape societies that lack cumulative culture), but did not require higher-level germline–soma differentiation as happened in the caste-differentiated colonies of social insects.

Neo-Darwinian synthesis Understanding based on the view that the most revolutionary insight established by Darwin's *Origin* was the omnipresence of naturally selected adaptation, not descent with modification; Alfred Russell Wallace and August Weismann held this view, and so did Ronald Fisher; neo-Darwinism developed into a synthesis more encompassing than the so-called modern synthesis after Tinbergen carved out his four complementary questions and Hamilton and Williams initiated the gene's eye view centered on maximizing inclusive fitness and on kin selection as dynamic process; the neo-Darwinian synthesis developed after the 1960s came to include the modern synthesis elements agreed upon in the 1940s and 1950s.

Organism versus society Societies always consist of organisms (cells or multicellular individuals) that interact to divide labor, but without being universally and irreversibly forced into fixed germline–soma arrangements, that is, into obligate reproductive division of labor; in contrast, organisms are never themselves societies, because they are defined by obligate and unconditional germline–soma differentiation, which cannot evolve from condition-dependent division of labor because it requires novel higher-level non-redundant closure; replicators can form societies at all levels (from viruses to memes), but there are only four domains of hierarchically organized life: prokaryote cell organisms, eukaryote cell organisms, multicellular organisms, and colonial superorganisms.

continued

Table 3.1 *Continued*

Organismal biology The branch of biology that is primarily interested in how whole organisms function when they interact with their abiotic, biotic, and social environments, how their ontogenetic, behavioral, and life-history traits are adapted to these environments, and how internal organismal homeostasis is maintained; organismal biologists are often inspired by Tinbergen's four complementary types of questions about adaptation, comparative phylogenetic history, physiological mechanism, and onto-genetic development; their approaches often, but not always, black-box molecular mechanisms because the reductionism intrinsic to molecular biology trades off with a heuristic understanding of overarching organismal function.

Organismality versus individuality These terms are often used interchangeably, but I use them here to relate to different aspects of life's organizational complexity; organisms come in four qualitative grades of individuality, that is, organismality levels in terms of nested organizational complexity: the prokaryote cell, the eukaryote cell, the cell-type-differentiated multicellular body, and the caste-differentiated colony; within each of these discrete domains of organismal organization, degree of individuality can be used as a continuous measure to describe the extent to which some lineages have reached higher organizational complexity, while other lineages experienced reductive evolution towards lower individuality and less independent agency.

Redundancy and social conflict Cooperation is maximally harmonious when informational closure is mediated by a functionally complementary pair (e.g., a female and a male gamete merging in a zygote; a single mitochondrial clone associating with the nucleus in a eukaryote cell; a lifetime monogamous parental pair founding an ant colony); I refer to these kinds of LTC closure as *non-redundant* because they contain a minimally required sample of the available genetic information in the population gene pool, which is preserved intact throughout the life of the organism; the origins of all nonhuman MTEs appear to have required such symmetrical, pairwise commitments; in contrast, any form of *redundant* partnership (e.g., polyspermy, multiple strains of symbiont associated with the same host, multiple or multiply inseminated colony foundresses) is a defining trait of societies; redundant partnerships induce competitive conflicts that should preclude future MTE potential and that required conflict regulation when they evolved secondarily in colonial superorganisms.

the logic in the final section of this chapter to set the stage for the remaining chapters of this book.

Before getting to these conjectures, I will use the penultimate section to challenge approaches that try to explain biological aspects of organizational complexity without including natural selection as a decisive force. As Griffiths (2009) explained from a philosophical perspective, Dobzhansky's (1973) dictum needs to be understood as *nothing in biology makes sense except in the light of adaptation*. Capturing this Darwinian principle requires a multiple causation perspective as provided by Tinbergen's four complementary angles (see Figure 1.1). Only this approach allows adaptations that evolved by natural selection in the past to be conceptually distinguished from adaptive traits that enhance differential reproductive success at present, and for mechanistic constraints to be evaluated in the wider context they deserve. When adaptation

by natural selection is purposefully left out (e.g., McShea, 2016), studying progressive tendencies cannot be more than a correlative description of arbitrarily defined criteria, not only in cladogenic comparisons but also in ontogenetic development. As Ruse (2003) summarized Williams' position, "the [isolated] notion of a Bauplan is not only 'misguided and dispensable' but cuts off further debate and inquiry. What right has one to say that homologies, or Baupläne, are so very nonadaptive?" Waddington's (1960) developmental canalization logic (see Chapter 1) is now increasingly being documented as depending on deeply integrated gene-regulatory networks (Table 3.1), shaped by selection to continue expressing complex optimized functions (e.g., Wagner, 2005; Uller et al., 2018). Formally including ontogenetic development is therefore required for understanding adaptation in a neo-Darwinian/Tinbergian manner (see Box 1.1, Box 1.3).

Wheeler's and Huxley's early understanding of organismality and individuality

In the very early 20th century, only two biologists wrote concisely and innovatively about the levels of organizational complexity across the existing forms of life, William Morton Wheeler in the US (Box 3.1) and Julian Huxley in England (Box 3.2). They pioneered perspectives on evolutionary progress that considered both gradual change, be it progressive or reductive, and stepwise discontinuous change, a distinction that had remained largely implicit in Darwin's time (Padian, 2009). The merits of their complementary approaches have been glossed over in later 20th-century appreciations, especially in comments on Huxley's (1912) book *The Individual in the Animal Kingdom* (e.g., Dawkins, 1982a, 1982b; Burkhardt, 1992; Ruse, 2003; Clarke, 2010; Strassmann and Queller, 2010). The book has recently been reissued (Huxley, 2022) with a foreword by Rick Gawne and myself, but I cover its contents here as well because they are indispensable for a theory of adaptation by natural selection that considers both continuous and discontinuous evolutionary change. Wheeler and Huxley (Figure 3.1) deserve to be acknowledged for jointly framing and defining the idea of MTEs more than eight decades before Maynard Smith and Szathmáry (1995) published *The Major Transitions in Evolution*. Their early neo-Darwinian thinking and the universal syndromes characterizing MTEs that they identified, largely in parallel, focused exclusively on abstract criteria of organizational complexity. Wheeler's (1910, 1911) arguments were primarily about the colonial social insects, while Huxley (1912) compared all levels of organizational complexity that had been studied at that time. As the logic of their analyses has never been formally refuted, it seems quite odd that their joint intellectual legacy was so long ignored.

Box 3.1 William Morton Wheeler: early North American neo-Darwinist studying social evolution

William Morton Wheeler (1865–1937) was the leading North American scholar on ant biology a century ago (Evans and Evans, 1970) and best known for proposing an ontogenetic superorganism concept that was tightly argued from first principles of comparative form and function. His original version was watered down when organicists began to question the distinctness between organisms and environments (Baedke, 2019), after which the Chicago School of ecology misappropriated the superorganism for conceptualizing putatively homeostatic ecosystem regulation. Wheeler's original was only recently reinstated by Boomsma and Gawne (2018). His scholarship deserves to be better known because it represented early ethology in the US, which petered out in the 1930s (Johnston, 1995), but not before it had significantly influenced the early European ethologists who would end up establishing the field, mostly under Tinbergen's leadership (see Chapters 1 and 2). In his teenage years Wheeler was deeply influenced by George and Elisabeth Peckham via his membership of the Wisconsin Natural History Society. The Peckhams were an entomologist/arachnologist couple who let him in on their pioneering studies of mating behavior in spiders during the 1880s, testing Darwin's and Wallace's opposing views about sexual selection. As Evans and Evans (1970) note, their "careful marshalling of descriptive data, their frequent references to published accounts, and their thoughtful argumentation: all of these came to be characteristic of Wheeler's writings." Tinbergen

Box 3.1 *Continued*

also read Peckham papers when starting on his digger wasp dissertation research (Tinbergen, 1932; Burkhardt, 2005).

After receiving a German-style secondary education (Melander and Carpenter, 1937), Wheeler received his PhD degree at Clark University in Massachusetts, one of the early research universities in the US shaped according to principles developed by Wilhelm von Humboldt (Watson, 2010, p. 741) and among the first to establish a graduate program in biology (Oleson and Voss, 1979; Pauly, 1984). Wheeler's interests in the social insects emerged during his five years as a zoology professor in Texas (1898–1903) and particularly during his curatorship of invertebrate zoology at the American Museum of Natural History in New York (1903–1908), that is, well before he moved to Harvard where he was to stay until his death. In his formative academic years, Wheeler had been mentored by Charles Otis Whitman, a student of Louis Agassiz. Whitman was the most outstanding pioneer of animal behavior research in North America and probably the most prominent thinker of this discipline in his time worldwide (Pauly, 1984; Burkhardt, 2005). This assessment is based on Whitman's advanced understanding of what we would now call the *phenotypic plasticity of instincts*, which compelled him to interpret the expression of instincts in a comparative phylogenetic context (Evans and Evans, 1970), ideas that have often been credited to Tinbergen. Whitman's meticulous observations of pigeons in captivity made him realize that instincts are evolved (not improvised) and that they need to be defined from the first time that behaviors are expressed, because learning, instruction, or imitation can then be excluded and added later to infer plasticity in instinctive behavior.

Whitman was never a field biologist, but his mentee Wheeler was. Evans and Evans (1970) mention that Wheeler noted in one of his first ant papers that "it can hardly be doubted that there is a phylogeny of instincts," a notion out of Whitman's book and an approach that would define his later comparative research on ants. Remarkably, he coined the term *ethology* for the systematic study of animal behavior (Wheeler, 1902), and he defined that field as part

of evolutionary biology shortly afterwards (Wheeler, 1905). This renders any claims that the term was introduced by German or British scholars incorrect (Evans and Evans, 1970; Burkhardt, 2005) and shows that North American ethology was born and initially nursed in a neo-Darwinian cradle going all the way back to the Peckhams. Wheeler's emphasis on the superiority of broad phylogenetic comparison versus intensive single-species study implies that he was the first to adopt the joint *ultimate* domain of always asking questions about adaptation in a phylogenetic context (Wheeler, 1905). This conceptual approach was made explicit by Huxley (1916), further specified by Mayr (1961), and completed by Tinbergen (1963) (see Figure 1.1 and Box 1.3). Wheeler's 1905 paper also proposed that taxonomic classification should include ethological data because "classification can be developed only by comprehensive, comparative study of behavior in a number of genera and families and not by any amount of intensive study of a few reactions in a few species" (see also Evans and Evans, 1970). In many respects, the ethology synthesis that Wheeler began in the US was completed by Tinbergen (1960, 1963, 1965) in the UK.

As a young man, Wheeler had obtained a superb knowledge of the classic and modern languages and was broadly read in a global scientific sense. His PhD training was in comparative embryology, and during the 1890s he met many internationally known scholars in this field, particularly during his stay in Europe to work with Boveri in Germany and Anton Dohrn at the Naples Marine Biological Station. He also became acquainted with Oxford zoologists after he had favorably reviewed E.B. Poulton's book on *The Colours of Animals* for *Science* in 1890. Wheeler shifted to studying ants in 1900 and spent the summer of 1907 in Europe, where he met almost all important myrmecologists. Around that time, he also played a key role in the foundation of the *Entomological Society of America* and became the society's second president after John H. Comstock. He moved to Harvard in 1908 as first director of the Bussey Institute where he would stay for the rest of his career. As Evans and Evans (1970) relate, his interests in animal behavior

Box 3.1 *Continued*

culminated in his "Instincts" lecture in 1917, which appeared in print a few years later (Wheeler, 1920–1921). He made explicit that Mendelian ratios fail to capture development and defined instinct in the simplest and broadest possible way as "an action performed by all the individuals of a species in a similar manner under like conditions," which delegated all additional behaviors to the explanatory realms of associative learning and plasticity. As Evans and Evans (1970) note, "Wheeler used the word *intelligence* (as was common in his time) for any departure from strict stereotypy, whether it be innate plasticity, physiological adaptation, or simple forms of learning and memory."

Writing for the ornithological journal *The Auk*, Wheeler (1904) put the obligations of the student of animal behavior on a rational footing, emphasizing the need to rebuild scattered popular knowledge of animal behavior "on the securer foundations of careful observation and experiment." He used the term *animal mind*, a common expression in those days consistent with animal behavior being referred to as comparative psychology, but asserted that inferences about animal movements in time and space always require critical comparative testing. He made clear that anthropomorphic interpretations are to be avoided as an eighth mortal sin of biologists (Sölch, 2016), and he may have been the first biologist to recommend Occam's razor: "Complicated explanations are inadmissible when simpler ones will suffice," concluding that we are not "to accept human reasoning as an explanation of any animal behavior, till simpler processes, like instinct and associative memory, have been tried and found wanting." This coherence of thinking pervaded his 1920-1921 *Instincts* paper where he also contemplated the significance of the Baldwin effect, calling it his "ninth mortal sin" because its Lamarckian taint seemed unorthodox for his neo-Darwinian training. Suggested two decades earlier (Baldwin, 1896, 1897), this concept had claimed that learned behavior can assimilate to become a genetically encoded adaptation. The Baldwin effect was later confirmed to be includable in neo-Darwinian theory without difficulty (Dawkins, 1982a;

Maynard Smith, 1987; Dennett, 1995; Frank, 2011a, 2011b). Finally, he was aware that organisms are in part irreducibly complex in their internal organization (Wheeler, 1926).

Wheeler's thinking about natural selection and adaptation was little affected by the turmoil among geneticists after Mendel's laws were rediscovered in 1900, and he avoided behaviorism when it emerged in the US (Evans and Evans, 1970). Wheeler met Thomas Hunt Morgan in the 1890s, and they remained lifetime personal friends even though Wheeler never showed any professional interest in Morgan's genetics because it failed to explain organismal development (Wheeler, 1920–1921; Sölch, 2016). Upon his retirement in 1934, he was "one of the most honored scientists of his generation" in the US (Evans and Evans, 1970), a well-deserved reputation because he continued to be an able synthesizer. In a late-career paper (Wheeler, 1934), he argued that "togetherness" behavior had now become documented in so many taxa and ecological settings that one should carefully categorize which of these were mere ecologically defined (aggregations, flocks, nutritive consociations, ecological communities) and which required evolutionary analysis (societies of insects, vertebrate societies, and humans, all separately). This suggests that Wheeler was aware of the confusion caused by the Chicago School of ecology trying to "environmentalize" his original superorganism concept. That some have accused Wheeler of crypto-Lamarckism (e.g., Ghiselin, 1974; Sleigh, 2004) or of being a metaphysical protosociobiologist (Sölch, 2016) reflects lack of understanding of his thinking as a neo-Darwinian (Evans and Evans, 1970; Mitchell, 1995) and of his motivation to understand adaptation holistically rather than by reductionist experiment.

Nonetheless, Wheeler's writing did occasionally become less incisive in his later years when he sometimes used poorly specified emergence terminology (e.g., Pauly, 1984; Gibson, 2013; Baedke, 2019), possibly under the influence of the philosophy of his Harvard colleague Whitehead (Sölch, 2016). This may have contributed to the superorganism becoming a moot concept in the hands of Alfred Emerson and

Box 3.1 *Continued*

the Chicago School of ecology after Wheeler's death (Banks, 1985), when all categorical social differences dissolved into smooth gradients similar to what Frederic Clements' school had pursued for plant communities (Baedke, 2019). However, that muddle should not affect our appreciation of Wheeler's original superorganism, which was clearly an evolutionary concept of ontogenetic colony development and not an ecological hypothesis. It was this original version (Wheeler, 1910; 1911) that Boomsma and Gawne (2018) recently reinstated as the correctly argued major transition view for caste-differentiated colonies. When Wilson (1967, 1971) dismissed the corrupted ecological superorganism idea (see e.g., Emerson, 1939), he should have gone back to Wheeler's source concept rather than encourage and develop new incomplete versions based on apparent homeostasis (e.g., Seeley, 1989; Moritz and Southwick, 1992; Tautz, 2008; Hölldobler and Wilson, 2009). Having coined *ethology* as a Darwinian discipline and having been the first to coherently define *instinct*, it seems astounding that Wheeler has not received more lasting credit by historians of biology. This neglect suggests that ethology had to develop excellence in the study of European vertebrates before it could be recognized as an incisive field of study.

Wheeler rarely used his original superorganism concept in his writing after World War I. This may have been related to his employment at the applied agricultural and horticultural Bussey Institute, but it also seems to be a form of self-censorship—in one of the rare occasions that he uses the term it is clear where it comes from for readers who know his early papers, but not why it is relevant (Wheeler, 1928a, pp. 41, 46). At the same time, he notes that biological theory has become very confused, and he reminds his readers that "an abiding interest in the more comprehensive theories [. . .] is all that prevents our sciences from lapsing into little more than empirical routines" and that "it is neither the observational and experimental data nor the investigational methods, but the theories that are the essential, vital constituents of a science" (Wheeler, 1929). As it appears, and in stark contrast with Huxley (Box 3.2), Wheeler had become a skeptical

scientist, the signs of which were already visible earlier when he wrote critical papers on the organization of research (Wheeler, 1920a), a satirical correspondence with the king of a huge fungus-growing termite colony (Wheeler, 1920b), and a blistering critique of academic governance (Wheeler, 1923a). His scholarly code of honor emerged early on, when he wrote that "whenever one does decide to publish, it is necessary to reckon with the great 'paper memory of mankind,' the conserved experience of other workers who have loved and investigated the same things. It then becomes a duty to study the 'literature of the subject,' if only for the purpose of bringing the new work into intelligible, organic relation with the old" (Wheeler, 1906).

It is important to note that Wheeler's superorganism concept always referred to full-sibling families, even though he was aware—as Darwin was when he coined family selection—that particularly ants may have colonies with multiple queens. He made this explicit when he wrote that "[i]t is now unanimously admitted that all insect societies worthy of the name, and no matter how populous, are families, that is, affiliations of the parents and, in most cases, of the mother alone, with the offspring" (Wheeler, 1928b, pp. 11–12). This notion was endorsed by Williams (1966a, pp. 93, 161–162, 197–201, 257) and earlier by Bonner (1955, p. 60), who wrote that "higher insect colonies are not only matriarchal but also monogamous. [. . .] [E]ven though the colonies may achieve a tremendous size and consist of thousands of individuals, they are in most cases but a family, one mother with her vast offspring." Bonner's assessment echoes conclusions by Wheeler (see also 1910, p. 3, 1928b, pp. 11–12), so it is difficult to understand how this agreed-upon naturalist consensus could later be ignored and replaced by a confusing mixture of colony selection and group selection (Wilson, 1971, 1975). Towards the end of his life, Wheeler once more demonstrated his brilliant analytical abilities when conjecturing that human societies remain unstable because they resolved their aggressive antisocial male problem only partly, while insect superorganisms achieved a complete solution

Box 3.1 *Continued*

by having "coerced" males into obligatorily monog-amous cooperation if not merely stored envelopes of their sperm (Wheeler, 1934). As we will see in Chapter 8, Wheeler's taming of bullying antisocial males was recently endorsed by comparative analy-sis of primate social systems, and highlighted as the key factor for human self-domestication (Wrangham, 2019).

Although many evolutionists and entomologists are accustomed to, or even fond of, the ecologi-cal convenience term "eusociality," it is essential to acknowledge Wheeler's earlier ontogenetic superor-ganism concept if we are to understand the parallels between obligate altruism by somatic cells in multi-cellular bodies and somatic individuals in caste dif-ferentiated colonies—something that eusociality ter-minology will not allow unless we would adopt the convention of calling our leucocytes eusocial cells. As we will see in Chapter 4, other natural history perspectives have also highlighted the weaknesses of Wilsonian *eusociality* as a scientific concept, and Ghiselin (2011) reached a similar conclusion from a

philosophical position. Semantic clarifications of this kind are important to keep asking incisive evolution-ary questions and to avoid the muddle introduced by ecological definitions of anthropomorphic conve-nience that make natural selection implicit or moot. This is not to negate the marvels of natural history and complex behavior disseminated by the eusociality liter-ature and by monographs that used later definitions of the superorganism (e.g., Seeley, 1989; Hölldobler and Wilson, 2009), but to make the point that these later versions were defined ad hoc and incorrectly focused on a mere subset of superorganisms that excluded the annual bumblebees and yellowjacket wasps with per-manent caste differentiation (Boomsma and Gawne, 2018). Using Wheeler's superorganism is straightfor-ward because it is precise, as were Batra's and Huxley's restrictive definitions of eusociality. Just using the term "social insects," as biologists did before the 1970s, and then stipulating the kind of colony organiza-tion one is concerned with, would greatly increase transparency as it would avoid using 'eusocial' in a throwaway manner (see also main text p. 70).

Box 3.2 Young Julian Huxley: understanding closure and discontinuous adaptive change

Julian Huxley (1887–1975) grew up as the grand-son of Thomas H. Huxley, and with Aldous Huxley—the author of *Brave New World* (Huxley, 1932)—as younger brother. He obtained a degree in zoology from the University of Oxford (1909) and held academic appointments at Rice University (1912–1916), the University of Oxford (1919–1925), and King's College London (1925–1927). He left academia to coauthor *The Science of Life*, quite likely the last scientific overview of all known biology (Wells et al., 1931). Huxley became secretary of the Zoological Society of London (1935–1942) and later the first gener-al director of UNESCO (1946–1948). In contrast to Wheeler (Box 3.1), whose contributions have been mainly underappreciated, Huxley's early legacy faces

the problem of first having been ignored by himself and then having been dismissed by others for unsci-entific progressionist tendencies. As I explain in the main text, this verdict is based on his later writing, and has no bearing on his first book, which is of particu-lar relevance here. Because Huxley's CV is accessible and relatively well known, I use this box to sum-marize the contents of *The Individual in the Animal Kingdom* (Huxley, 1912) and thereby show that ignor-ing the insights of this book has been detrimental to the development of thinking about the MTEs during the past few decades. The paragraphs below sum-marize, chapter by chapter, what Huxley had to say and aim to clarify why the inferences he drew more than a century ago are still relevant. A more elaborate

Box 3.2 *Continued*

complementary account has recently been given in the foreword of the reissued book (Huxley, 2022).

In the preface, Huxley states that his scope will be wide and general, posing that "living matter always tends to group itself into [...] 'closed independent systems with harmonious parts'"—it reflected his awareness that organizational closure and independent agency are key characteristics of all organismal individuals. As he notes in Chapter 1, entitled "The idea of individuality," "nothing homogeneous can be an individual," setting crystals of inanimate nature apart from biological organization as being mere aggregations without division of labor. He makes clear that his closure and division of labor principle will continue to apply throughout the subsequent levels of organizational complexity that he will come to address, observing that "difference of function always implies difference of structure, so that the more independence—the more individuality—an individual is to possess depends very closely on the amount of heterogeneity of its parts. Look for instance at such an individual as a colony of Termites [...], its defence delegated to one caste, its nutrition to another, its reproduction to another; the various castes are specifically adapted in their structure for their various functions." He further recognized the requirement that the "diverse parts all work together in such a way as to ensure the whole's continuance, or, as the evolutionist would say, whose structure and working have 'survival value,'" to which he added that survival's sole purpose is reproduction. This may be the first explicit use of the term survival value for which Tinbergen would be credited decades later when he specified adaptation in his four-quadrat scheme (see Figure 1.1; Box 1.3; Chapter 2).

Young Huxley showed himself to be a true Weismannian evolutionist when he wrote that "bodies are only 'cradles for our germ cells,'" inferring that complex life-cycles of liver flukes and standard ant colonies have analogous forms of individuality even though their parts move around. As he writes, being "separated in space" as in an ant colony "is no bar to individuality." In just this single chapter, Huxley succeeded in characterizing a unitary individual with a degree of coherence that I do not think has been improved upon since.

In the next chapter, "The biological foundations of individuality," Huxley turns to what we would now call the loss of totipotency of the subordinate units of an individual, starting again with an illustrative example provided by social insect colonies. After first having referred to such colonies as families, Darwin later called them "communities" to allow for colonies sometimes having multiple queens (Dugatkin, 2006). Huxley therefore wrote: "There are communities, such as those of bees and ants, where, though no continuity of substance exists between the members, yet all work for the whole and not for themselves, and each is doomed to death if separated from the society of the rest." Somewhere here, he shifted to using organismality terminology, and concluded that "[a]n organism is more definite [than a crystal] because its size is defined as well as its form. A crystal will continue to grow without limit if only the appropriate mother-liquor in which it hangs is kept saturated: its form is definite, its size indefinite. It is a fact of common observation, however, that each organism has a typical size—not invariable, but fixed within certain not very wide limits."

In the third chapter, "Some other definitions of animal individuality," it becomes clear that Huxley is handicapped by the dearth of knowledge about bacteria at the start of the 20th century, but he acknowledges their existence as a "large group of organisms [that] seem not to possess [sexuality] at all, while here and there in higher groups it [sexuality] has been lost." Given there were no microscopes yet that allowed bacteria to be seen, this is a sound assessment, because it delegated the origin of sex and meiosis to the eukaryote protists and acknowledged the recurrent later loss of sex in multiple lineages. Huxley knew that subsequent meiotic events define individuality from generation to generation in the higher Metazoa, but noted that clonal propagation by "bud-sports" in plants is an alternative that modifies this rule. He appears to have had a solid

Box 3.2 *Continued*

Weismannian grasp of meiotic recombination and the difference between zygotic and somatic mutations. As he wrote, "[a] plant will often appear showing a mutation in all its parts, so that the change inducing the mutation must certainly have affected the single sexually produced cell from which the whole plant has sprung. Once formed, mutations will persist in cuttings or slips of the parent plant, but will usually be lost when the sexual chromosome-shuffling is allowed to take place and offspring are raised from seed." He thus concluded that "we must recognize that [the] connection [between sex and individuality] is not obligatory, that in origin the two are entirely distinct, and that therefore the one cannot possibly be used as the basis for the definition of the other"—once more, a correct assessment today.

In chapter 4, "The second grade of individuality and its attainment," Huxley reaffirmed his neo-Darwinian schooling by specifying how he understood levels of complexity, while at the same time making clear that discontinuous hierarchical transitions in organizational complexity are real. As he wrote, "[m]any Neo-Darwinians [. . .] argue that adaptation is the great reality gained by organisms through natural selection, and that, therefore, no one species now alive has preference over any other—for to be alive both must be adapted to their surroundings. But to exist and nothing more, to vegetate merely, is not the fate of all organisms. There *is* a higher and a lower, for some are more independent, more powerful than others." This statement refers particularly to the fact that the specific levels of individuality, that is, protists, multicellular organisms, and caste-differentiated social insect colonies, show distinct degrees of control over their environment and have concomitant levels of organizational complexity. Huxley illustrated this point noting that Protozoa are amazingly complex but cannot exploit division of labor beyond a single cell, and that metazoans cannot exploit division of labor beyond a single individual. The key principle for uniting the three distinct levels of individuality that he recognized was that the "whole is greater than the sum of its parts; for the problem is one of combination,

not of mere addition," a principle that would later be beautifully developed by Weiss (1967). Thinking as a neo-Darwinian biologist trained in comparative embryology made it easy for Huxley to see the non-linear nature of organismal development across these distinct levels, as did Wheeler (1926) (Box 3.1).

Well into chapter 4 on second-grade individuality, Huxley noted that synergy is achieved only in obligatorily integrated organisms such as *Volvox* and the higher Metazoa, and not in sedentary colonies of cells, which are "merely a device for making the fullest use of a place with good food supply"—here he inferred that the composing parts merely mind their own business. In Chapter 4 of this book, I will use a similar argument when proposing a general explanation for the origins of MTEs, arguing that synergy in resource acquisition, not suppression of conflict, was the key condition and only when relatedness among offspring cells or individuals remained maximal. Huxley highlighted that the *Volvox* transition, where "we can positively affirm that the step from an aggregate to a higher individual has actually taken place, is one of the most important in biology." He emphasized that germline–soma division of labor normally followed after the evolution of obligate multicellularity, but he also presented the Dinoflagellate parasite of polychaete worms, *Haplozoon*, as a contrasting example where division of labor evolved before multicellularity (see Rueckert and Leander, 2008, for a recent study). While explaining these cases, Huxley recognized the trade-off in free-living eukaryote cells between division, nutrition, and motility, and he concluded that it is in this way that division of labor creates synergy by lower-level specialization. Many of these topics were later taken up by Buss (1987) in his book *The Evolution of Individuality*, which criticized the field for its Weismann-zoology bias, but without citing Huxley (1912) who thought broadly about similar issues many decades earlier.

In chapter 5, "The later progress of individuality," Huxley focused on hydroid colonies that have some division of labor for defense or nutrition while the rest of the colony consists of totipotent polyps. Thus

Box 3.2 *Continued*

characterized, such colonies appear analogous to later discovered gall-forming thrips and aphids with a few differentiated soldier individuals (Aoki, 1982; Crespi, 1992). He then considered the related, more complex Siphonophora (e.g., Portuguese man o' war), showing that they have achieved third-grade individuality because the collective of differentiated polyps forms a single individual with extensive reproductive and somatic division of labor without any remaining totipotent parts. Huxley compared the third-grade individuality of Siphonophora with obligate interspecific mutualisms such as lichens and yucca trees and moths, and concluded that these mutualisms barely qualify as third graders because their division of labor was given from the start and little new became added later. He illustrated the less advanced nature of such interactions by a marine flatworm that hosts symbiotic green algae for photosynthetic nitrogen acquisition, but later digests them while turning mutualism into parasitism (for recent studies of similar kind, see Douglas, 1983; and for unicellular protist hosts, Lowe et al., 2016 and Kiers and West, 2016). Finally, he noted that "[t]he communities of ants and bees are undoubted individuals," referring to Wheeler (1911). Although Huxley did not make very explicit in 1912 that these colonies, which Wheeler came to refer to as superorganismal, always fitted his third grade of individuality, his later writing clarified that he understood third-grade individuality as being particularly defined by the advanced caste-differentiated social insect colonies (Huxley, 1923a).

As I explain in more detail in the main text, Huxley's concept of evolutionary progress in 1912 emphasized adaptive design, as could be expected from an early neo-Darwinian who acknowledged the all-sufficiency of natural selection (Weismann, 1893). In this view, natural selection is a continuous creative force allowing some lineages to progress to more complex and versatile phenotypes when random mutations provide novel phenotypic traits with higher reproductive success in specific environments. Young Huxley and Wheeler (Box 3.1) agreed on these principles, and they jointly

inspired the early organicist movement, whose most prominent figure heads had no problem seeing that adaptive design was not about mere gene frequency change but about natural selection maintaining the complex design of ontogenetic development. Particularly Waddington (1960) continued to stand firm on this principle at a time when most later organicists had lost focus on proper differentiation between organism and environment (Baedke, 2019).

Despite the softening of his neo-Darwinism over time, Huxley's inferential imagination remained superb. Huxley (1923b) appears to have been the first essay to compare the human MTE with the MTEs of the colonial superorganisms in exquisite biological and cultural detail. He later lyrically summarized the difference when he wrote that "[t]here is [. . .] no reason to suppose that man is destined to sterilize nurses or manual workers, to breed armoured or gas-resistant soldiers, communal parents the size of whales, or an intelligentsia all head and no body" (Huxley, 1930, pp. 120–121). Similarly, his Darwin Centennial Thanksgiving Convocation address ably summarized his then modern insight in human evolutionary history: "For their [our prehuman ancestors] transformation into man a series of steps was needed. Descent from the trees; erect posture; some enlargement of brain; more carnivorous habits; the use and then the making of tools; further enlargement of brain; the discovery of fire; true speech and language; elaboration of tools and rituals. These steps took the better part of half a million years: it was not until less than a hundred thousand years ago that man could begin to deserve the title of dominant type and not until less than ten thousand years ago that he became fully dominant. After man's emergence as truly man, the same sort of thing continued to happen, but with an important difference. Man's evolution is not biological but psychosocial; it operates by the mechanism of cultural tradition, which involves the cumulative self-reproduction and self-variation of mental activities and their products" (Huxley, 1960b, pp. 251–252). Also this text remains largely correct today (see Chapter 8).

Wheeler and Huxley were among the few well-trained and broadly read neo-Darwinians of their time (Boxes 3.1 and 3.2). They had both studied comparative anatomy and embryology in their early years, which made it natural for them to interpret comparative observations hierarchically (Brigandt, 2006). They also both knew that Darwin had struggled to explain the neuters (workers and soldiers) of ants, honeybees, and other advanced social insects as products of adaptation through natural selection (Costa, 2006, 2013; Dugatkin, 2006). Richards (1987) relates in considerable detail that the well-known quote from the *Origin* that the problem of explaining these neuters "at first appeared to me insuperable, and actually fatal to my whole theory," was resolved by Darwin only just before his book went to print in 1859. As already mentioned in Chapter 1, Darwin's explanation that family selection could generally account for sterile workers was extremely powerful, as he illustrated with several convincing analogies. One of them highlighted how cattle breeders artificially select when the visual quality and taste of meat can only be assessed after an animal has been slaughtered by breeding with the closest possible relatives. The social insects and their neuters retained center stage as role models for evolutionary thought experiments in the late 19th century. They featured, for example, prominently in the *all-sufficiency of selection* exchange between Herbert Spencer and August Weismann (1893). Weismann's work and particularly family selection remained a generally endorsed concept for Haldane, Fisher, Huxley, and many other leading evolutionary biologists and geneticists in the first half of the 20th century (see Boomsma and Gawne (2018) for a series of representative quotations documenting the continuing intellectual prevalence of the concept of family selection).

Apart from three encyclopedic monographs (Wheeler, 1910, 1923b, 1928b), Wheeler is best known for the idea that colonies of ants, bees, wasps, and termites are organismal when they only consist of individuals with physically differentiated caste phenotypes (Box 3.1). This perspective allowed him to formalize Darwin's idea of higher-level family selection, so that neuters could be conjectured to have evolved according to the same logic that Weismann had used for his germ-line and soma segregation. As Wheeler wrote in the introduction of his *Ants* monograph (1910, p. 7): "There is [...] a striking analogy [...] between the ant colony and the cell colony which constitutes the body of a Metazoan animal; and many of the laws that control the cellular origin, development, growth, reproduction and decay of the individual Metazoon, are seen to hold good also of the ant society regarded as an individual of a higher order." This phrase appears to directly elaborate on Weismann's (1893, pp. 326–327) statement that "in the case of the ant colony, or rather state, the barren individuals or organs are metamorphosed only by selection of the germ-plasm from which the whole state proceeds. In respect of selection the whole state behaves as a single animal; the state is selected, not the single individuals; and the various forms behave exactly like the parts of one individual in the course of ordinary selection." Wheeler thus concluded that each colony has its own "germ-plasm" and "somatic tissue," is coordinated and individualized for optimal nutrition, self-protection, and reproduction, and has its own parasites and diseases (Evans and Evans, 1970). He elaborated this concept in a separate paper (Wheeler, 1911), and one year later we find his "individual of a higher order" phrasing back in Huxley (1912). Another account of the conceptual proximity of Huxley and Wheeler has recently been provided by Baedke (2019).

Like Wheeler, Huxley was a prolific writer from early age, covering a wide range of topics in biology, medicine, politics, ethics, and humanism in scientific papers, monographs, and essays for the lay public. The two early monographs of interest here are his first book *The Individual in the Animal Kingdom* (Huxley, 1912) and his concise later book *Ants* (Huxley, 1930). Towards the end of *The Individual in the Animal Kingdom* (2022, p. 108), Huxley pays tribute to Wheeler, writing: "The communities of ants and bees are undoubted individuals. Wheeler in a recent paper (1911) has abundantly justified this view from a somewhat different standpoint. Here I can only say that if the ideas and definitions put forward in Chap. 1 [a.o. that individual functional design always implies non-totipotent parts working for the preservation and reproduction of the whole] are accepted, their individuality is beyond dispute." This endorsement came after a wide-ranging analysis of the major discontinuous transitions in organizational complexity, captured for the first time in an overarching perspective that was significantly broader than Wheeler's: the emergence of the free-living eukaryote cell (grade 1), the evolution of the major multicellular lineages (grade 2), and the origins of the caste-differentiated social insect colonies (grade 3). In his own words, "Suppose that instead of separating from each other after each division, the cells remain connected. The result will be a colony of cells each one like all its fellows. If division of labour sets in later among the cells, they are rendered mutually dependent, and the colony is transformed into a true individual, which is obviously of a higher order than the cell" (Huxley, 2022, pp. 46–47). This is how he moved from first grade, to second grade, and to third grade without needing to adjust the parsimonious coherence of his arguments (Box 3.2).

In *The Individual in the Animal Kingdom*, Huxley formulated a composite definition of individuality that made it natural to switch to organismality language not too far into the book and to identify not just key characteristics but entire syndromes that organismality

grades needed to match. He argued that an individual always has (1) agency, complexity, and adaptability; (2) closure and indivisibility, such that when one splits an individual in two, it will always imply at least temporary loss of function; (3) heterogeneity in structure and function emanating from division of labor between parts; and (4) recognizable adaptive design of these cooperative parts to ensure the whole's continuance, defined as its survival and reproduction. In listing these defining characteristics, he showed a keen awareness of the irreversible reductions in totipotency of cells or multicellular individuals whenever a higher level of organismality arose. In later chapters he acknowledged that the cell, as the most fundamental unit of life, must represent the first grade of individuality, so it then logically follows that multicellular organisms with division of labor between germ line and soma have individuality of the second grade. He also noted that "the sexual act stood originally in no relation to the life of the cell, or of the multicellular organism, or of the race [i.e., species or lineage], so that any conclusions with regard to individuality based on the periodical recurrence of sexual fusion cannot be fundamentally true" (Huxley, 2022, p. 54). This is consistent with sex not being an MTE in its own right (Box 3.2). Huxley further emphasized the crucial significance of adaptive function when evaluating the different grades of organizational complexity, always having in mind the obligate division of labor that renders the organismal whole more than the sum of its parts, as Williams (1966a) would reiterate in multiple contexts more than 50 years later.

Well before biologists started to think in terms of ecological and evolutionary dynamics from the 1930s onwards (Elton, 1927, may have been the first such ecology book; written while he was Huxley's assistant), young Huxley had already shown that he understood the logic of evolutionary arms races where "[e]ach advance in attack has brought forth, as if by magic, a corresponding advance in defence" (Huxley, 2022, p. 88). He appreciated that lichens and other obligate symbioses such as *Yucca* and its pollinating moths evolved mutual dependences of varying depth, and he recognized that benefit/cost ratios always need to remain positive for both parties because "such a system appears like a double parasitism, and twice the evil that parasitism brings should be its portion" (Huxley, 2022, p. 103). The insight that mutualisms are really forms of mutual exploitation would not be reformulated until the very end of the 20th century (Herre et al., 1999) (see Chapter 2).

Huxley argued that organisms come exclusively in grades or levels (a discrete hierarchical measure), but that individuality can also vary in degree, a continuous measure within these grades. As he wrote (Huxley, 2022, p. 100), "[t]he necessity for effort—the 'struggle for existence' in the most general sense—has from age to age raised the average level of independence, the measure of individuality's perfection in living beings. In spite of this general rise of levels, there has been in every age a falling away, a decline in perfection of individuality in certain species. This decreased independence reveals itself not only as structural degeneration, but also in degeneration's opposite, structural specialization. There is, however, a common cause beneath these opposite effects and that is overclose adaptation, adaptation to very narrow conditions." Although we would not use the term perfection today, it is clear that Huxley understood that such declines are complexity reductions, not reversals to a previous domain of organizational complexity, consistent with Wheeler's (1910) arguments for socially parasitic ants that lost the worker caste.

The agreement between Huxley's and Wheeler's analyses is striking. As Wheeler (1910, p. 7) wrote: "we may say that since the

different castes of the ant colony are morphologically specialized for the performance of different functions, they are truly comparable with the differentiated tissues of the Metazoan body." In *The Social Insects* that appeared almost two decades later, Wheeler (1928b, p. 23) reiterated that "the insect society, or colony as a whole, [. . . is,] as I have shown in another place (1911) [. . .] so strikingly analogous to the Metazoan body regarded as a colony of cells, or indeed to any living organism as a whole, that the same very general laws must be involved." Conceptual alignment continued in Huxley's *Ants* book (1930), which offered, in his typically transparent terms, a concise account of the types of social insect colonies that deserve to be kept apart for their distinct traits that can respond to natural selection (see also Chapter 6). As he wrote: "Three main grades of social habit may be distinguished. In the lowest there is some sort of a family life, either the mother or both parents living with and helping the developing young. This may be called the subsocial, or family, grade [possibly the first use of subsocial as term]. The second is the true social, or colonial, grade, in which the young, when fully grown, stay with their parents and co-operate with them in building the nest and caring for further broods of young. The highest grade is that of the caste-society, in which some of the young are transformed into unsexed 'neuters,' who take off the shoulders of the fertile caste all the duties of the colony, save only that of reproduction" (Huxley, 1930, p. 10).

My positive appraisal of Wheeler's and young Huxley's consistent stepwise grade (rather than continuum) thinking (Boxes 3.1 and 3.2) appears to run counter to assessments by non-biologists who have downgraded particularly Huxley's writing as unduly progressionist (Divall, 1992; Waters and Van Helden, 1992; Herring, 2018). These assessments are unjustified as far as the biological facts and the synthesis of his 1912 book are concerned, as explained in detail elsewhere (foreword to Huxley, 2022). Non-biologist reviewers appear to have confused Huxley's and Wheeler's careful thinking about levels of organizational complexity in a comparative phylogenetic context with progressionist orthogenesis arguments like those of Jean Baptiste Lamarck, Ernst Haeckel, and Henri Bergson. In particular, many have taken issue with Huxley's (1912) acknowledgment of Bergson (1911), one of the last monographs defending an internal *élan vital* drive of organisms. However, careful reading of Huxley's (1912) text reveals that he only cites Bergson for the concept of closure and that he does that along with a quote of Nietzsche on independence being crucial for individual agency. For understanding Huxley's legacy, we need to distinguish between the anthropomorphic inspiration that increasingly shaped his later thinking and the clarity of his early materialistic neo-Darwinian logic. His reasoning on organismal agency and his adoption of the closure concept were very incisive and ahead of his time. As Baedke (2019) concluded recently, the joint efforts by Wheeler and Huxley were essential for initiating a much-needed anti-vitalist focus on the organism as distinct level of organization. Their argumentation is clearly incompatible with Bergson's philosophical position, but they may have felt misunderstood in this respect, as both made explicit statements dismissing Bergson later on (Huxley, 1923b; Wheeler, 1926).

Much of the mediocre press that Huxley received after his death in 1975 appears to be based on retrospective extrapolation of the reputation that he started to develop in the 1920s (Huxley, 1923a, 1923b), which was to gradually become an explicit public dissemination framework (e.g., Huxley, 1936, 1940, 1960b). His 1923 essays still confirm the scientific appreciations of his 1912 book, but there is also text on the human condition that we would now consider misplaced. Just a few years later he published a less coherent individuality essay devoid of neo-Darwinian arguments (Huxley, 1926), and his editorial introduction of Elton (1927) shows that he now embraced a reductionist view of biology that was incompatible with the conceptually more abstract book of his youth that had focused on comparative natural history. Huxley's foundational contributions to ethology in Europe have likewise been dismissed as less significant than originally claimed, and he has been accused of seeing natural selection working mostly for the good of the species (Burkhardt, 1992). However, good-for-the-species thinking was rampant throughout the 1940s and 1950s (but see Simpson, 1941; Nicholson, 1960; see also Chapter 2), and young Huxley remained among the few supporters of the sufficiency of individual selection at that time (Provine, 1998). Moreover, dismissing Huxley's early work ignores that he was the first to understand the fundamental difference between proximate and ultimate explanations in biology. This distinction profoundly influenced the Oxford zoology school (Provine, 1998) and contributed decisively to the international influence of his mentees, particularly David Lack and Niko Tinbergen, who would later translate some of Huxley's ideas into first-principle neo-Darwinian conjectures (see Chapter 2) enabled by more incisive phrasing than Mayr (1961, 1974) ever achieved (see Figure 1.1; Box 1.3).

As it appears, young Julian Huxley did not have an unusual humanistic agenda when he published his 1912 book; at that time he seems to have been primarily driven by the pursuit of comparative studies that could professionalize ornithology (Zuckerman, 1992; Morrell, 1997), once more an interest he shared with Wheeler who published about this topic in the ornithological journal *The Auk* (Box 3.1). As detailed elsewhere (foreword to Huxley, 2022), it is wrong to dismiss his early book just because his later contributions tried to reconcile purposeless evolution by natural selection with progress towards a humanistic future for our own species (Provine, 1988; Smocovitis, 2016). Huxley's reluctance to accept that natural selection "is a poor guide to the living of a human life, let alone to the conduct of a society" (Durant, 1992) remains an ambiguity that persists until his Darwin centennial writing. However, his conclusion that "purposes in [human] life are made, not found" (Huxley, 2010, p. 576) does show that he distinguished between the human condition and all other organisms, in contrast to later sociobiology that all but erased this distinction by suggesting that genes can ultimately explain morality (Wilson, 1975). The connection between Huxley and Wilson was made by Smocovitis (2016), who noted that that both authors combined progressionism with a devotion to reductionism (Huxley dedicated his 1942 book to "T.H. Morgan many-sided leader in biology's advance"). Her analysis helps to explain why Huxley came to neglect his early ethology research and *The Individual in the Animal Kingdom* in his *Modern Synthesis* monograph (see also Burkhardt, 1992). However, while confirming the strong anthropomorphic connotations of Huxley's and Wilson's views in the middle of the 20th century, Smocovitis failed to note that Wilson's sociobiology views lacked a solid foundation in natural selection theory. Throughout his writing, he ignored the earlier work by Pittendrigh (1958), Tinbergen (1963), and Williams (1966a), and he covered Hamilton's (1964a) insights without in depth understanding.

The problems of pursuing organismal biology without first-principle Darwinism

As we saw in Chapter 1, a cultural belief in progress reemerged in the 1960s (Hull, 1988b), so it could be argued that Huxley (1942,

1960b) anticipated this sociological trend and that E.O. Wilson's (1975) *Sociobiology* was an attempt to finalize it (Ruse, 1988). In a subsequent contribution, Ruse (2003) cited and evaluated a paper by Wilson (1992, p. 187) which offered the conclusion that "[p]rogress [...] is a property of the evolution of life as a whole by almost any conceivable intuitive standard, including the acquisition of goals and intentions in the behavior of animals"—a much stronger and less nuanced statement than Huxley ever made. I mentioned in Chapter 2 how Wilsonian sociobiology was criticized for its loose argumentation and its attempt to subjugate the more precisely defined discipline of ethology (e.g. Figure 1.1), while offering a mix of ecological and genetic group thinking that made the natural selection specificities of Darwinism disappear. And as far as valid scientific synthesis points were made, they were drowned in the dialectical arguments of sociobiology's most vocal critics (Gould and Lewontin, 1979; Gould, 1988; but see Queller, 1995, and Box 2.3), who were all too happy that adaptation had disappeared from the Darwinism they experienced among many of their peers. All that was lacking was making the point that natural progress of any kind was just an example of the wrong kind of cultural or political bias (Segerstråle, 2000). It is interesting that the journal *Behavioural Ecology and Sociobiology* came to capture both academic cultures, a gene's eye approach that was, as Dennett (1995, p. 228) put it, about reverse engineering in biology to figure out what *Mother Nature* had in mind in terms of adaptive design, and a descriptive mere phenological and non-hypothesis-testing approach that reflected sociobiology's anthropomorphic narrative of conflict-blind optimism on nature's behalf.

Towards the end of the 20th century, thinking about evolutionary progress became re-conceptualized as *The Major Transitions in Evolution*, the title of a monograph by Maynard

Smith and Szathmáry (1995). Part of the book's rationale was to rebut central tenets of Buss (1987) (Folse and Roughgarden, 2010), but neither of them referred to Huxley (1912). One therefore misses Huxley's clear hierarchical understanding of the structural and functional aspects of organizational complexity—an approach that has recently been incorrectly credited to modern philosophers of biology (e.g., by McShea, 2016). As summarized above and in Box 3.2, young Huxley proceeded with unambiguous logic from eukaryote unicellularity (complexity level 1) to eukaryote multicellularity (complexity level 2) to colonial superorganismality (complexity level 3), with the prokaryote domain as the unspecified start of organized life (level 0 by implication; Huxley, 2022). One also misses Simpson's (1941) version of hierarchical individuality. His printed address to the Paleontological Society of Washington lacks references, but uses logic similar to Huxley's (1912, 1923b) completer narrative. Simpson emphasized that transitions to higher organizational complexity followed the auction chant *going, going, gone*, which cannot be read other than meaning that MTEs should be irreversible ratchet clicks. At the same time he understood that "[o]ne level of individuality can arise only to the degree that the subordinate level is suppressed" [in the sense of being somaticized]. He then contrasted this appreciation of irreversibility with the conclusion that "[m]ost animal and all human social groups are collectives the members of which retain complete metazoan individuality", noting that "[t]his distinction between group and individual is fundamental and has implications of the greatest importance, extending even into the political sphere" (Simpson, 1941). However, there is almost no text on irreversibility of MTEs in Maynard Smith and Szathmáry (1995). Those authors feared that reference to irreversibility could be associated with belief

in evolutionary progress, as they state in a brief section entitled "The fallacy of progress," arguing that progress has a bad Lamarckian name because it suggests belief in "ladder of life" orthogenesis. Huxley (1912) had resolved that issue by emphasizing the naturally selected adaptive design of subsequent hierarchical levels of organization, but none of that was mentioned.

The clearest statement on evolutionary progress is in a book chapter preceding the 1995 major transitions monograph, where Maynard Smith (Figure 3.1) agnostically suggests that life's complexity has simply drifted upwards over time: "It is true that, on the largest scale, evolution does seem to have given rise to increasingly complex organisms. But since the first living things were necessarily simple, it is not surprising that the most complex things alive today are more complex than their first progenitors" (Maynard Smith, 1988). A similar "mere diversification" idea was more recently argued to represent biology's first "zero-force" law (McShea and Brandon, 2010; McShea, 2016), suggesting that selection is a passive fringe phenomenon that will somehow reduce non-neutral variation when phenotypes diverge by random mutation. Such an inference turns things on their head because organizational complexity can only arise and be maintained by the continuous scrutiny of selection, a natural order of events that was clear to Dobzhansky (1958) as a geneticist, Pittendrigh (1958) as a chronobiologist, and Williams (1966a) as a phenotypic organismal biologist. In the monograph that denies the adaptive nature of the stepwise increases in life's organizational complexity, Maynard Smith and Szathmáry (1995) contradict the 1988 narrative when they write that "transitions must be explained in terms of immediate selective advantage to individual replicators" because "we are committed to the gene-centred approach

William Morton Wheeler   Julian Sorell Huxley

John Tyler Bonner   John Maynard Smith

Figure 3.1 The four major evolutionary transition pioneers of the 20th century. The photograph of Wheeler is from 1910 and the image of Huxley from ca. 1915, that is, from around the time they wrote their first monographs on ants (Wheeler, 1910) and individuality/organismality (Huxley, 1912). Both developed a largely coinciding, and mutually inspired, framework for understanding the distinct nested complexity levels that can be recognized in the overall organization of life. Bonner (image 2009) offered developmental perspectives on complexity transitions through the 1950s–1970s, and later adopted a more evolutionary approach (Bonner, 1988). Maynard Smith (image ca. 2000) addressed the topic primarily from a genetic conflict perspective, partly prompted by Buss (1987), but without making reference to insights by Wheeler, Huxley, or Bonner (Maynard Smith, 1988). He and Eörs Szathmáry later coined the term "major transitions in evolution" in a coauthored monograph (Maynard Smith and Szathmáry, 1995). The Wheeler image is from the public domain and the Huxley image is a courtesy of the Woodson Research Center, Fondren Library, Rice University. The Bonner image is reproduced with permission from Princeton University, and the image of Maynard Smith with permission from the AAAS.

outlined by Williams (1966), and made still more explicit by Dawkins (1976)." Sitting-on-the-fence attitudes towards adaptation have continued in recent MTE papers, echoing Maynard Smith (1988), who agreed with Buss (1987) that all one could say was that selection tends to operate at different levels simultaneously. In opposition to this ambiguity, I will maintain (below and in Chapters 4 and 5), that

the simultaneous operation of different levels of selection was irrelevant for MTE origins.

In North America, John Tyler Bonner (1974, 1988) (Figure 3.1) was among the first to consider questions of evolutionary progress. Like later evaluations by Dennett (1995), Bonner had no reservations in acknowledging progress, albeit with a critically investigative mind. As if he anticipated the fallacy of progress section in Maynard Smith and Sza-thmáry (1995—who do not cite Bonner's 1988 book or any of his previous books), he noted that "[t]here is an interesting blind spot among biologists. While we readily admit that the first organisms were bacteria-like and that the most complex organism of all is our own kind, it is considered bad form to take this as any kind of progression. [...] There is a subconscious desire among us to be democratic even about our position in the great scale of being. [...] In an early notebook Darwin cautioned that we should never use the terms 'lower' and 'higher' (although he was sensible enough not to follow his own advice), and I have been reprimanded in the past for doing just this" (Bonner, 1988, pp. 5–6). Denying that obvious progress needs a first-principle explanation has handicapped MTE research ever since. Valuable empirical details have been added, but the field made little progress in understanding how MTEs originated before their adaptive radiations took off (e.g., Calcott and Sterelny, 2011; Ruiz-Trillo and Nedelcu, 2015; Szathmáry, 2015), in spite of extensive mathematical modeling (e.g., Michod, 2000). Students of MTEs have continued to work from the general assumption that any progress in organizational and functional complexity, if that term could be lawfully used, has been an accidental emergent property—the result of a process that follows no general rules and does not need naturally selected (i.e. functional) adaptive design.

Reluctant to think in unambiguous terms about stepwise increases in complexity, all Maynard Smith and Szathmáry (1995) could offer was hand-waving "contingent irreversibility," consistent with the implicit hypothesis of bottom-up emergence presented in another book chapter by Maynard Smith (1991) that predated Maynard Smith and Szathmáry (1995). Here he envisaged that "[a]fter the higher-level entity has existed for a long time, its components may no longer be capable of independent existence. [...] There is, then, a kind of irreversibility associated with the transitions. This may help to explain the long-term stability of higher levels, but cannot help to explain their origins." This conclusion is deeply unsatisfactory, as Bonner (1995) noted in his review of Maynard Smith and Szathmáry (1995), writing that he missed a concluding section on "comparative transitionology." Maynard Smith and Szathmáry (1995) made the correct point, later elaborated by Bourke (2011, 2019), Clarke (2013, 2014), and West et al. (2015), that different mechanisms must be involved in the origin and secondary maintenance or elaboration of new levels of organizational complexity, but the persistent, often implicit, assumption of contingent emergence has precluded deeper understanding. Nor have the necessary and sufficient conditions for MTEs ever been reevaluated after Huxley's (1912) admirably insightful first attempt (details in Box 3.2). Although many intuitively endorsed a framework reminiscent of Huxley's levels 0–3 (e.g., Bonner, 1988; McShea, 2001a, 2016), the lack of clear qualifying criteria has meant that additional MTEs could be conjectured as a matter of taste. Authors proposing such additions, explicitly or implicitly, assumed that any form of non-continuous evolutionary change might well qualify as an MTE (e.g., Barrett, 2008), even though they would never satisfy Huxley's (1912) syndrome of necessary criteria that should apply across all MTEs (see page 63).

The unfortunate consequence of accepting what Bonner referred to as democratic inclusiveness is that theory becomes either untestable or that testing becomes circular. For example, McShea (2001b) recognized the ratchet-like nature of the fundamental Huxley-type MTE grades, but he then invoked continuity language by characterizing them as having stronger directional bias in building new higher levels in the hierarchy of life. This is not a mere semantic issue, for it muddles transition criteria that should remain the same no matter what level of organizational complexity is addressed. While Huxley (1912) consistently maintained such first-principle coherence (Box 3.2), the McShea approach allows a haphazard set of other discontinuous events of evolutionary change to apply for admission to a kind of premier MTE league where teams have become too grand to fail. In this view, all that matters in the lottery of life is upward drift to higher complexity and downward drift towards simplification. Other kinds of discontinuous change, sometimes referred to as "minor transitions" with "different degrees of individuation," may then reveal correlations among complexity traits, but such changes can have no bearing on the MTEs if we define them with Huxley's strictly hierarchical criteria. These recent developments make clear that evolutionary theory without natural selection as a driving and scrutinizing force becomes a vacuous descriptive exercise, unable to produce testable predictions because one is comparing "Bonner-democratized" apples and oranges. This is not to say that it would not be of interest to explore whether some obligate "minor transitions" became as irreversible as the Huxley MTEs and whether they achieved that state by the same set of closure rules, but such comparisons would require explicit focus on the necessary and sufficient conditions across all known MTE origins (to be elaborated in Chapters 4, 5 and 8).

The semantic confusion of major and minor transitions has been particularly acute for what Maynard Smith and Szathmáry (1995) called the origin of social groups, what McShea (2001b) called the origin of colonies, and what Michod (2000) called the origin of societies. All these authors referred to E.O. Wilson's (1971, 1975) definition of eusociality. Wilson chose to define that term in an all-inclusive manner, dismissing an earlier history of precise meaning by Suzanne Batra (1966), who defined the term in the same way as did Julian Huxley (1930) when coining the second *true social or colonial grade* of social insects (see previous section). Both referred to subsocial insects where joint nesting between mother and offspring had become obligate but where physically distinct castes had not evolved (see Chapters 4 and 6) – a colonial state clearly distinct from superorganisms sensu Wheeler, which consist only of morphologically distinct caste phenotypes. In contrast, the all-inclusive Wilsonian definition of eusociality was a semantic maneuver never properly explained or defended that concealed the irreversible, evolutionary convergent MTEs to superorganismality in the ants, corbiculate bees, vespine wasps, and higher termites that had been obvious not only to Darwin, Huxley, and Wheeler, but to most evolutionary biologists until the 1960s (Boomsma and Gawne, 2018). By using imprecise sociobiology terms as their frame of reference (already in Maynard Smith, 1978a), Maynard Smith and Szathmáry could not discriminate among types of social organization that differed fundamentally in organizational complexity. And, as expected for an "inclusively defined" concept that had unjustifiably merged societies and irreversibly more complex superorganisms, it was no surprise that "reversals of eusociality" could indeed be found (Wcislo and Danforth, 1997). Imprecise definitions thus breed imprecise science—confusion emanating from semantic relaxation of definitions has haunted research for decades by turning unambiguous natural history facts into artificial human constructs that cannot have a Darwinian explanation.

My current analysis agrees with McShea (2016) that Maynard Smith and Szathmáry (1995) did not provide consistent criteria for MTEs (see also McShea and Simpson, 2011), but he did not distinguish either between obligate forms of social life representing deeply ancestral MTEs and facultative forms of coming or staying together that usually did not proceed very far, were reversible, and never achieved MTEs. As already noted, McShea's approach seems to represent a full surrender to the idea that complexity only originates by upward drift, a null model that does neither invite contemplation *sensu* Williams (1966a) of what type of adaptive states natural selection can in fact produce, nor awareness of the principle that no organizational complexity can have arisen without natural selection (Pittendrigh, 1958). When we know that a colonial superorganism sensu Wheeler has subordinated somatic workers, whose bodies consist mostly of somaticized cells, and whose cells have formerly independent but deeply somaticized mitochondria, it seems puzzling that a comparative analysis can lead to the conclusion that "no trend has been demonstrated in sophistication, advancement, or excellence" that could justify progressive advance in complexity in a broad "colloquial sense" (McShea, 2016). This phrasing underlines how anthropomorphic human constructs preclude understanding of the hierarchical structure of life's fundamental levels of organization. It is interesting to note that McShea (2016) acknowledges the mid-20th-century "systems biology" school for their views on organismal autonomy, while their insights were virtually identical to what Huxley (1912) wrote down from an adaptation-focused perspective. We will encounter another such example of incorrect crediting in Chapter 8, underlining that the disappearance of Huxley's first book into oblivion had a considerable negative effect on current understanding of the MTEs.

These incoherences appear to have a deeper cause in the rarely made explicit confusion between parsimonious Darwinian *understanding* of adaptive design and the *interpretation* of life's patterns of diversity.

The latter cannot explain adaptation but often pretends to have a form of encompassing feel-good inclusiveness—a contrast that I likened to pursuing least common multiples in calculus (see Box 1.3). Least common multiple approaches always suggest that environments are drivers rather than just modifiers of the outcome of natural selection, a form of disorientation that greatest common divider approaches would normally avoid. Comparing *Adaptation and Environment* by Brandon (1990) and *Adaptation and Natural Selection* by Williams (1966a) illustrates the difference very clearly. Written a quarter of a century later, Brandon's monograph implicitly denies Williams' first-principle "greatest common divider" Darwinism that aims for general understanding vulnerable to Popperian refutation. Closer inspection reveals that the arguments in Brandon's narrative do not meet the criteria of good scientific theory—there are no new predictions and no evaluation of how contentions would be an advance relative to existing understanding. In this broad-brush environmentalized approach, every new data set pushes the least common multiple further upwards as a zoomed-out descriptor, while new data tend to zoom in as meaningful tests of established theory in a Darwinian greatest common divider approach. A more recent monograph by Godfrey-Smith (2009) took a similar approach, first defining new terms without explaining why they are helpful or even necessary, and then producing a narrative that complicates descriptions without giving new answers—while carefully avoiding if not dismissing the topic of adaptation by natural selection. In the next section, I will use greatest common divider inclusive fitness arguments to try carve nature's organizational complexity at its most fundamental joints while offering a general testable framework for the origins of all nonhuman MTEs.

Conceptualizing the major transitions and their adaptive origins

To conceptualize MTEs from first principles, Huxley's notion of closure needs to be formulated in terms of genetic information and commitment shared by non-redundant co-replicons (see Box 2.2; Table 3.1). It is also necessary to define MTEs such that they always jump from one domain of social evolution to the next higher domain of organizational complexity. This implies that MTEs are, as organismality phenomena, very different from societies that emerge recurrently within existing domains of social evolution. When a novel event of informational closure that initiates an MTE has no relation to forms of society in the ancestral domain, it must be a functional exaptation (rather than a structural one as in Gould and Vrba, 1982), reminiscent of what Mayr (1960) wrote about the emergence of evolutionary novelties in his Darwin centennial contribution (see Chapter 1). Conceptualizing life's nested levels of organizational complexity (prokaryote cell, eukaryote cell, obligate multicellularity, superorganismality) as sequential events of informational closure means that a transition was only an MTE when it both preserved already present organization and added a new layer to the hierarchy. Every MTE's domestication (Table 3.1) of a previous level of independent agency therefore jumped levels of adaptation rather than combining levels of selection as societies do. Defining closure in terms of a minimally necessary set of lifetime committed carriers of genetic information means that the division of labor process could be realized either internally or externally. Early MTEs were based on internal domestication when the first cell membrane physically closed around some ribosomal structures and when the eukaryote cell originated via higher-level membrane closure, but later MTEs were based on external somatization of clonal cell copies

or full-sibling offspring. However, no matter which model applied, subsequent division of labor needed to evolve in order to cement an MTE's potential for adaptive radiation.

As Huxley (1912) realized, an MTE has become irreversible when its novel naturally selected division of labor has made all parts obligatorily interdependent. Combining this insight with understanding that "[n]atural selection is the only acceptable explanation for the genesis and maintenance of adaptation" (Williams, 1966a, p. vii; p. v in the second 1996 edition), it follows that every MTE origin must have produced an adaptive innovation syndrome that enabled niche space expansion in competition with other life forms—in particular its sister lineage that maintained and elaborated less complex and more redundant forms of society (Table 3.1). The MTEs as I define them (Table 3.1) thus represent the most unambiguous examples of progress in evolution, a principle that can be visualized in a phylogenetic MTE tree (Figure 3.2a) to make explicit the domains of organizational complexity that social evolution could produce in the wake of MTEs. This figure is the backbone of this book, for it communicates what I mean by clicks of the MTE ratchet (every line drawn between ellipses) and by domains (the groupings of convergently evolved MTEs fitting within same-color ellipses). I have drawn the figure so that progress in organizational complexity across domains goes from left to right rather than from bottom to top to avoid kneejerk responses of orthogenesis. The diagram illustrates that some MTEs, such as the start of cellular life (last universal common ancestor (LUCA)), the last eukaryote common ancestor (LECA), and the human domain, were unique, while other MTEs happened a handful of times independently. The latter events are known as fraternal MTEs (Queller, 1997, 2000) because they involved obligate cooperation between individuals from the same gene pool during external domestication of previously independent agency, in contrast to the unique egalitarian MTEs that merged elements from different pools of information during internal domestication. The latter includes the human MTE which can be conceptualized as a symbiosis between algorithmic symbionts and individual host brains, as I will elaborate in Chapter 8.

I have made the discreteness of the nonhuman domains of social evolution explicit in Figure 3.2b by plotting the stacked nestedness of present and previous (domesticated) levels of organizational complexity on the horizontal axis. With minor modification, these are identical to what Huxley (1912, 2022) referred to as grades of individuality or levels of organismality (Table 3.1). The y-axis plots degrees of individuality as a continuous variable representing variation in complexity within adaptive radiations produced by separate MTE origins. Each "brick" defines a nested level of phenotypic agency, empowered by the latest advance in stepwise organizational complexity (asterisks) that domesticated the previous level (now gray punctuated). These metaphorical bricks are internally heterogeneous, not only at their top-of-the-stack level of agency, but also at the lower levels that were preserved as functional remnants of closed, previously domesticated organizational complexity. The white bands, drawn across the differently colored bricks, represent irreducible and functionally self-organized complexity that organismal biologists aim to interpret from above, using both ultimate (adaptive and phylogenetic) and proximate (physiological and developmental) understanding. The total metaphorical white-zone cascade thus pragmatically represents the dividing line between organismal biology pursuing heuristic top-down understanding while mostly avoiding deterministic reductionism and molecular biology approaching unknown complexity

with bottom-up reductionism (terms given in the bottom right brick of Figure 3.2b apply to all bricks). The reductionist molecular branches of biology can, up to a point, pretend to only need the laws of physics and chemistry but organismal biology approaches always need Darwinian natural selection theory to explain how adaptive complexity originated and is maintained.

The stylized red trees in Figure 3.2b illustrate how variation in degree of individuality has likely evolved during adaptive radiation following an MTE origin. First, organizational complexity will have increased beyond the MTE baseline, both solitarily and when societies were formed. Second, lineages may experience later reductions in degree of individuality due to over-specialization in parasitic or mutualistic niches, as Huxley (1912) already noted (Box 3.2), implying that organizational complexity can become eroded to levels below the MTE origin baseline (downward red branches). However, closer inspection always reveals that these reductions are never reversals to a previous domain of social evolution, for the level of obligate cooperation that defined the MTE origin will always be retained even when elements are lost. Third, no branch tips of the red adaptive radiation trees ever produced a new MTE, no matter what size of society was achieved or what symbionts were involved. This is because society partnerships remain redundant by definition, so that selection for new obligate reproductive division of labor cannot be sufficiently consistent, irrespective of partners being cytoplasmatic elements, neighboring cells, or family members. Instead, MTEs are hypothesized always to have emerged from generalist ancestors with low degrees of individuality in their ancestral domain (relative to the domain's average) and to have evolved a novel type of closure as an exaptation that happened to prevent partnership promiscuity or redundancy (see Table 3.1 for definitions). As Huxley (1912) already noted, ectosymbiotic mutualisms also have the potential to increase in combined degree of individuality (Figure 3.2c), but I conjecture that known symbioses of this kind have not led to MTEs, in contrast to the endosymbiotic origin of LECA. This in spite of some adaptive radiations based on pairwise lifetime commitment between ectosymbiotic partners offering compelling analogies to how endosymbiotic egalitarian MTEs could have arisen (see Chapters 5 and 8).

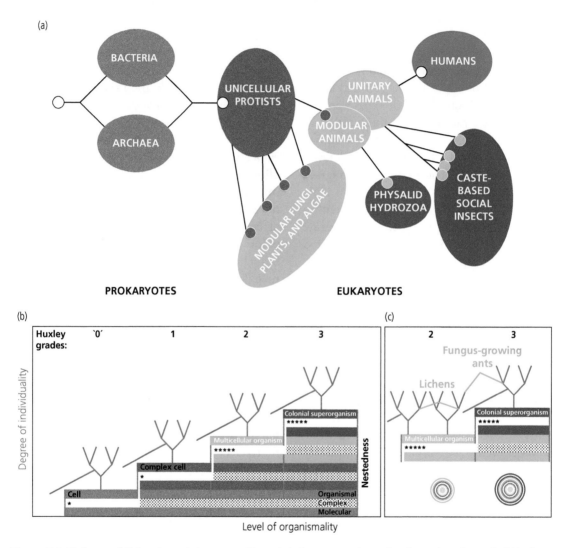

Figure 3.2 Phylogeny of life's major evolutionary transitions, their discrete nested organismality, and continuous individuality. (a) The major transitions in evolution (MTEs) defined as irreversible transitions in organizational complexity that realized global adaptive radiations with (from left to right) the origin of cellular life (last universal common ancestor (LUCA)), the origin of the complex cell of the last eukaryote common ancestor (LECA), the multiple transitions to obligate organismal multicellularity (green algae and plants; red algae, brown algae, some fungi, animals), the multiple transitions to colonial superorganismality (ants, bees, *(Continued)*

When I maintain that the white irreducible complexity zones of Figure 3.2b are home to what is known as self-organized complexity, that contention is not a metaphysical form of reductionism denial but acknowledgment of the fact that subordinate agents with an aligned interest in the best performance of their whole have collective agency with very limited proximate predictability. For example, the individual qualifications of players in two soccer teams offer statistical odds for the outcome of a match but give essentially no information on how they will achieve a final result via the huge number of feedback loops that their interactions set in motion. Much of the impressive collective performance of a colony of ants or bees is known to be self-organized (Bourke and Franks, 1995; Camazine et al., 2001), and the same is true for the way in which leukocytes patrol a metazoan body and for the assembly and disassembly of cytoplasmatic proteins (Nicholson, 2013, 2019). As a phenomenon, self-organization has produced mixed responses among organismal biologists, because it originated in physics (Prigogine and Nicolis, 1971) and has been conjectured to create order for free (Kauffman, 1993). However, the belief among many system biologists that natural selection is not a universal requirement for adaptive organization is untenable from a

---

Figure 3.2 (*Continued*)

wasps, termites, and physalid Hydrozoa), and the major transition that produced *Homo sapiens*. The first prokaryote and eukaryote MTEs represent egalitarian transitions (white dots) and the convergent transitions to multicellular organismality and colonial superorganismality (solid gray and blue dots) represent fraternal transitions (Queller, 2000). The transition to the human condition is special because it was driven by the accumulation of cultural traits, but its symbiotic character makes it formally an egalitarian transition. (b) Huxley (1912) distinguished between organizational complexity of grade 1 (eukaryote cellular), grade 2 (somatic eukaryote multicellular), and grade 3 (eukaryote caste-differentiated insect colonies and physaliid Hydrozoa), with bacteria (largely unknown at that time) as unspecified base-level "grade 0," giving the four levels of organizational complexity that would remain canonical until Bonner (1988). The diagram illustrates the stepwise domains of social evolution with their nested organizational complexity as stacked bricks, each defined by a hierarchically progressive level of non-redundant genetic-informational closure (see Table 3.1 for definitions of terms). As a consequence, domain-specific units with independent agency became "somaticized" subunits in a next domain of social evolution, with the core family as the highest level at which adaptive somatic design can evolve for the exclusive higher-level benefit (Williams, 1966a; Gardner and Welch, 2011). Each MTE that made a lineage enter a new domain of social evolution produced its own adaptive radiation tree (only one such red tree is drawn here for each Huxley grade; multicellularity and superorganismality have a handful each as shown in panel (a)). These red trees have mostly ascending branches because organizational complexity usually continued to increase, but also some hanging branches representing secondary reductions (but not reversals—see main text) of organizational complexity after specialization in parasitic or mutualistic niches (Huxley, 1912). The MTEs that emerged from each new level of closure (asterisks; single for egalitarian MTEs and five for fraternal MTEs) also created new independent levels of irreducibly complex self-organization (white zones) that are maintained throughout the nested hierarchy of organizational complexity while previous self-organization layers became more deeply somaticized at each subsequent MTE origin (gray dotted white zones). Each metaphorical brick thus has a central irreducibly complex layer that is accessible from the top down by organismal understanding and from the bottom up by reductionist molecular understanding. (c) Interspecific mutualistic dependencies can be included in this type of diagram by connecting branch nodes or tips within or across domains (green wedges), here represented by lichens (merging a multicellular algal and fungal partner) and attine fungus-growing ants (merging a superorganismal ant colony and a multicellular fungal cultivar partner) as examples. In Huxley's (1912) view, such mutualisms have combined individualities that reflect the symbiotic sum when that amounts to more than the partners' individualities separately, a minor increase in lichens and a steeper green wedge in the fungus-growing ants because the ant and multicellular fungal lineages coadapted in bilateral symmetry. The concentric circle icons (that can also be drawn for the cellular domains where they have only one or two gray "Russian doll" layers) characterize domain-specific lineages in schematic cross section showing the subsequent concentric levels (grades) of organizational complexity, a type of imagery that I will reuse in Figure 8.3 to evaluate the MTEs in terms of their discontinuous increase in control information.

neo-Darwinian perspective, for organization only becomes adaptive when the variants produced for free by self-organization have been recurrently subjected to natural selection (see Box 1.2). Denying that natural selection must have shaped the characteristics of lower-level units is as misleading as maintaining that individual talent, training, and selection are unimportant for soccer teams. In Chapter 8, I will return to the idea that the irreducible white- and gray-punctuated complexity layers in Figure 3.2b represent level-specific capacities to self-organize control information under the auspices of natural selection (Corning and Kline, 1998a, 1998b).

The nested hierarchical complexity of my MTE definition (Table 3.1) can be illustrated by the naturally selected traits of a honeybee colony, which include (1) a self-organized adaptive colony ontogeny mediating traits like the dance language, queen replacement procedures, and collective social immune defenses; (2) a self-organized individual ontogeny producing different caste phenotypes from initially totipotent larvae; (3) a self-organized patrolling of immune cells in single bee bodies having retained phagocytosis capacity reminiscent of free-living protist ancestors; and (4) self-organized interactions within each body cell when prokaryotic mitochondrial symbionts help to metabolize nutrients and repair damage in a carefully controlled manner. This entire cascade is coordinated by the honeybee nuclear genome and expressed via interactions with the multiple, and likewise nested, levels of environmental exposure that Williams (1966a, p. 259) specified (see Box 2.3). A sample of the entire organizational history of life is thus represented in the genomes of superorganismal bees, ants, wasps, and termites, giving them four levels of adaptive biological complexity (Figure 3.2b) where humans have only three (see Chapter 8). Closed family is the highest level for specific

adaptation syndromes to evolve, as Williams (1966a, pp. viii, 161, 197–201, 257) emphasized repeatedly. However, this condition needs to be the *full-sibling* family, consistent with the necessary condition for MTE origins being non-redundant genetic information closure— this strict condition secures there is only a single higher level of adaptation during an MTE origin rather than the usual coexistence of two main levels of selection that characterizes all forms of society. Williams' (1966a, pp. vii, 4–5) principle that "adaptation should be attributed to no higher a level of organization than is demanded by the evidence" thus refers up-front to a single higher level when MTEs originate and mostly to the lower-level in the sister lineages that retained society life with various forms of partnership redundancy.

Another way of phrasing Williams' level-of-organization principle is that parsimony demands rejection of higher-level adaptation unless lower-level explanations fail. The honeybee dance language is a good example of such failure, because it is adaptive only at the superorganismal family level and meaningless at the individual bee level. It could evolve by family selection that first produced the MTE origin of the corbiculate bee lineage, establishing the lifetime celibacy and obligate altruism of entire worker castes in the common ancestor of the bumblebees, stingless bees, and honeybees. The next steps, during the adaptive radiation that produced the extant branches of this superorganismal clade, must have been additional selection for improved collective performance of these worker castes. Dance language subsequently evolved only in the evolutionarily derived honeybees and could not possibly evolve in a non-superorganismal social insect even when its societies reached similar colony sizes as, for example, in some epiponine wasps (see Chapter 4). The reason is that dance language represents a self-organized higher-level form of collective agency, analogous to but beyond self-organized agency among the cell types in a metazoan body (see Table 3.1 for definitions). Being precise about the level at which naturally selected agency operates is therefore essential and related to the level of environment that such agency interacts with. As stipulated by Williams (1966a, pp. 61–62), it is important not to use "environment" in a hand-waving manner, but to distinguish between the *genetic*, the *somatic*, the *social*, and the *ecological* environment, where the genetic environment refers to other genes in the same genome and the somatic environment to other (organelle) genomes in the same cell (see Box 2.3). The ecological environment represents the final orbit and comes in different layers itself: (1) the biotic environment defined by mutualistic, competitive, predatory, and parasitic interactions; and (2) the abiotic environment.

My approach represents a firm break with recent practices in which no attempts were made to concisely differentiate between levels and degrees of organismality and individuality, in contrast to Wheeler's and Huxley's clarity of understanding in the early 20th century (see Folse and Roughgarden, 2010; Clarke, 2010, 2013, 2014; Lidgard and Nyhart, 2017, for partial exceptions). The century-long perspective across the three first chapters of this book has allowed me to explicitly connect to the origin of neo-Darwinian animal behavior studies (see also Johnston, 1995). This neglected history gives the emergence of the gene's eye view

around 1960 a deeper perspective, starting with Pittendrigh's (1958) "legalization" of teleonomic agency, an approach that helped Williams (1966a) to formulate his general gene's eye view of evolution, focusing on the parsimonious albeit heuristic explanation of adaptive design (Table 3.1). Williams' book amplified Tinbergen's efforts to turn the study of animal behavior into a quantitative field-orientated science, and provided a firm platform on which Hamilton's inclusive fitness theory could be further developed and tested, as first reviewed, elaborated and disseminated by Dawkins (1976, 1982a). In the chapters to follow, I will explore the validity of Figure 3.2, by confronting its conjectures with existing gene's eye theory about the evolution of altruism and mutualism (see Chapters 4 and 5) and by evaluating the empirical support (see Chapters 6 and 7). A neo-Darwinian synthesis appeared to be, by and large, agreed upon around the turn of the millennium (e.g., Seger and Stubblefield, 1996; Eshel and Feldman, 2001; Godfrey-Smith, 2001; Leigh, 2001; Grafen, 2002). However, some unexpected erosion emerged during the last two decades, owing to the kind of confusion that I highlighted in the previous section. In Chapter 8, I will revisit the root cause of the most prevailing manifestation of such misunderstanding, the improper separation between genetic information itself and the environmental conditions shaping that naturally selected information as it is stored in population gene pools.

CHAPTER 4

Necessary and sufficient conditions for major evolutionary transitions

Summary

From an individual perspective, there is a fundamental difference between cooperation for mutual benefit and self-sacrificing altruism. However, there is no such difference for the gene's eye view of social evolution, which stipulates that both types of cooperation are equally self-serving. Gene's eye explanations of altruism started with the pedigree version of Hamilton's rule, and were later generalized when Price equation logic produced a statistical and fully general genetic theory of social evolution, and an approximate phenotypic theory amenable to empirical testing. However, this generalization applied to societies with redundant partnerships where relatedness is variable and social adaptations are mediated by condition-dependent altruism of cellular or multicellular individuals. Under such conditions, adaptations for the exclusive benefit of the collective higher-level of organizational complexity cannot evolve. Yet, such adaptations of unconditional somatic altruism universally characterize the convergent major evolutionary transitions (MTEs) towards multicellular organismality and colonial superorganismality. I show that the origins of (super)organismal MTEs can be conjectured to have required invariably maximal relatedness among cell copies or siblings, owing to lifetime commitment between a pair of gametes or monogamous parents. Such pairwise functional closure in terms of genetic information partitions Hamilton's rule in an invariant necessary condition of relatedness equivalence between dispersing and adhering offspring, and a contingent sufficiency condition ($b/c > 1$) that is additionally required to forge an MTE origin. MTEs are thus expected to evolve by directional kin selection for unconditional, obligate altruism by entire offspring cohorts, which is different from individual selection for conditional altruism in societies. I contrast the predictions of the classical continuous version of Hamilton's rule for societies and the partitioned discontinuous version applicable to MTE origins, and argue that conflict reduction stabilizes societies but played no role in the origins of multicellular organisms and colonial superorganisms.

> *Hamilton's theory, now often labelled 'kin selection' (Maynard Smith's name, not Hamilton's own), follows directly from the neo-Darwinian 'Modern Synthesis'—directly, in the sense that kin selection is not an extra, not an addition bolted on to the neo-Darwinian synthesis: it is a necessary part of the synthesis. You cannot divorce kin selection from neo-Darwinism, any more than you can divorce the Pythagorean theorem from Euclidean geometry.*

Dawkins, *An Appetite for Wonder* (2013, p. 197)

In the preceding chapters, I gave historical and conceptual context for Dawkins' quote. The completion of an encompassing neo-Darwinian synthesis was framed by Niko Tinbergen and initiated, within this frame, by George Williams' and William Hamilton's

Domains and Major Transitions of Social Evolution. Jacobus J. Boomsma, Oxford University Press. © Jacobus J. Boomsma (2022).
DOI: 10.1093/oso/9780198746171.003.0004

gene's eye view of adaptative evolution. Their neo-Darwinian theory is rooted in Darwin, Wallace, and Weismann and was championed by Fisher—the only synthesis architect who coherently integrated all three necessary elements of evolutionary theory—Darwinism, Mendelism, and biometry (Provine, 1971). Building on that foundation, Williams and Hamilton established that genes are both the drivers and archives of adaptation, the only units permanent enough to store information about the adaptive achievements of populations of unicellular and multicellular organisms across generations (Williams, 1966a; Dawkins, 1976, 1982; Cronin, 1991; Frank, 2012a, 2012b, 2012c, 2012d). I showed that the early neo-Darwinian synthesis was primarily a British achievement, for first-principle thinking about adaptation continued to mature in the UK during the decades in which North Americans developed an adaptation-agnostic version of population genetics to achieve a pragmatic and proximate "modern synthesis" around 1960. Moreover, I argued that the punctuated and geographically divergent history of neo-Darwinian understanding muddled discussions about the fundamentally progressive nature of adaptation. Even after removing political and anthropomorphic bias, discussions often failed to distinguish between gradual phylogenetic "progress or regress" and stepwise advances in organizational complexity, known as the major transitions in evolution (MTEs) (see Table 3.1). I also showed that neo-Darwinian thinking about MTEs was transparent until World War I, then faded away and did not return to that same clarity in postmodern organismal biology.

In this chapter, I return to inclusive fitness theory to elaborate on the text boxes of Chapter 2, updating where the theory stands

and which aspects still divide theoreticians. I will conclude that inclusive fitness, as a scientific theory in the sense of Popper (1975), remains the only viable framework for understanding adaptation in Tinbergen's ultimate sense, for no significant empirical anomalies have been identified that would threaten the underlying view of how adaptive phenotypic agency is shaped by the information stored in genes. However, this achievement—what I call the continuous version of inclusive fitness theory—is incomplete. If Hamilton's rule deserves the foundational status claimed in the opening quotes of Chapter 2 and this chapter, it should also be able to explain the dilemma of progress expressed in the opening quote of Chapter 3. Moreover, it should offer that understanding while meeting the parsimonious causal sufficiency principle expressed in the opening quote of Chapter 1. My main point will be that we need an explicit discontinuous version of Hamilton's rule to explain the stepwise advances in level of organismality (see Table 3.1) that were first identified as grades by Huxley (1912) (see also Figure 3.2b). I will outline the contours of that kin selection theory for organismality here and in Chapter 5, and I will review the evidence in Chapters 6 and 7. I will argue that the idea of gradual emergence via size increase and conflict reduction is inadequate as an explanation of MTE origins, and that continuous (Pricean) inclusive fitness or levels-of-selection models fail to explain MTE origins. The rationale is that only lifetime commitment upfront between two gametes or two parents giving rise to clones of cells or families with maximally related full-sibling offspring should offer a universal necessary condition for achieving MTEs. My theory implies that the MTEs that did evolve originated according to the same set of functional rules.

Families, groups, and the gap between them

We have already seen that cooperation can be for both mutual and altruistic benefits (see Box 2.1). It is important to keep these forms of social evolution apart, because altruism requires positive relatedness between interacting agents, while mutualism can evolve between nonrelatives, either of the same species, as in a prisoner's dilemma or a tragedy of the commons setting (Foster, 2004; Bourke, 2011a), or of different species, as in host–symbiont interactions (Leigh, 2010a). Figure 4.1 presents a nested diagram of the ultimate contexts and proximate mechanisms involved; it shows that the two types of cooperation only have overlap in some proximate mechanisms. Here I focus on the right-hand side of the diagram, that is, on how inclusive fitness theory allows understanding the different ways in which altruism can be expressed. I will address interspecific mutualisms in Chapter 5. Figure 4.1 is a simplification of an earlier scheme by West et al. (2007a, 2007b), designed to relate the diagram's keywords more parsimoniously to Tinbergen's ultimate and proximate scheme of asking complementary questions in organismal biology (see Figure 1.1; Box 1.3). Hamilton's (1963, 1964a) original theory elucidated altruism among family members—hence Maynard Smith (1964) coined the term kin selection. Already at that early stage, Hamilton (1964a) abstracted the theory to also include the hypothetical thought experiment that not just individuals but also single genes could obtain phenotypic agency to propagate their interests, a concept that was later named green-beard dynamics (Dawkins, 1976). Altruism and green beards are analogous: kin selection needs kin discrimination, and green beards need gene discrimination. However,

single-gene agency (see Table 3.1) is very rare, so inclusive fitness is primarily about complex organismal phenotypes and quantitative genetics (Gardner and West, 2009; West and Gardner, 2010; Gardner, 2019; Madgwick et al., 2019; Scott and West, 2019)(see Box 2.3).

Inclusive fitness theory was generalized after Hamilton was inspired by George Price's (1970) covariance equation (Hamilton, 1970, 1975). This produced a conceptual framework that accommodates selection and transmission of abstract genes across nested levels of social structure, allowing the theory to handle different levels of selection without the stringent population structure that explicit group selection models need (Frank, 2013b). A practical advantage, foreseen by Hamilton (1964a), is that social behaviors in viscous (limited dispersal) populations of microbes could also be captured in inclusive fitness terms, so the theory became independent of pedigree information (Crespi, 2001; West et al., 2007c). The logic of the Price equation made it possible to conceptualize family life as a form of limited dispersal that affects the scale of competition (West et al., 2001, 2007b; Frank, 2012a). However, this generalization implicitly suggested that *family* was merely another case of *group* and thus no longer worthy of separate consideration. This is not a problem as long as one is interested in determining how gene frequencies change in response to selection at different levels, but the two approaches are not equivalent if one asks how complex adaptations can arise. As Williams (1966a) pointed out repeatedly, adaptations for the sole benefit of a group can evolve in core families but not in other kinds of groups. For example, he wrote that "group-related adaptations do not, in fact, exist" and that "[a] group in this discussion should be understood to mean something other than a family and to be composed of individuals that need not be closely related" (Williams

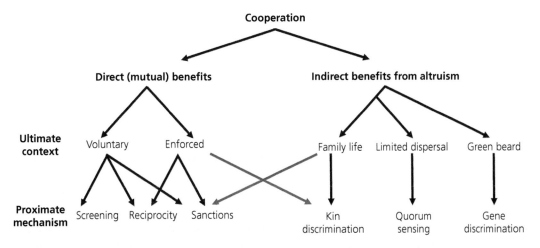

Figure 4.1 Partitioning cooperation for direct and indirect benefits into ultimate and proximate components. A hierarchy modified and simplified from West et al. (2007a, 2007b), underlining that without indirect fitness as a driving force towards altruism (right hand side), cooperation must have direct mutual benefits for all parties involved (left hand side). Cooperation is then either voluntary, in which case a priori partner screening is an important maintenance mechanism, or it needs to be a posteriori reinforced by sanctioning free-riding symbionts. If cooperation is top-down enforced by vertical symbiont transmission, reciprocity happens by default unless secondary symbiont invasion breaches functional closure and induces selection for sanctioning responses. Working with these categories updates earlier classifications restricted to non-mutually exclusive symbiont choice and partner-fidelity feedback (Bull and Rice, 1991; Sachs et al., 2004), and allows incorporation of recent empirical data while mostly retaining the parsimonious terminology of West et al. (2007b). A detailed discussion of cooperation for mutual benefit will be provided in Chapter 5. The topic of the present chapter, altruism, concerns behavioral or life-history traits that can only be maintained if the indirect reproductive benefits exceed the direct reproductive costs for the individuals expressing altruism (see Box 2.1). If that condition is fulfilled, altruism is voluntary by definition. For altruism to be expressed, it is necessary that social interactants are positively related, that is, that they share more genes by common descent than two random members of a population. Relatedness can also be expressed at the level of a single gene or a non-recombining supergene, in which case the green-beard metaphor applies (West and Gardner, 2013). Altruism by gene-level discrimination is normally detrimental to the fitness interests of the remaining genome (West and Gardner, 2013), so green-beard phenomena are of great proof-of-concept interest but remain rare (Scott and West, 2019) unless one conceptualizes non-additive genetic effects as behaving in a green beard like manner (Queller, 2011, see also Box 5.1). Alternatively, and much more commonly, genetic relatedness is relevant across the genomes of cellular or multicellular individuals (Batty et al., 2014). In such situations, proximate cell-level agency can be mediated by quorum sensing in viscous populations, usually for assessing density as a proxy for the percentage of clone mates in the vicinity. However, when social traits are expressed in families of multicellular individuals, variation in pedigree relatedness is key and may give rise to proximate recognition mechanisms of direct kin discrimination, when pedigree connections are known to agents, or to indirect kin discrimination, when agents can infer them from context. Although acts of cooperation for mutual or altruistic benefits mostly involve separate ultimate contexts and proximate mechanisms, the latter are not always mutually exclusive, that is, sanctions (left-pointing blue arrow) may be used to resolve conflicts in both genetically heterogeneous families and in mutualistic partnerships with chimeric (i.e., redundant; see Table 3.1) cohorts of symbionts. In addition, enforced vertically acquired symbionts may (with or without their host) practice a combination of sanctions and kin discrimination (right-pointing blue arrow) when additional (redundant) symbionts can secondarily invade (e.g., Poulsen and Boomsma, 2005).

1966a, p. 93). Separating selection and adaptation is essential here, for it implies that MTE origins, which are about higher-level adaptive syndromes, need an explanation that traditional levels-of-selection theory cannot provide.

Box 4.1 summarizes the differences between family selection and group selection as they historically unfolded. If Williams is right, it must be family selection (i.e., Hamilton's original pedigree kin selection) that can produce higher-level altruistic adaptations when Hamilton's rule is fulfilled (bottom right Figure 4.2a), whereas population viscosity or other kinds of group structuring (top left Figure 4.2a) cannot. This distinction relates to the necessity to use an "others only" definition of altruism as Hamilton (1964a) had in mind when focusing on pedigree relatedness (see Box 2.1). This definition cannot be used in population viscosity scenarios (e.g., Gardner, 2015a) where one has to accept a less specific all-inclusive definition. Such cases are often referred to as trait group altruism, where altruism includes benefits to self and refers to selection for "weak altruism" (reviewed by Pepper, 2000; Kerr et al., 2004), even though that would fail to be altruism proper in the scheme of Figure 4.1. It is interesting, therefore, that kin selection and trait-group selection could be reconciled via the Price equation (Figure 4.3), but only with neighbor-joining notation that removes the explicit causation that is obvious when working with pedigree relationships. The opposite red arrows in Figure 4.2a illustrate that only the bottom-right scenario allows for the consistent adaptive phenotypic improvement (Birch, 2019) that could ultimately establish obligate somatic altruism, an insight that goes back to Williams and Williams (1957) and ultimately to Fisher (1918, 1930). Without the Williams/Fisher conceptualization, the investigator needs to work with demes as locally inbred subpopulations *sensu* Wright (1949)—referred to as metapopulations by Wade (2016). However, it seems impossible that complex higher-level adaptations can evolve that way (Maynard Smith, 1982b; although Wade (2016) suggested otherwise; see also Williams, 1985).

All this implies that there are good reasons to distinguish between groups in viscous populations and families where there is immediate

---

Box 4.1 Group selection versus family selection and the levels of adaptation

As explained in Chapter 1, good-for-the-species thinking was common in the first half of the 20th century, but came under attack when Wynne-Edwards (1962) made its implications explicit. However, even after Wynne-Edwards' "old style group selection" was refuted, Maynard Smith (1982b) emphasized that considerable semantic confusion remained almost two decades after he had defined the difference between kin selection and group selection (Maynard Smith, 1964; see also Leigh, 2010b). In an era before Price equation formality had become the standard, Maynard Smith argued that kin selection for altruistic traits is always based on interactions between family members in randomly mating populations, whereas group selection for altruism *sensu* Wynne-Edwards or Sewall Wright always relies on reproductively isolated randomly composed groups (demes). This confusion continued after David Sloan Wilson (1975) introduced so-called synergistic selection or trait-group selection models operating in ephemeral non-reproductive demes—"new group selection" now meant differential deme-wise survival and propagule contribution to a population-wide mating pool. Maynard Smith (1982b) concluded that old style group selection could only contribute to gene-frequency change when demes are small and genetically isolated (see also Williams, 1966a; Leigh, 2010b). He then showed that synergistic trait-group selection also requires small groups and argued that it matters whether alleles induce their bearers to be unconditional or conditional altruists because the latter would evolve much more easily. Since then, it has generally been acknowledged that Hamiltonian inclusive fitness logic is about the evolution of conditional altruism (Charlesworth, 1978; Bourke, 2011a).

Box 4.1 *Continued*

It is important to note that trait-group selection models have been reconciled with the most general forms of kin selection theory via the Price equation (e.g., Keller, 1999; Lehmann et al., 2007; Foster, 2009; Leigh, 2010b; Gardner et al., 2011) (Figure 4.3), even though Maynard Smith considered these models to address mutualistic cooperation between nonrelatives, a logical inference when altruism is conceptualized by the Hamiltonian 'others only' definition (see Box 2.1). However, for the purpose of this book, I am not interested in gene-frequency change per se, but in this overarching question: can group selection of any kind produce adaptations that evolved only for the benefit of the group? In that respect, group selection approaches, based ultimately on Sewell Wright's (1932) shifting balance theory, have not fared well (Coyne et al., 2000) (but see Goodnight and Wade, 2000, for a response). Even though single genes coding for social traits can go to fixation by some combination of selection and genetic drift in abstract models, there is no way of knowing whether such derived states are evolutionarily stable, that is, whether they would be resistant to invasion by other mutants encoding novel, quantitatively different variants of the same social trait. Experimental work in the lab has indicated that Wright's scenario might work (reviewed by Wade, 2016) but, to my knowledge, no field study has proved that group selection can produce important adaptive change without being challenged by a simpler alternative explanations based on individual kin selection. Discussions about group selection scenarios have carried on at low frequencies, but continue to be handicapped by levels of selection and levels of adaptation not being properly separated (West et al., 2011; West and Gardner, 2013).

As it appears, the extra assumption of deme structure makes explaining the adaptive landscape less parsimonious, as Goodnight and Wade (2000) admit. Group selection scenarios based on drift are thus stronger claims that should require stronger empirical evidence, which has not been provided for any trait that could be considered to be a complex social adaptation. Additional arguments based on epistasis have also not produced theoretical models with agreed-upon conclusions (e.g., Wade, 2002; Turelli and Barton, 2006; Barton and Turelli, 2007). This strongly suggests that the Fisherian assumption of panmixis as a default setting of nature—often an implicit but not necessary assumption of inclusive fitness theory—is a reasonable generalization, similar to Fisher's emphasis on genes that code for complex adaptations normally being part of what he called "modifier complexes" (Provine, 1971)—we would now use the term gene regulatory networks (see Table 3.1). The individual contributions of allelic variants are then likely to be minute so that breeding value approaches are better suited for understanding adaptation than population genetics models (see Box 2.3). As recently argued by Birch (2019), it is important to keep qualitative arguments of adaptive improvement separate from genetic mechanism arguments that can be more exact but at the significant price of losing generality and overall accuracy (see also Frank, 2013b; Queller, 2017). In other words, the rate of evolution will be affected when populations are fragmented or somewhat inbred, but not the phenotypic outcome when the same directional selection pressure is maintained.

Neglecting possible complications of genetic mechanism is justified when considering that MTEs have arisen via directional kin selection, as I will argue in the final section of this chapter. In such scenarios, the challenge is to explain why MTEs originated, not how long it took. It is often assumed that the trait-group selection controversy was resolved when Hamilton's rule was re-derived as a levels-of-selection principle using the Price equation (Hamilton, 1970, 1975), but this is only true in the abstract genetic sense (Frank, 2013b; West and Gardner, 2013; Grafen, 2015; Queller, 2017). In practice, approaches emanating from the new group selection approach by David Sloan Wilson (1975, 1990) have often continued to be pursued as if they offer insights that cannot

Box 4.1 *Continued*

be obtained from inclusive fitness theory (Figure 4.3). I will cover some of these alternative approaches in the middle section of this chapter, mostly to emphasize that inclusive fitness language, and particularly the direct fitness (or neighbor-modulated) version of that theory, remains superior for explaining social adaptations (Gardner, 2015a). The key argument is that first-principle neighbor-modulated theory can always be switched over to an analogous inclusive fitness formulation that makes testable predictions based on agency of a focal altruist, which is impossible to do with trait-group models (West et al., 2007a, 2008; West and Gardner, 2013). There continues to be some controversy on this issue (Gardner, 2015b, 2015c; Goodnight, 2015), but this seems unlikely to affect empirical testing under natural conditions, for which only the direct usefulness of the theory matters (West et al., 2008).

(a) **Social dynamics have two distinct starting points precluding a relatedness continuum**

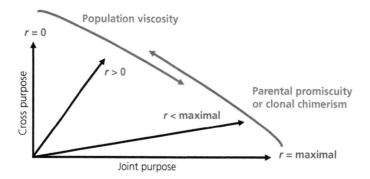

(b) **Partitioning the relatedness parameter space in Hamilton's rule: $br_x > cr_o$**

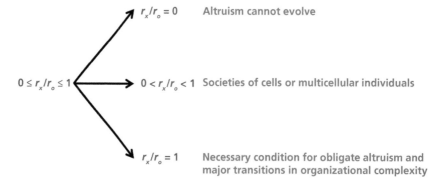

Figure 4.2 The difference between family groups and non-family groups. (a) Relatedness (r) is not a smooth continuum between zero and one, but two opposite gradients defined by vectors that combine into degrees of joint and cross purpose. When relatedness is zero, interacting cells or multicellular individuals invariably compete for resource acquisition, so they only have a cross-purpose vector and no form of altruism can evolve, consistent with zero-relatedness reducing Hamilton's rule to the impossible condition that the product of two positive numbers needs to be negative ($cr_o < 0$). In contrast, when relatedness

(*Continued*)

and tractable common descent. I use family here in its broadest possible sense, applying it not only to parents and sibling-offspring but also to clonal cell lineages. This analogy is valid because family relatedness can become diluted by parental promiscuity while cell line relatedness can become reduced by chimeric coexistence of multiple clones (Figure 4.2a). To explain the origins of MTEs, I conjecture that we need to focus on clones of cells or families of multicellular individuals that stay together because of maximal relatedness enforced

by non-redundant closure in terms of genetic information (see Table 3.1), which has implications for how we look at Hamilton's rule (Figure 4.2b). It is customary to apply that rule, and inclusive fitness theory more generally, in a continuous manner provided relatedness is positive (West and Gardner, 2010). That version of Hamilton's rule explains societies of cells or multicellular individuals where levels of condition-dependent altruism correlate with relatedness in local kin groups, but never perfectly so because all three parameters (r_x, b,

---

Figure 4.2 (*Continued*)

is maximal, interacting individuals are always full-sibling offspring or clonal cells, so that $r_x = r_o$ (see Box 2.1). In such cases, both relatedness terms cancel, and Hamilton's rule reduces to $b > c$, which implies that there is always a joint purpose vector for resource acquisition as long as $b > c$. From these two extreme starting positions, population viscosity may create positive relatedness from a zero-relatedness ancestral state, and parental promiscuity or chimerism may create less than maximal relatedness among interacting siblings or cell copies (red opposite arrows). Hamilton's rule applies in both cases (because relatedness $\neq 0$) and so will the Price equation, but the types of altruistic social systems that selection can produce along these alternative pathways are different; the two relatedness gradients overlap, but they will not reach the opposite extreme. In other words, viscous populations never gradually move on to structures that invariably secure clonality or full sibling families, and starting points with maximal relatedness never become reduced to randomly distributed populations of zero relatedness. The top-left scenario leads to social adaptations that involve condition-dependent forms of altruism, as for example in microbial biofilms, cellular *Dictyostelium* slime molds, and societies of families nested within communal non-kin aggregations as in human societies. The bottom-right scenario concerns parent–offspring associations or clonal cell lines where relatedness can be maximal but where dilution through promiscuity or chimerism occurs, albeit not to the point that family structure disappears. This applies in many cooperatively breeding vertebrates, *Polistes* paper wasps, and wood-dwelling lower termites, where expressions of altruism can be impressive when relatedness is high, and where some individuals may evolve to be fully "somatic," although this never becomes the norm for all because relatedness remains variable. The bottom-right domain represents what Darwin (1859) and Williams (1966a)—and by and large every other biologist in between their times—called family selection, and what Maynard Smith (1964) had in mind when coining the term kin-selection. (b) Using the general notation of Hamilton's rule $br_x > cr_o$ (see Box 2.1), it is of interest to explore the rule's boundary conditions. This is straightforward after rearranging the expression so the relatedness ratio and the cost/benefit ratio can be directly compared. The expression $r_x/r_o > c/b$ then has two extreme states and an intermediate range corresponding to the axes and arrows of panel (a). This is consistent with altruism being impossible to evolve when r_x equals zero, independent of whether the alternative relatedness (r_o) is 0.5 (as with outbred sexual offspring) or 1.0 (as with clonal cell copies). When $r_x > 0$, Hamilton's rule becomes meaningful as a general expression for the evolution of condition-dependent altruistic traits under the assumption that relatedness is a variable ($0 < r_x/r_o < 1$). However, the other extreme state of $r_x/r_o = 1$ has rarely been considered as a special case, as I do in this chapter and Chapter 5 to develop an inclusive fitness explanation for MTE origins, i.e. a kin selection theory for organismality. It is this situation of mathematical equivalence between relatedness to siblings and offspring, represented by the horizontal vector in panel (a), that makes the relatedness terms cancel out of Hamilton's rule. When that happens, Hamilton's rule becomes the product of two separate conditions, one that is fulfilled generally and a priori ($r_x = r_o$) and the other one ($b/c > 1$) that remains to be fulfilled contingently and a posteriori. However, when both conditions are fulfilled, kin selection will allow obligate unconditional altruism to evolve without facultative, conditional altruism as transitional stage. As we will see in the final section of this chapter, such a partitioning gives us, in an ultimate evolutionary sense, the hypothetical necessary and sufficient conditions for the origins of all fraternal MTEs. Note that the diagram does not explicitly consider situations that may give rise to spiteful behavior when $r_x/r_o < 0$ (see Box 2.1). This type of negative sociality also requires forms of family life or limited dispersal as ultimate settings, and kin discrimination or quorum sensing as proximate mechanisms.

(a)   (b)

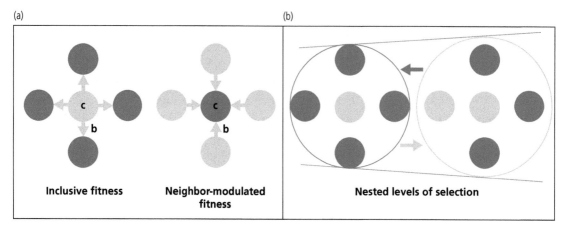

Figure 4.3 Alternative approaches for analyzing social evolution dynamics. (a) A summary diagram of the conceptual difference between the inclusive fitness approach and the neighbor-modulated fitness approach for understanding the dynamics of social evolution. Both notations consider the propagation of genes coding for altruistic traits, but, in reality, conceptualize entire gene regulatory networks in which altruistic genes contribute to complex phenotypic agency, consistent with the expression of altruism being a quantitative, i.e. biometric trait (see Box 2.3). Inclusive fitness notation emphasizes that such genes in a focal altruist (gray) pay a cost (c) in direct reproduction, scaled by relatedness to own offspring (r_o), to be more than compensated by indirect reproductive benefits (b) represented by extra offspring that could only be raised because of the altruist's help to the neighbors or nestmates producing them, scaled by the focal altruist's relatedness (r_x) to these non-own offspring (blue). Inclusive fitness thus separates direct and indirect fitness of the same individual so that these fitness components add up to the total fitness of an agent (see Box 2.1). Neighbor-modulated fitness only considers direct fitness and emphasizes that the average altruist pays a cost (c) in direct reproduction scaled by relatedness to its offspring (as in inclusive fitness), but then focuses on the help received from neighboring altruists rather than on help given, so that the compensating benefits that ensure altruism (when $b > c$) are expressed as another component of direct fitness. In this notation, the relatedness scaling of fitness gains induced by others only applies at the population level in the background. In contrast, the inclusive fitness approach works with focal agency of a single representative cell or multicellular organism, which yields theoretical predictions that allow direct validation by fieldwork or laboratory experiment. The neighbor-modulated fitness notation is mathematically more general and therefore often preferred for theoretical modeling, but does not normally generate testable predictions because populations have no agency (see Table 3.1). It is thus less useful for empirical biologists. Both inclusive fitness and neighbor-modulated fitness approaches work without the need to specify a rigid group structure and conclusions phrased in the notation of one version can normally be converted to the other notation without ambiguities. Note that in the genetic versions of these complementary approaches, r_x, b, and c are always partial regression coefficients, whereas r_o is assumed to be constant. (b) When considering genes encoding altruistic traits, the nested levels-of-selection approach is equivalent to neighbor-modulated fitness logic when the latter considers the same discrete group structure. Levels of selection (trait-group) notation entails that altruists (gray) realize lower reproductive success within groups because non-altruists (blue) avoid paying the cost of altruism. However, groups with more altruists are assumed to waste fewer resources as they avoid within-group competition so that group-level productivity is higher (i.e. the circular envelope is larger). Altruists are then favored by selection at the between-group level because more productive groups produce more dispersing propagules carrying altruistic genes that will populate a next generation of groups (also called demes). Reproduced with minor modification from Social evolution theory: a review of methods and approaches, T. Wenseleers, A. Gardner and K. R. Foster (2010). In: *Social Behaviour: Genes, Ecology and Evolution*, edited by T. Székely, A. J. Moore and J. Komdeur, with permission of Cambridge University Press through PLSclear. See also Frank (1997b, 2012a, 2012d), West et al. (2011), and West and Gardner (2013).

and c) vary independently (see Box 2.1). In contrast, the discontinuous version of Hamilton's rule assumes that relatedness is

a maximal-value constant decoupled from b and c. To capture the distinction between these versions, it is useful to partition the total parameter space in which Hamilton's rule is mathematically valid as an explanation for altruism ($0 < r_x/r_0 \leq 1$) into a quantitative domain of dynamic adaptive change in societies of cells or multicellular individuals ($0 < r_x/r_0 < 1$), and a qualitative (static) domain where relatedness to clonal cells or full sibling individuals is always maximal and equal to relatedness to offspring ($r_x/r_0 = 1$). In the latter case, Hamilton's rule reduces to $b > c$, as a sufficient condition for the evolution of obligate, i.e. unconditional altruism (Figure 4.2b).

Since the 1980s, research has concentrated on Hamilton's rule in variable relatedness settings ($0 < r_x/r_0 < 1$). In this continuous version, the product $r_x b$ needs to exceed the cost in lost personal reproduction ($r_0 c$) for the inequality to be fulfilled. The inequality will therefore differ across individuals even when b is constant because r_x is a variable. It is this continuous version of Hamilton's rule that can be captured as a levels-of-selection concept (Figure 4.3), because it may pay some individuals to be altruists helping others reproduce while others in the same family or group achieve higher inclusive fitness by personal reproduction. However, in the discontinuous version, it either benefits all or none of the individuals to be altruists depending on whether b is larger or smaller than c. The discontinuous version therefore has no "levels of selection" analogy because there is only one level—all respond to individual selection as long as $b < c$ and all respond to family selection when $b > c$. Thus, when $b/c > 1$ applies consistently for a sufficiently long time, family selection can forge an irreversible change towards higher organizational complexity via obligate somatization of entire offspring cohorts simultaneously.

In what follows, I will continue to segregate arguments of continuity versus discontinuity (see Table 3.1) as two sides of the same inclusive fitness coin, for the former are known to explain social behavior while the latter can be

expected to explain somatization during MTE origins, and both do so from first kin-selection principles. In the next section, I briefly review the state of the art of inclusive fitness theory working with continuous parameter settings; in the final section I will then continue to develop the discontinuous version of Hamilton's rule and explore its historical roots. The conclusion will be that complex societies of cells or multicellular individuals may gradually "emerge," but that it is not compelling to assume that MTEs have originated that way.

I am not aware that others have explicitly emphasized the importance of keeping arguments of high but variable relatedness and invariably maximal relatedness separate in an a priori manner. Gardner and Grafen (2009) and West et al. (2015) separated cooperation and conflict reduction but continued to assume that conflict reduction must be necessary for MTE origins. In my approach, conflict reduction is relevant only for allowing societies to gain complexity, but not for stepwise MTE origins (see Figure 3.2). In retrospect, Williams (1966a) was one of the last conceptual thinkers to be inspired by Darwin's argument of family having been decisive for the evolution of physically differentiated "neuter" castes in social insects (see Boomsma and Gawne, 2018 for quotations illustrating this pre-sociobiology understanding). The generalization of inclusive fitness theory via the Price equation (Hamilton, 1970, 1975) had the side effect that transparent terms such as family selection and Hamilton's original pedigree version of kin selection became mere special cases of a more general inclusive fitness paradigm. The insight that relatedness can only be invariably maximal in cell clones or full-sibling families thus became implicit. This had unfortunate consequences for deeper understanding of the origins of MTEs, which I aim to rectify in the final section of this chapter and in Chapter 5. As we saw in Chapter 3, these developments also biased theorizing about evolutionary progress into an unproductive form of fortuitous emergence thinking, devoid of first-principle arguments of adaptation through natural selection. The extremes of invariably zero and invariably maximal relatedness (Figure 4.2b) both invalidate the continuous version of Hamilton's rule, and the invariably maximal relatedness case appears to be as important as the invariably zero relatedness case.

How the Price equation captured Hamilton's rule

This section focuses on the conceptual developments in inclusive fitness theory that have occurred during the last ca. 25 years. These advances extended and generalized the gene's eye theory of social evolution reviewed in Chapter 1 and 2, building particularly on pioneering work by Queller (1992a, 1992b), Taylor and Frank (1996), and Frank (1998). They gave us both an explicit quantitative genetics perspective on inclusive fitness theory and an alternative, more general "direct fitness" notation based on the Price equation (Figure 4.3). The conceptually most comprehensive review of the present status was provided by Gardner et al.'s (2011) genetical theory of kin selection, which developed an explicitly Fisherian view that explains adaptation by only considering the additive genetic variation (i.e., breeding value) of the social traits of interest. As shown by Gardner et al. (2007) and later summarized by Queller (2017), the Price equation is fundamental for the quantitative genetics of social evolution because it captures not only genic and phenotypic selection but also Fisher's fundamental theorem. Further, Queller (2017) showed that the Price equation is the pinnacle of a hierarchy of fundamental equations of evolutionary change (Figure 4.4) and that two other key mathematical expressions rank just below these

five: Hamilton's (1964a) original pedigree version of inclusive fitness theory and Lande and Arnold's (1983) multiple regression approach to analyze selection on correlated characters. Gardner et al. (2011) also demonstrated that the genetical theory produces a fully general form of Hamilton's rule that requires no special assumptions because the parameters b, c, and r are partial regression coefficients that accurately predict the direction of selection for any social trait of interest (Box 4.2; Figure 4.5).

In their further elaboration of the theory, Gardner et al. (2011) discussed 12 misunderstandings that have often confused biologists and showed that they are all groundless. They concluded that "Hamilton's rule does not break down in response to factors such as gene interactions, strong selection, frequency dependence, pure strategies, mechanisms that favor altruism between non-genealogical relatives, facultative strategies, greenbeards, evolutionary bifurcations, indirect genetic effects, fitness consequences over multiple generations, group selection or non-pairwise interactions." The quantitative genetics version of Hamilton's rule implicitly allows for these factors, consistent with Grafen's (1984) phenotypic gambit based on breeding values and the optimality principles associated with that approach (see Box 2.3). All conclusions emanating from the general genetical theory thus relate to evolutionary changes in social traits that are shaped by natural selection, for this version of kin selection theory remains agnostic about additional stochastic evolutionary change, that is, nonadditive or random change. Having evolutionary theory at this level of conceptual generality is gratifying but not necessarily of direct practical value because empirical research can only proceed with predictions based on concrete variables that represent real-life causality. The phenotypic gambit black-boxes all proximate detail while applying general genetical theory (what Gardner et al. (2011) called a streamlined

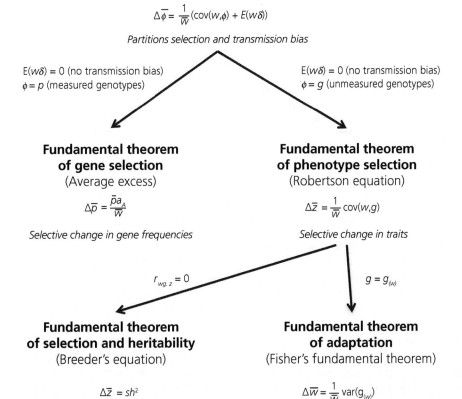

Figure 4.4 The hierarchy of fundamental equations for evolution by natural selection from Queller (2017) where details can be found. Several of the connections between equations were also derived by Frank (2012b, 2013b) and Gardner et al. (2011) but without making the entire hierarchy explicit. Given that Fisher (1930) was the only author who named his equation a fundamental theorem, it seems surprising that Fisher's equation is preceded by two even more fundamental mathematical expressions. The Price equation occupies the top position because it not only captures natural selection but also all other forces that affect evolutionary change, and because all lower-level equations can be derived from the Price equation (arrows) either directly (Fisher's average excess equation) or indirectly (via Robertson's equation; sometimes referred to as the secondary theorem of natural selection; Gardner et al. (2011)). The well-known breeder's equation that multiplies responses to selection with the heritability of a phenotypic trait (see Box 2.3) appears towards the bottom left. Queller also mentioned two theorems that rank just below the breeder's equation and Fisher's fundamental theorem, which are Lande and Arnold's (1983) multivariate selection equation and Hamilton's (1964a) original pedigree version of inclusive fitness theory, consistent with Hamilton's appreciation that Fisher's work was a fundamental source of inspiration for developing what Maynard Smith (1964) would coin kin selection theory. As Queller argues, this entire hierarchy is a strong vindication of Fisher's emphasis on adaptation by natural selection, the principle that identifies the fundamental difference between biology and physics/chemistry as we have seen in Chapter 2 (see also Chapter 8). Fisher's work plays such a central role because neither Haldane nor Wright shared Fisher's wide-ranging ambition to prove that Mendelism could account for phenotypic correlations of continuous traits between relatives (Fisher, 1918; Provine, 1971, p. 143; Lande and Arnold, 1983; Frank, 2012d). Fisher's insights remained the hallmark of all approaches to social evolution that are (often implicitly) based on quantitative genetics (see Box 2.3). Grafen (2019) and Levin and Grafen (2019) offer formal Darwinism updates relating to the same insights. Symbols: the key variable ϕ can either refer to the frequency of genes, or to phenotypes, or to breeding values—in fact to any trait value. δ is the change in ϕ from parent to offspring, w is fitness, p is allele frequency, a_A is average excess, g is breeding value, z is phenotypic value, r is partial correlation, s is selection differential, and h^2 is heritability. Republished with permission of The University of Chicago Press – Journals, from Fundamental theorems of evolution, D.C. Queller, *The American Naturalist* volume 189, 2017; permission conveyed through Copyright Clearance Center, Inc.

Box 4.2 The genetic and phenotypic versions of Hamilton's rule

There is a substantial literature on the accuracy of what recently has been coined the genetic theory of kin selection. Gardner et al. (2011) and a series of seven review papers by Frank (2011a, 2011b, 2012a, 2012b, 2012c, 2013a, 2013b) made it clear that Hamilton's rule is completely general when phrased in terms of genes or the heritable (additive) component of a phenotypic trait, also known as the individual's genetic or breeding value for that trait (see Box 2.3). This implies that the Hamiltonian parameters (c, b, r) are all defined as partial regression slopes of a fitness component on a genetic or breeding value predictor variable, as illustrated in Figure 4.5. This approach gives us precise estimates of the direction of selection for or against conditionally expressed altruism, as derived for the first time by Queller (1985). It is also precise in its mutually exclusive partitioning of direct fitness (to self) and indirect fitness referring only to others than self. The regression method of estimating relatedness as a parameter that quantifies indirect fitness benefits thus produces a measure of statistical association between the heritable inclinations of social partners for a trait of interest such as altruism, that is, the product ($r_x b$) that needs to exceed the cost ($r_o c$) as illustrated in Figure 4.5b,c,d. Using fitted linear regression slopes makes this fully general form of Hamilton's rule insensitive to essentially all conceivable genetic complications (Gardner et al., 2011; Frank, 2012b). The robustness of the genetic theory of kin selection is a remarkable achievement, although the original pedigree term kin selection may be perceived as somewhat confusing when used in this general statistical theory (Leigh, 2010b), as Gardner et al. (2011) concede.

It is important to note that the scatter of individual data points around the regression lines (Figure 4.5) does not affect the predictive power of Hamilton's rule because natural selection only "sees" the average additive effects (Gardner et al., 2011). The same is true for strength of selection and the presence/absence of frequency dependence as long as we consider the fully general version of Hamilton's rule. In statistical terms, a regression slope is a ratio, with the covariance (Cov) between a response variable and a predictor variable in the numerator and the variance (Var) of the predictor variable in the denominator, so the Price equation expresses Hamilton's rule as a sum of statistical (co)variance terms. However, as discussed in detail by Gardner et al. (2011) and Frank (2013b), this abstract genetic information version of Hamilton's rule does not address formal causation because we do not know the genes for altruism or the gene expression networks behind the heritable components of the social traits of interest. This is not because we have not looked hard enough, but because it will always remain very difficult to obtain knowledge of the complex genetic mechanisms and their variation across social interactants and environments. This is why we separate proximate mechanisms and ultimate adaptive explanations (see Chapter 1) and why we use Grafen's (1984) phenotypic gambit (see Chapter 2) to formulate testable predictions based on abstract genetic information. This is also why we derive predictions from the notion that selection will tend to both optimize phenotypes in terms of benefits and costs (Grafen, 2002) and maximize inclusive fitness (Grafen, 2006) (see Crewe et al. (2018) and the seven 2011–2013 Frank papers for a detailed discussion of these principles from several complementary angles).

The abstract genetic theory of kin selection is useful because it produces general predictions about social behaviors of interacting phenotypes that are sufficiently accurate to have often been validated by empirical evidence (e.g., Abbot et al., 2011). This justifies confidence in the theory being correctly and adequately argued, albeit with the usual Popperian caveat of an even more general theory possibly emerging in the future (see Chapter 8). However, these predictions cannot explain all the variance in actual social phenotypes when we study field populations or analyze the results of lab experiments. This is not a weakness of inclusive fitness theory but a general characteristic of the theory of adaptation through natural selection (Lande and Arnold, 1983), as Queller (1992a, 1992b) and Frank (1997c, 2013a, 2013b) discuss in detail while focusing on social traits. To supplement abstract general understanding with insight into concrete causation, these authors proposed a multiple regression

Box 4.2 *Continued*

approach where a specific fitness response variable becomes a function of many predictor variables, after which a statistical technique called path analysis can rank the empirically measured predictors in order of their importance for explaining proportions of the variance in fitness. Taking this approach means that one is now working with a phenotypic version of inclusive fitness theory that is concrete on one hand (assuming, for example, one can measure predictors such as fertility and age-specific reproductive value), but less general because one only considers a concrete population of a particular organism in a specific habitat setting.

Any phenotypic version of Hamilton's rule is a tool for applying the predictions of the fully general genetical theory of kin selection (Gardner et al., 2011; Marshall, 2015) to real-world situations (Frank, 1997c). Its approximate nature therefore never compromises the principal validity of the overall inclusive fitness concept, but merely confirms the truism that in biology and economy all concrete models of reality are wrong in the sense of being incomplete but that some are more useful than others (Box, 1976). The first incisive general attempt to use Hamilton's rule as an instrument for understanding phenotypic causation was developed by Queller (1985), who showed that the genetic version of Hamilton's rule Cov $(G_A, G_O)b >$ Var $(G_A)c$ can (at least for interacting pairs) be replaced by a phenotypic version Cov $(G_A, P_O)b >$ Cov $(G_A, P_A)c$, where G and P stand for genotype and phenotype, respectively, and subscripts A and O for actor and other. As summarized by Fletcher and Zwick (2006), this notation makes Hamilton's rule emphasize how natural selection may change the frequency of social genes in the actor (G_A) as a function of being surrounded by altruistic phenotypes of others (P_O), independent of the genetic underpinning of the social traits expressed by these others, that is, there is no G_O term in the second equation. This approach sacrifices genetic detail by assuming that certain complications do not occur (Grafen, 1985b), but the general gain in power of predicting the direction of change in social adaptations is substantial, and the match of these predictions with

observable behavior then becomes the final arbiter of whether the simplifying assumptions made are likely to match reality in field populations that would justify adaptive design interpretations.

The phenotypic version of Hamilton's rule is formulated in direct neighbor-modulated fitness terms and thus lacks the directional altruistic agency that characterized Hamilton's original inclusive fitness approach. One can normally easily swap between inclusive fitness and neighbor-modulated notation (Gardner et al., 2011; Queller, 2011; Frank, 2013b; West and Gardner, 2013) but it works best to use the direct fitness notation when we want to maintain a connection with formal population genetics and levels-of-selection approaches (Figure 4.3). Queller (1985) established this connection by showing that an interaction term Cov $(G_A, P_A P_O)d$ needs to be added to the Price equation to account for the fact that fitness of phenotypes will also have a nonadditive component. This follows from quantitative genetics (see Box 2.3), where heritability has both an additive (narrow sense) component and a nonadditive component such that the sum of the two is defined as the broad sense heritability of a trait (Falconer and Mackay, 1996). We could ignore nonadditivity of costs and benefits in the abstract genetic form of Hamilton's rule because natural selection only responds to the additive genetic part of the variation expressed by individuals that pay the costs (c) or reap the benefits (b) of social interactions. However, we can no longer do that for the phenotypic version of Hamilton's rule because these nonadditive genetic components affect the phenotypes of individual agents. The proposed interaction term is indeed zero when there is only additive genetic variation $(d = 0)$, but when there are nonadditive effects (i.e., $d \neq 0$), P_A and P_O either take the value of 0 (no altruism expressed) or 1 (altruism expressed).

The inclusion of nonadditivity implies that each two-by-two interaction scheme with four possible outcomes produces a single 1 (when $P_A = P_O = 1$) for every three zeros (when either P_A or P_O or both are zero), and these 0 or 1 predictors differentially affect G_A.

Box 4.2 *Continued*

The $P_A P_O$ interaction will normally produce a positive conditional synergy effect similar to the TIT-FOR-TAT strategy for reciprocity (Trivers, 1971; Axelrod and Hamilton, 1981) because, as Queller (1985) phrased it, the following rule of thumb will apply: "be altruistic to an untested partner; otherwise, behave towards the partner as he behaves towards you." Queller's phenotypic version of Hamilton's rule thus became Cov $(G_A, P_O)b$ + Cov $(G_A, P_A P_O)d$ > Cov $(G_A, P_A)c$. It applies to both conditional and unconditional altruism and it also captures reciprocal mutual cooperation between unrelated players, as was later re-derived, tested, and further elaborated by Fletcher and Zwick (2006). When both sides of the equation are divided by Cov (G_A, P_A) it becomes transparent that the average benefits of altruism or reciprocity (b) and the nonadditivity deviation effect (d) are both scaled by regression slopes, and that d is essentially also a relatedness coefficient but one that comes into play (with a value of 1) only when two altruists randomly end up in the same pair. How often that happens depends on how common the altruistic phenotype P_A is in the breeding population. The phenotypic version of Hamilton's rule can therefore capture adaptive design across many different social evolution settings within and between species, but it remains agnostic about the origin of altruism from very rare mutant beginnings, whose squared frequency is essentially zero.

It is here that Queller's (1985) approach appears to meet models by Frank (1994, 1995a, 1997a) that I will discuss in Chapter 5 and where some minimal threshold value of P_A is required (i.e., altruism is already an established trait) for synergism to emerge, as Queller (1985) also indicated. His model established that when we use unrelated collaborator phenotypes to predict the fate of genes coding for cooperative traits, it appears possible to at least qualitatively explain reciprocity between nonrelatives (see Fletcher and Zwick, 2006, for a worked-out example), a conclusion that the genetic version of Hamilton's rule does not allow (Gardner et al. 2011). This in turn inspired other papers (e.g., Fletcher and Doebeli, 2006, 2009) arguing that assortment and synergy are fundamental necessary conditions for the evolution of cooperation (either mutualistic or altruistic; cf. Figure 4.1), albeit with disagreement about the mathematical details

(Bijma and Aanen, 2010; Fletcher and Doebeli, 2010). As it appears, the crucial point remains related to the definition of altruism, because the stringent "others only" definition for who receives indirect benefits excludes that social interactions between nonrelatives can be altruistic. Only this original Hamiltonian definition gives a conceptually clean 100% partitioning between direct and indirect fitness. Replacing it by an alternative definition that somehow blends direct and indirect fitness benefits has always caused confusion, as for example with the term reciprocal altruism (Trivers, 1971), which is a misnomer according to the "others only" definition (see e.g., West et al., 2007a).

In the end, the validity of phenotypic models of kin selection will depend on the extent to which they correctly predict altruistic and mutualistic interactions in populations of real unicellular or multicellular individuals. Working with arbitrary predictors rather than with breeding values as in the general genetic theory of kin selection will always restrict the concrete validity of models; this is also true when using single-gene loci as predictors. This is nothing new as Hamilton's rule was originally developed as a population genetics model for the spread of alleles that encode altruistic phenotypes (Hamilton, 1964a, 1964b), and Hamilton was the first to state that such an approach requires simplifying assumptions, but nothing of a kind that population geneticist modelers are not generally happy to accept, as an early review by Michod (1982) made clear. Gardner et al. (2011) showed that genetic nonadditivity, indirect genetic effects, and other complications are implicitly accommodated in the general genetic theory of kin selection, so the actual research questions asked determine how interesting or relevant it is to make these factors explicit (see e.g. McGlothlin et al., 2014). Just as in estimates of heritability from parent–offspring or sibling comparisons (Falconer and Mackay, 1996), only the slopes of the partial regressions matter—not the scatter, which represents a fundamental difference from the causal multiple regression approach (Lande and Arnold, 1983; Queller, 1992b; Frank, 2013a) where the objective is to maximize the proportion of fitness variance explained. However, as we saw already, that approach produces specific models of particular cases, not an overarching conceptual principle.

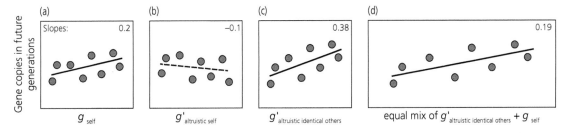

Figure 4.5 The genetic version of Hamilton's rule in graphs. Natural selection will make any heritable trait (g), for example, the frequency of a gene or the average breeding value of a quantitative trait, increase in frequency or value when, respectively, it realizes more copies of itself in future generations than alternative alleles at the same locus, or there is a positive correlation between breeding value and that heritable trait. Depending on the costs and benefits of solitary reproduction (scaled by r_o, i.e., relatedness to offspring cells or offspring individuals) or social reproduction (scaled by r_x), either allele may be favored by selection as the arbitrary examples in these four panels illustrate—always referring to g as (quantitative) trait even though it could also be an allelic contribution to that trait. (a) Solitary reproduction: a positive regression slope (0.2) of fitness on a trait mediating direct reproduction (g_{self}) will ensure that the trait increases in frequency. (b) An alternative trait (g') arises that expresses altruistic behavior such that a higher value of g' implies a lower number of gene copies in future generations, producing a slope of −0.1. This means that g' should be selected against and gradually disappear from the population because it pays a direct fitness cost (c) for altruism. (c) However, being surrounded by altruists that carry copies of the g' allele for altruism gives benefits (b) in the transmission of gene copies of the focal altruist to future generations that more than compensate its cost (c), that is, the positive slope of 0.38 combined with the negative slope of −0.1 in panel (b) gives a combined slope of +0.28, which exceeds the slope of 0.2 in panel (a). The altruistic trait g' will therefore increase in frequency because altruism towards carriers of identical gene copies enhances the total reproductive success of the g' allele. An imaginary example of this scenario is the evolution of social fruiting bodies with sterile somatic stalk cells in cellular *Dictyostelium* slime mold populations where they would only make fruiting bodies with clonal cell copies (Gilbert et al., 2007). Note that the argument, so far expressed in direct neighbor-modulated fitness terms (Figure 4.3a), could also have been phrased in inclusive fitness terms, in which case each altruist would be 1.4 times as efficient in transmitting gene copies to future generations by helping others reproduce than by reproducing herself. Being surrounded only by clonal copies with independent agency is therefore crucial for the increase of the g' allele at the expense of the g allele with the particular costs and benefits specified. (d) In contrast, if we consider a scenario where the focal altruist is always surrounded by an equal mix of g' altruists and g non-altruists, the horizontal axis becomes twice as long to realize the same increase in copies of the g' allele and this 50% dilution reduces the slope by 50%. So now the net reproductive benefit of altruism is still positive (0.19 − 0.1 = 0.09), but that gain rate is well below the 0.2 slope of solitary reproduction in panel (a). Thus, if this 50/50 self/non-self chimerism between clones persists in the population, the altruistic trait g' should gradually disappear consistent, for example, with chimerism slowing down slug migration in *Dictyostelium* (Foster et al., 2002; Castillo et al., 2005). This numerical example illustrates that partitioning reproductive success into direct and indirect fitness components involves regression slopes for costs and benefits and that the benefits of altruism need to be scaled by variable relatedness (r_x) towards other cells in the social environment. The genetical theory of kin selection (Gardner et al., 2011) therefore writes Hamilton's rule as a set of three regression slopes, one each for r_x, b, and c (see Box 4.2 for details and Gardner (2015d) for a three-dimensional version of the figure panels given here). In the general genetical theory of kin selection, both direct and indirect fitness components are always evaluated as partial regression slopes holding the effect of g or g' on the other component constant.

methodology), but the alternative approach, based on Queller (1985), has been to work with phenotypic versions of the Price equation that require one or several extra terms (e.g., Frank, 2013a; Queller, 2014). My summary of the characteristics of the genetic and phenotypic theory of kin selection in Box 4.2 emphasizes

that such phenotypic Price equation models have several concrete advantages but are also less generally valid.

Gardner et al. (2011) made explicit that general kin selection theory needs to be framed in terms of genic change, not genotypic or phenotypic change, because quantitative genetics tells us that natural selection only elicits a response in the additive genetic variation for social traits (see Box 2.3). Phenotypic theory or genetically explicit single-locus models that track genes and genotypes thus each have their concreteness advantages and non-generality disadvantages. The conceptual generality of first-principle theory versus the practical causality in specific models is an issue of general, but often misunderstood, importance (Lehmann and Keller, 2006; West et al., 2007a; Gardner et al., 2011; Frank, 2012d; West and Gardner, 2013). The extent to which one feels the need to supplement general theory with more specific modelling is an empirical question. Dual approaches usually reinforce each other as, for example, Hamilton (1963, 1964a) pursued when he developed both phenotypic inclusive fitness logic and a population genetics model to show that the two gave similar results—albeit with simplifying assumptions for the latter. It is therefore not appropriate to deny the importance of first-principle theory because an arbitrary model can produce another outcome (e.g., Nowak et al., 2010). Frank (2012b) quoted Boulding (1956) to argue that "[s]omewhere [...] between the specific that has no meaning and the general that has no content there must be, for each purpose and at each level of abstraction, an optimum degree of generality." Overall, Gardner et al.'s (2011) conclusions match those of Queller (2017), of Frank's series of seven reviews in the *Journal of Evolutionary Biology* (particularly Frank, 2012a, 2012b, 2013a, 2013b), and of Rousset (2004, 2015). Their consensus is that specific models can give useful causal understanding as long as the caveat of reduced generality is made clear (Box 4.2).

Queller and Frank have also tried to clarify what Hamilton's rule would look like if one used concrete but arbitrary costs/benefits predictors (e.g., fecundity or another parameter that strikes a researcher as potentially relevant)

that take the generality out of the genetic theory of kin selection (Gardner et al., 2011). Similar issues were addressed earlier by Grafen (1985), Queller (1992a), Frank (1997c, 1998), Rousset (2004), Lehmann et al. (2007), and Gardner et al. (2007), and later by Marshall (2015, 2016). If one imagines single material predictor genes coding for specific arbitrary traits, then one either has to assume weak selection to arrive at good approximations of dynamic evolution (Lehtonen, 2020) or one needs to define, as Queller (1985) proposed, an extra synergism term. This can be useful, as Gardner et al. (2011) concede, as long as one realizes that adding such extra relatedness-like terms is a self-inflicted compromise with generality and exactness for the sake of obtaining insight in concrete causation and applicability in empirical research (see also Gardner et al., 2007). There can be clear rewards in this approach because it allows analysis of social evolution between non-kin, such as reciprocity, interspecific mutualism, and even parasitism (Queller, 1985; Frank, 1998; Queller, 2011; McGlothlin et al., 2014; Queller, 2014), types of questions that I will discuss in more detail in Chapter 5. Of particular interest is the analysis of indirect genetic effects between social partners (Wolf et al., 1998; Linksvayer and Wade, 2005; Linksvayer, 2006; McGlothlin and Brodie, 2009), not because it invalidates the genetic version of Hamilton's rule, but because it allows the additive genetic variation of social traits to be partitioned into a direct and an indirect component (Bijma, 2010, 2014).

Several other aspects of social evolution have been illuminated by phenotypic covariance models of inclusive fitness. First, verbal notions about the importance of local competition between relatives were made explicit (Griffin and West, 2002; West et al., 2002), which showed that global rather than local competition is a requirement for indirect fitness benefits to affect interactions as diverse as aggression (West et al., 2001) and virulence evolution (Frank, 1996b). This led to novel insights into budding as a joint-dispersal mechanism that maintains indirect fitness benefits in viscous populations (Gardner and West, 2006; Gardner et al., 2009). In retrospect, it seems surprising that it took so long for this complication to be made explicit, for history and mythology have illustrative examples of brother princes cooperating to conquer new territory but killing each other when competing for zero-sum opportunities at home. Second, as already mentioned, theoretical generalizations have emphasized that it is often useful to compare the predictive efficiency of open, ultimate models where relatedness is an independent variable with that of closed models where proximate specifics emanate from multilocus genetics, life cycles, or demography (Gardner et al., 2007; Cooper et al., 2018). Third, Hamiltonian logic has shed light on Hardin's (1968) "Tragedy of the Commons," the principle that public goods are liable to be destroyed by exploitation unless stable reinforcement mechanisms (Figure 4.1) can evolve (Leigh, 1977). Finally, Foster (2004) showed that the diminishing returns in social groups where individuals reap benefits from both reproductive competition and cooperation discourage competition more than cooperation, suggesting that societies based on condition dependent altruism or mutualism can remain stable in spite of considerable levels of reproductive competition.

As it turns out, the *Genetical Theory of Kin Selection* by Gardner et al. (2011), Frank (2012b, 2013b) and Queller (2017) pervades the domain of Tinbergen's adaptation quadrat (see Figure 1.1), which is meant to explain adaptive changes mediated by natural selection on additive genetic variation. It is appropriate therefore to assert that Tinbergen's adaptation quadrat is home to the entire hierarchy of equations of Figure 4.4 (see also Gardner, 2015a). Now, more than half a century after Tinbergen's (1963) seminal paper, we can conclude with reasonable confidence that any gene's eye theory of adaptation must always be based on the Price equation, but that the general theory formulated this way gives only partial answers. It leaves open that not all of life is adaptive, but claims at the same time that the fractions that are adaptive can be understood from a single principle that is so comprehensive across the entire variety of concrete biological systems that it may look tautological (Gardner, 2015a). This remains consistent with Tinbergen's principle that no major question in biology will be adequately understood until mechanisms of proximate causation, developmental facilitation or constraint, and historical (phylogenetic) contingency have been addressed as well—that is, the three other quadrats provide context for adaptation or lack thereof (see Box 1.3). As Queller (2017) noted, it is remarkable that to achieve this generality, "the unmeasured-genotype approach of quantitative genetics has been more productive of fundamental theorems than the measured-genotype approach of population genetics." To which he added that "[t]o have captured [the] essence [of adaptation] in a simple [Price] equation is a tremendous feat" (see Bull and Wang, 2010 for a similar conclusion based on optimality arguments). Conceptually, the codex information form of genetics therefore trumps material population genetics in making sense of social adaptations. As Frank (2012b) emphasized, axiomatic theory has no units, so that "[t]he enduring power of the Price equation arises from the discovery of essential invariances in natural selection."

Since the late 1990s, Hamiltonian social evolution theory has compared very well with alternative theoretical approaches. Gardner et al. (2007) showed that multilocus population genetics methods are analogous to

Price-equation-based social evolution theory, including both levels-of-selection and neighbor modulated fitness approaches. Further comparisons demonstrated that a multilocus genetic version of Hamilton's rule has the dynamic sufficiency that Price equation covariance arguments lack, confirming that the two approaches are most explicitly equivalent when genetically specific models explore invasion conditions of mutants coding for slightly different social traits. However, although multilevel selection theory is often equivalent to kin selection theory (see also Lehtonen, 2016), Gardner (2015a) showed that the multilevel genetic approach also has important limitations when compared to a fully general theory of multilevel selection based on the Price equation (see also West and Gardner, 2013). This is because group traits and group fitness are often poorly defined and there is no obvious link between individual and group selection, that is, between what Okasha (2006) termed MLS-1 and MLS-2. This lack of connection is consistent with the opposite red-arrow argument that I presented in Figure 4.2, with critique by Rainey and Kerr (2011), and with Okasha's appreciation of the gap between MLS-1 and MLS-2. As Gardner (2015a) observed, genetic multilevel selection approaches also have difficulties handling within-population class structure, such as the coexistence of males and females, age groups, and ploidy variation with different Fisherian reproductive values. More explicit reviews of the interaction between arguments of reproductive value and direct or inclusive fitness have recently appeared (Grafen, 2020; Rodrigues and Gardner, 2021).

Finally, Gardner et al. (2007) pointed out that what appears to be the most powerful modelling tool, Hamilton's original neighbor-modulated (direct) fitness version of inclusive fitness theory (Figures 4.4 and 4.5), turned out to combine marginal value approaches in evolutionary game theory with stability analyses of equilibria (Christiansen, 1991; Taylor, 1996; Taylor and Frank, 1996) that helped to produce a generality on which many subsequent theory papers would successfully build (West and Gardner, 2013). The usual assumption in this approach is that the variation of breeding values for the focal traits considered is negligible, which implies that effectively only one gene with small effect is adapting at any particular time. This notion has been criticized (see Birch, 2019 for a recent review) but gained momentum from an argument by Fromhage and Jennions (2019), who maintained that low-penetrance genes of small effect are the only genes that are cumulatively able to build the complex polygenic adaptations that represent the interest of the entire genome (see also Queller, 2019 for a commentary). This reminds us that classic Hamiltonian inclusive fitness is only a maximand in the sense of Grafen (2002) when interactions are additive. As Fromhage and Jennions (2019) noted, it is useful to apply the concept of an abstract reference gene of low penetrance because such a gene automatically meets the phenotypic gambit assumptions and because its rare conditional expression simplifies the indirect fitness definition. Their paper revisited Hammerstein (1996), who proposed that any gene coding for a social phenotype that could be promoted by natural selection, but "suffers" from additivity and weak selection conditions not being met, will always be transient and be replaced by alleles of small additive effect.

Given the coherence and consensus summarized above, it seems surprising that a few mathematically inclined theoreticians remain critical of the general validity of Hamilton's rule and the Price equation approach, particularly because they have no alternative conceptual framework or testable predictions to offer, neither previously (West and Gardner, 2013) (see Box 2.3) nor at present. As aptly described by Welch (2017), these critics merely argue that established terms need redefinitions because "life is complex" and that what they call the "adaptationist paradigm" is "wanting" while ignoring its explanatory power. Welch's (2017) rhetorical question *What's wrong with evolutionary biology?* shows that these critics shun interdisciplinarity *sensu* Tinbergen (see Figure 1.1; Box 1.3) and prefer to draw up "geneticist," "developmentalist," or "phylogeneticist" laundry lists of what they feel the presumed "adaptationist" approach is missing. Welch noted that much of this appears to be driven by emotional dislike of the neo-Darwinian theory of adaptation as it was developed in particular by ethologists and behavioral ecologists during the second half of the 20th century (see Chapter 2). This unscientific attitude towards first-principle concepts remains an unproductive element in current evolutionary biology discourse (Welch, 2017), particularly because exchanges have become increasingly acrimonious (e.g., Nowak et al., 2010; Abbot et al., 2011) after an initial period of constructive interaction between genetic modelers and inclusive fitness theoreticians (Wade, 1979, 1980, 1985; Michod, 1982). Given that science is ultimately a hypothesis-testing enterprise, it is puzzling that critics ignore both the conceptual coherence of inclusive fitness theory and the many nontrivial falsifiable predictions that have been tested and found to match the logic of Hamilton's rule (Abbot et al., 2011; Bourke, 2011b, 2014, 2021).

Partitioning Hamilton's rule to explain the origins of major transitions in organismality

The previous section has been about Hamiltonian theory to explain conditional altruism in societies but, as it appears, that approach has no direct bearing on the evolutionary origin of unconditional altruism, as required for obligate somatic functionality and thus for MTE origins. To develop the discontinuous version of Hamilton's rule as a conceptual

instrument for explaining the origins of MTEs, I use kin selection theory as it was formulated by the 1980s. This excursion to the past reflects that, for all the mathematical sophistication via the Price equation that I summarized in the previous section, it is now appropriate to focus on core families and pedigree relatedness (Figure 4.2). Given the occasional controversy that has transpired, it is important to restate some of the conclusions of a review on "The theory of kin selection" by Michod (1982) because his summary account, formulated four decades ago, is still not always correctly understood even though no Price equation reformulation has rendered them incorrect. First, Michod concluded that "[t]o evolve by kin selection a genetic trait expressed by one individual (termed the actor) must affect the genotypic fitness of one or more other individuals who are genetically related to the actor in a nonrandom way at the loci determining the trait. These conditions are necessary and sufficient for kin selection to occur and embody the essential characteristics of the models presented [. . .] as well as preserving the original intent of Maynard Smith (1964) when he coined the term kin selection." Second, he confirmed that even then the "others only" version of altruism was already superior, writing that "it is the inclusive fitness *effect* that controls the outcome of selection in [. . .] simple genetical models and not the average fitness of relatives (self included) weighted by genetic relatedness, which is often used in optimization models of kin selection (e.g. Oster and Wilson (1978)." Third, he concluded that "[t]he results of [. . .] quantitative genetic models [of kin selection] parallel in several important respects the single locus results [. . .] in that conditions analogous to Hamilton's rule can be shown to apply to the increase of altruism."

Michod (1982) continued with further important points: fourth, he confirmed that "[i]f either there is no fitness interaction or the interactants are randomly related, then kin selection cannot be an explanation of the observed [cooperative] behavior." This may now seem to be a truism, but it highlights that invariably zero relatedness is fundamentally different from low positive relatedness, similar to my argument that invariably maximal relatedness is an extreme state of special importance (Figure 4.2). Fifth, Michod concluded that "it is more accurate to view kin selection and parental manipulation as two aspects of one phenomenon, population structure into families, rather than as two competing and mutually exclusive hypotheses," an issue that Bourke and Franks (1995) confirmed but that remains occasionally misunderstood (e.g., Nonacs, 2019). Sixth, although Michod's review mostly covered insights that predate the generalization of inclusive fitness theory via the Price equation, he recognized the potential of partitioning "within- and between-group components of gene frequency change [...] to show, by using Price's covariance form [...], that the rate of change of social genes depends upon the between-group variance in fitness." Seventh, Michod concluded that "[a]ll analyses to date of the family structured model with random mating have confirmed Hamilton's rule [...], so long as selection is weak and there is no overdominance," and that "the central heuristics of kinship theory receive adequate justification from explicit population genetic models. They apply most appropriately when selection is weak, requiring, in addition, a host of other assumptions, most of which are 'usual' in population genetics theory." Any modern challenge of the inclusive fitness paradigm (e.g., Nowak et al., 2010) would thus need to explain how all these early insights were already mistaken.

To explain the multiple fraternal (Queller, 1997, 2000) MTE origins (the gray and blue dots in Figure 3.2a), I will use both this classical kin selection logic and comparative empirical data organized by Hughes et al. (2008) and Fisher et al. (2013) that I will present in some detail in Chapter 6. The key conjecture is that MTEs towards obligate somatic altruism by cells or multicellular bodies have not arisen by a gradual increase of operational scale and/or conflict reduction in societies, but by lifetime commitment (LTC) of non-redundant founding units (see Table 3.1). This implies that all MTEs to colonial superorganismality in ants, bees, wasps, and termites originated from ancestors where parents were always strictly monogamous for their entire life (Figure 4.6a), whereas all MTEs to permanent multicellular organismality in plants, animals, fungi, and clades of algae originated from ancestors that initiated individuals by LTC between a pair of non-redundant complementary gametes (see Table 3.1)(Figure 4.6b). I have developed elements of this LTC concept in previous reviews (Boomsma, 2007, 2009, 2013, 2022; Boomsma et al., 2014; Boomsma and Gawne, 2018) and expand them further here. My arguments pinpoint the importance of a single zygote or (fungal) dikaryon bottleneck between generations of multicellular organisms (so far considered to be helpful rather than necessary; Maynard Smith and Szathmáry, 1995; Michod & Roze, 2000) and highlight with equal force the higher-level monogamy bottleneck in social insects (see also Davies and Gardner, 2018). The LTC principle is general for it uses relatedness ratios (r_x/r_o) (see Box 2.1; Figure 4.2) which, independent of ploidy, cancel out when r_x among cohorts of unicellular or multicellular offspring is maximal. It is this cancellation that provides the invariant necessary condition for irreversible loss of reproductive totipotency, that is, for evolving unconditionally altruistic celibacy across somatic cell copies or colonial neuter offspring.

If one assumes that kin selection must always have played some role in forging fraternal MTEs, it would follow that MTEs could in principle have happened under any conditions that satisfied Hamilton's rule for the number of generations needed to rewire developmental pathways towards permanent differentiation of somatic cell or neuter caste phenotypes. This null model would imply that MTEs could have happened by crossing the diagonals in Figure 4.6 at any point to enter the gray triangle where Hamilton's inequality is satisfied. If this null model were true, all we would need is the continuous version of Hamilton's rule. However, the alternative LTC hypothesis maintains that MTEs could only happen via the black circles in the center which would, by implication, reject that MTEs could have originated by gradual emergence. In the null-model scenario, normally implicitly assumed to be the only relevant trajectory, insect societies would first secure benefits of facultative altruism and then increase in scale of operations and regulation of potential social conflicts to make the Hamiltonian b/c benefits large enough to pass the diagonal even when nestmate relatedness was not maximal. My contrasting conjecture, based on the discontinuous version of Hamilton's rule, is that facultative conditional altruism of cells or multicellular individuals was not ancestral to any form of obligate somatization (Box 4.2; Figure 4.6a, b). Rejecting these null-model scenarios in favor of the LTC hypothesis would imply that (1) there exists a universal set of necessary and sufficient conditions for all fraternal MTEs to explain the origins of both obligate multicellular organismality and colonial superorganismality, and (2) irreversible MTE innovations to these higher levels of organizational complexity likely originated from inconspicuous and barely social ancestry, rather than from societies of cells or multicellular animals that were already complex.

This conjectured universal LTC explanation for the origin of all fraternal MTEs is parsimonious and conceptually compelling, for LTC ensures that relatedness (r_x) to sibling cells or individuals is always maximal and mathematically identical on average to offspring relatedness. No matter whether cells disperse or adhere after division, both are clonal copies so $r_x/r_o = 1$. The same unity ratio applies in diploid termites where full siblings are related by 0.5, identical to parent-offspring relatedness should any such sibling mate and reproduce herself. In the Hymenoptera with their asymmetric haplodiploid relatednesses, the average relatedness to a full sibling is—assuming a 50/50 Fisherian sex ratio—the mean of 0.75 relatedness to a full sister and 0.25 to a brother and thus again, but now only on average, the same as relatedness to sons and daughters. With LTC, all these relatedness terms thus cancel out of Hamilton's rule, which implies that r_x becomes irrelevant when it is always maximal as in the horizontal bottom vector of Figure 4.2a and the bottom arrow in Figure 4.2b. We have thus obtained a discontinuous version of Hamilton's rule based on the combined effect of LTC at founding and the ensuing maximal offspring relatedness as an invariant necessary condition, but remaining without effect unless the Hamiltonian $b > c$ inequality also remains satisfied for enough evolutionary time to irreversibly rewire developmental pathways into specialized germ and soma cell lines or into specialized queen and worker phenotypes. Hamilton's rule has now become partitioned into an ultimate general necessary condition and a contingent benefit/cost sufficiency condition that captures all relevant proximate factors. This contrasts with the continuous version of Hamilton's rule where the entire inequality ($br > c$) is *both necessary and sufficient* for making conditional altruism evolve (Michod, 1982)(see his first point cited at the start of this section).

There have been multiple, albeit rather few fraternal MTEs (see Figure 3.2a), suggesting

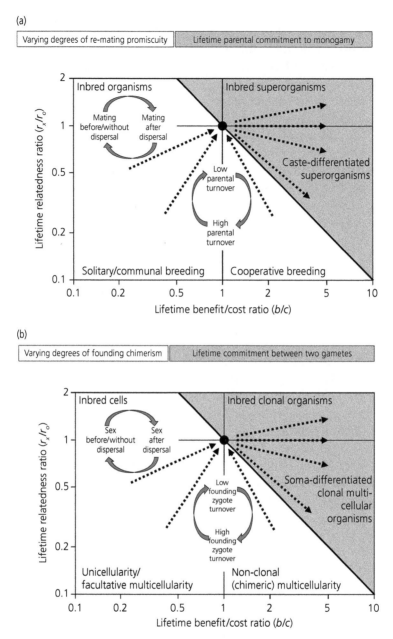

Figure 4.6 The LTC principle as necessary condition for fraternal MTEs. (a) I have proposed that strict lifetime monogamy is the universal necessary condition (Boomsma, 2007, 2009, 2013) for MTEs from solitary or family-based cooperative breeding with totipotent phenotypes (white area) to superorganismal colonies (*sensu* Wheeler)(Chapter 3) consisting only of differentiated complementary caste phenotypes (gray area)—see Davies and Gardner (2018) for a modeling study confirming this conjecture. When plotted on log-transformed axes in Hamilton's rule parameter space, the so-called monogamy window appears as a black circle in the center of the diagram, representing the situation where both $r_x/r_o = 1$ and $b/c = 1$. The gray triangle represents all parameter combinations for which $br_x > cr_o$, here plotted as $\log(r_x/r_o) > -\log(b/c)$, is fulfilled across the lifetime of families of multicellular individuals. Assuming outbreeding, relatedness equivalence implies that the r-terms cancel out of Hamilton's rule, so that $b/c > 1$ remains as sufficiency condition to be met before an MTE towards superorganismality can be realized. This is a

(*Continued*)

that the necessary and sufficiency conditions for MTE origins have rarely been fulfilled jointly, quite possibly because both are exceedingly unlikely to be met when they do not tolerate any exceptions or interruptions until an MTE is achieved. Any form of parental promiscuity or foundational chimerism—however rare—will annihilate the necessary condition and any temporary reversal into $b < c$ will quickly erode already partially evolved developmental pathways towards permanent segregation of germline and soma. As in the continuous version, the discontinuous version of Hamilton's rule also works with black-boxed genetic detail, for we are only considering what the qualitatively necessary and sufficient conditions for MTE origins are, not how long it takes directional kin selection to forge irreversible

transitions. In this perspective, LTC and $r_x/r_0 = 1$ offer an ultimate condition by default, while $b > c$ represents a conglomerate of proximate factors that will facilitate or constrain selection efficiency and which can be analyzed for causal effects via the Lande and Arnold (1983) multiple regression approach (Box 4.2), consistent with the ultimate/proximate complementarity of Tinbergen's four-quadrat scheme (see Figure 1.1). However, relative to the continuous version of Hamilton's rule, constraints emanating from proximate genetic mechanisms seem less likely in a discontinuous scenario for MTE origins—uninterrupted $b > c$ will ensure that directional selection continues to push in the same direction, so that whatever overdominant or strongly selected alleles appear, such constraints can just be

---

Figure 4.6 (*Continued*)
stringent condition because (together with the necessary condition) it has to remain fulfilled without interruptions for many generations before directional kin selection will have irreversibly "somaticized" the developmental pathways of entire cohorts of non-reproductive offspring into lifetime unmated worker castes, that is, neuters in the sense of Darwin (1859), only to produce specialized "reproducticized" gynes later in the colony cycle that will normally disperse to mate and found new colonies as queens (and kings in termites). Any form of parental promiscuity will ensure that the necessary relatedness equivalence condition remains unfulfilled and would thus make the sufficiency condition moot, so kin selection loses its consistent directionality towards producing obligatorily altruistic castes and can only produce more "reluctant" condition-dependent altruistic phenotypes. Corresponding social breeding systems are written into the white and gray areas, and the promiscuity conditions are given in the horizontal bar at the top. Straight arrows illustrate that both inclusive fitness logic and available evidence (Hughes et al., 2008) indicate that transitions from subsocial nesting with parental care to morphologically caste-differentiated colonies were not made by crossing the borderline between white and gray anywhere other than passing through the central monogamy window. Bent arrows illustrate that ancestral or sister-clade breeding systems will tend to be solitary when parental turnover (promiscuity) is high and that various forms of family-based cooperative breeding can evolve in lineages where this form of promiscuity (from the perspective of older offspring helping to raise younger siblings) is reduced, as indeed appears to apply in cooperatively breeding birds and mammals (Cornwallis et al., 2010; Lukas and Clutton-Brock, 2012) (see Chapter 6). The other pair of bent arrows illustrates that mating before (possible) dispersal or after dispersal will determine whether a breeding system is inbred or outbred, but no social systems with significant inbreeding are known to be sister clades to superorganismal lineages, so this $r_x/r_0 > 1$ scenario is not considered further. Reprinted with minor modifications from Kin selection versus sexual selection: Why the ends do not meet, J.J. Boomsma, *Current Biology* 16, R673-R683 (2007), with permission from Elsevier, and from Beyond promiscuity: mate-choice commitments in social breeding, J.J. Boomsma, *Philosophical Transactions of the Royal Society B* 368: 20120050 (2013), with permission of the Royal Society conveyed through Copyright Clearance Center, Inc. (b) The same diagram but now applied to the MTEs from unicellularity to permanent multicellularity with cell-differentiated soma. The inclusive fitness logic is identical to the details given for panel (a); supporting data have been provided by Fisher et al. (2013) (see Chapter 6). As far as eukaryotes are concerned, two lifetime committed gametes (usually in a zygote) are analogous to two lifetime monogamous parents in panel (a), and chimerism at foundation is prohibitive here, similar to parental promiscuity in panel (a)—both eliminate the necessary condition for an MTE origin (see also Chapter 6).

waited out to be overcome by the emergence of smaller effect alleles that are weakly selected. This would likely permit the gradual accumulation of new gene regulation networks for obligate germline-soma differentiation by cells or multicellular bodies early in development according to the scenarios suggested by Hammerstein (1996) and Fromhage and Jennions (2019) that were discussed in the previous section.

In Chapter 6, I will consider the empirical evidence for the scenarios outlined in Figure 4.6. In spite of their parsimony, they may seem daring conjectures, so I drew up Figure 4.7 to make explicit which phylogenetic patterns empirical data should significantly deviate from in order to refute, in a Popperian sense, the LTC hypothesis for fraternal transitions to permanent multicellularity and colonial superorganismality. As far as extant clades have lineages that practice forms of non-obligatorily committed multicellularity or cooperative breeding, that is, have retained at least some lower-level totipotent agency, I expect them to lack the invariably maximal relatedness condition, so none of them can have been a viable precursor towards an irreversibly higher level of organizational complexity. MTEs (red branches) can thus be expected to have had ancestral states that evolved along the outer "edges" of both trees where $r_x/r_o = 1$ was

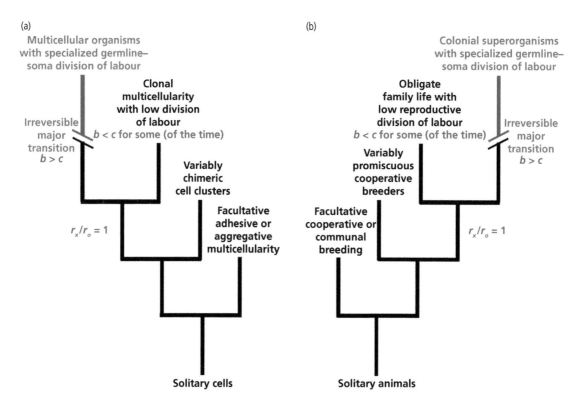

Figure 4.7 Phylogeny patterns by which societies and (super)organisms are expected to evolve. Societies of cells or multicellular animals have black text and branches and permanently multicellular organisms and caste-differentiated superorganisms have red text and branches while Hamiltonian variables are given in blue. (a) The four levels of multicellularity with higher organizational complexity than unicellularity, of which only a single terminal branch represents an MTE origin towards irreversibly more complex

(Continued)

retained, which required no more of ancestors than being solitary with parental care (i.e., subsocial breeding) or being unicellular with optional clonal cell adhesion, for only the long-term $b/c > 1$ sufficiency condition remained to be met. When that second condition was satisfied, continuously for a long enough time, completion of an MTE could potentially have proceeded relatively rapidly on an evolutionary timescale. Relatedness equivalence ($r_x = r_o$) implies that selection for resource acquisition always remained at the competitive lower cell or individual level as long as $b < c$ but unambiguously shifted to the cooperative multicellular or family level when $b > c$. Variable r and b values, as we typically find in societies with redundant partnership (see Table 3.1) will typically mean

---

Figure 4.7 (*Continued*)

permanent multicellularity. In the first branch, multicellularity itself is facultative, that is, unicellular life remains the default and only under some conditions can cell clustering be observed, either by chimeric aggregation or by daughter cell adhesion, as, for example, in many yeasts and unicellular algae (Brückner and Mösch, 2012; Kapsetaki and West, 2019. In the next branch, unicellularity is no longer a reproductively viable lifestyle, so multicellular aggregation is required, as, for example, for asexual reproduction in cellular slime molds (Strassmann et al., 2000). In the third branch, LTC by two complementary gametes (usually merged in a single zygote) has become compulsory for the foundation of new individuals so that daughter cells may commit to adhering to each other because the necessary MTE condition ($r_x/r_o = 1$) is fulfilled. However, somatic differentiation will be stalled when there are no consistent b/c benefits across all possible environments, so selection for increased somatic division of labor remains ineffective ($b < c$). FLO11-mediated cell adhesion in yeast may represent this situation (Regenberg et al., 2016) similar to unicellular *Chlamydomonas* algae that can be experimentally induced to express higher-level heritable variation for division of labor among cells, but without having become naturally multicellular (Bernardes et al., 2021). Alternatively, forms of chimerism originate later in the life cycle, so the necessary (clonality) condition fails and selection for later cell copies to remain adhered is ambiguous. Only in the other (red) terminal branch, where there is consistent directional kin selection for unconditional altruism ($b > c$), will the benefits of division of labor always surpass the costs and can developmental pathways become rewired so an irreversible MTE towards permanent multicellularity may result. (b) The analogous putative branching of organization in multicellular colonial breeders. In the first branch, retaining helpers at the nest is facultative, that is, most parents reproduce without helpers but some nests have helpers, either relatives or non-relatives, as, for example, in many cooperatively breeding birds and halictid bees (e.g., Cornwallis et al., 2009; Cardinal and Danforth, 2011; Quiñones and Pen, 2017). In the second branch, reproduction has become impossible unless individuals are part of a social nest or den, but turnover of breeders and promiscuity between them persists as in, for example, polistine paper wasps (e.g., Reeve, 1991; Hughes et al., 1993; Gadagkar, 2001; Leadbeater et al., 2011; Kennedy et al., 2018) and naked mole rats (e.g., Sherman et al., 1991). In the third branch, lifetime parental monogamy has become compulsory, but the productivity of nests with helpers is too variable ($b < c$ occurs frequently) to select for universal loss of reproductive totipotency in all offspring that stay as helpers, so that at least some and usually many helpers will retain phenotypic plasticity to become a local replacement reproductive or to disperse and become an independent breeder elsewhere. This is the stage that Batra (1966) originally defined as "eusocial," exclusively for halictid bees, while implicitly assuming that founding females always nest alone and are always singly inseminated, similar to Huxley's (1930) definition of "true sociality" (see Chapter 3). This branch likely applies also to the log-dwelling lower termites where progression towards superorganismality was stalled because strict lifetime monogamy often fails when colonies merge towards the end of the colony cycle (similar to gall-making social thrips and aphids whose colonies accumulate unrelated intruders; see Box 3.2) so that selection cannot remove reproductive totipotency in later offspring cohorts. Hence, essentially all termite lineages produce some morphologically differentiated soldiers (sometimes invariably just one; Lüscher, 1953) early in the colony cycle when all offspring are always full siblings, but only evolutionarily derived higher termites became superorganismal with colonies consisting only of specialized reproductive and permanently sterile somaticized castes. Only the other (red) terminal branch represents the MTEs to superorganismality, allowed by consistent directional kin selection for permanent caste differentiation among all nestmates. Note that all branches of the right hand tree fit Wilson's (1971, 1975) all-inclusive definition of eusociality and that all branches of the left hand tree match similarly unspecified multicellularity. Such hand-waving definitions are therefore meaningless and should no longer be used in evolutionary arguments because they muddle transparency of what selection can target (see Chapter 3).

that individual kin selection may induce some individuals to become altruists, consistent with condition dependence but likely constrained by genetic correlations due to caste load (Holman, 2014). However, when both relatedness equivalence and the $b > c$ condition are fulfilled, entire cohorts will be simultaneously selected to become unconditional altruists, a process of directional kin selection for higher-level organismality.

In a section entitled "Kin selection arises from patterns of variance, not genealogy," Frank (2012a) wrote that "[s]ince Hamilton's (1970) work, no fundamentally derived theory of kin selection based on genealogy has existed. However, it is often convenient to use the fact that genealogy typically associates with the underlying causal processes of variance and correlation." This characterization of kin selection explicitly refers to Price equation dynamics that are consistent with the dynamically continuous version of Hamilton's rule because Frank did not consider the static extreme of invariably maximal relatedness ($r_x = r_o$) (Figure 4.2b). This is logical because LTC removes the within-group term from the Price equation, whereas that term always remains of some importance in society-type sister lineages that did not realize MTEs. As noted at the start of this chapter,

there are two levels of selection in the latter case but only one in the former. In other words, when pairwise LTC is universal during (super)organismal founding and the $b > c$ condition applies, all variance that selection for separate or joint resource acquisition can work on is restricted to the family-level. Frank's (2012a) coefficient of individual within-group competitiveness (z^*) then reduces to zero up front rather than just decreasing after a long process of gradual conflict reduction over evolutionary time (see Chapter 5 for details). It is here that my analysis differs fundamentally from previous MTE concepts, which all go back to Maynard Smith (1988) who wrote that "[o]ne can recognize in the evolution of life several revolutions in the way in which genetic information is organized. In each of these revolutions, there has been a conflict between selection at several levels. The achievement of individuality at the higher level has required that the disruptive effects of selection at the lower level be suppressed." All this is true for societies as they increased in size and complexity while reducing but not eliminating Frank's z-coefficient. However, I maintain that the scenario outlined by Maynard Smith is irrelevant for the evolution of unconditional collective altruism that defined the fraternal MTEs.

CHAPTER 5

Inclusive fitness as driver of cooperation for mutual benefit

Summary

Mutual aid cannot evolve unless it offers compensating direct benefits to each actor or indirect benefits for her relatives. This phrase is a gene's eye version of Darwin's statement that no adaptation can arise for the exclusive good of another species. Such thinking has inspired studies to explore how Hamilton's rule can explain adaptation in symbiotic mutualisms. When studies tracked genetics, they converged on the conclusion that interspecific cooperation is always driven by clonal or family altruism within the partner species, which confirmed Darwin's insight and underlined earlier conclusions that "altruism between species" is a misnomer. I review the dynamics of cooperation and conflict in symbioses where a multicellular host associates with mostly unicellular partners, either by voluntary horizontal acquisition or in a coerced setting of vertical transmission. I then consider symmetry and redundancy of partnerships and their degree of genetic closure, showing that hosts either maintain societies of symbionts or have a lifetime non-redundant partnerships with a single symbiont. This suggests that the egalitarian MTE origin of LECA can be understood by the same non-redundant closure principle that I used in Chapter 4 to explain the recurrent MTEs towards organismal multicellularity and colonial superorganismality. This explanation includes the origin of meiotic sex and can possibly be extended to the origin of the first prokaryote cell. In general, the arguments in this and the previous chapter imply the conjectures that: (1) higher levels of organismality did not just emerge, they were naturally selected adaptive syndromes; (2) their ancestral properties had little or nothing to do with being big or complex; and (3) non-redundant partnership was essential to avoid conflict over resource acquisition, while reproductive allocation conflicts appear to have been unimportant for explaining the origins of MTEs.

> *Competition brings forth mutualistic interdependence in human society (Smith, 1776). Likewise, the competitive process of natural selection drives the evolution of mutualism in ecological communities (Darwin, 1862).*

Leigh, The evolution of mutualism
(*Journal of Evolutionary Biology*, 2010)

Asking how reproductive altruism evolves and is maintained by natural selection was not considered to be a relevant question before the seminal work by William D. Hamilton (1963, 1964a, 1964b). Any individuals or cells were supposed to work unambiguously for the interest of the conspecific collective that they were part of, and mutual aid was often considered to be the default setting of nature also between allospecific individuals (e.g., Kropotkin, 1902). As I mentioned in Chapter 1, these views were held in particular

Domains and Major Transitions of Social Evolution. Jacobus J. Boomsma, Oxford University Press. © Jacobus J. Boomsma (2022).
DOI: 10.1093/oso/9780198746171.003.0005

by those who believed that accepting Darwin's principle of differential survival and reproduction driven by competitive adaptations would be morally corrupting (Dugatkin, 2006; Harman, 2010). However, in Chapter 4 we saw that the evolution of reproductive altruism is now well understood as a gene's eye form of competitive multiplication, contradicting the anthropomorphic notion that nature has moral lessons to offer. In this chapter, I explore our conceptual understanding of social interactions between allospecific nonrelatives, arguing that symmetric lifetime commitment (LTC) was also necessary for the origin of the first eukaryote cell and its somaticized mitochondria, the best-known example of an egalitarian major transition in evolution (MTE) (Queller, 2000). I claim that there is no difference in principle between two unrelated gametes committing to a zygote or two unrelated parents committing to lifetime monogamy on the one hand, and a nuclear and mitochondrial genome committing to joint resource acquisition and cell division on the other. Non-redundant informational closure (see Table 3.1) always guarantees the immediate establishment of a single higher level of selection and possible adaptation.

The significance of uncompromising LTC between founding units at the origin of discrete levels of organismal organization (see Figure 3.2) can hardly be overstated, but this condition is rarely made explicit because most research has focused on cooperation and altruism in lineages where partner redundancy is the norm. This contrasts with understanding by early evolutionists who had a clear notion that open cooperation for mutual benefits requires a different explanation than closed cooperation between family members. As mentioned by Gibson (2013), Herbert Spencer (1881) acknowledged "super-organic evolution" in the social insects before Weismann (1893). The momentous disagreements between Weismann

and Spencer were about the inheritance of acquired characters in social insects and the significance of proximate (Spencer) versus ultimate (Weismann) explanations that were not yet recognized as complementary (Tinbergen, 1963). However, they agreed that natural selection can target the colony level, echoing Darwin's family selection argument. For Spencer, advanced insect colonies were not mere social aggregates—they differed fundamentally from human society, which he defined as "a union among like individuals substantially independent of one another in parentage, and approximately equal in their capacities." This distinction, echoed by Wheeler (1928b), Huxley (1930), and Simpson (1941), reflects the key difference between organismal and society organization that was unjustifiably abandoned when E.O. Wilson's version of sociobiology merged all social systems into a "eusociality gradient" (see Chapter 3). This chapter will show that also in mutualisms there is a crucial difference between symbiont–society-like arrangements that involve chimeric redundancy, and higher-level organismal commitments between different species' propagules as strong as those between single-gene-pool zygote-forming gametes or monogamous parents. Based on these conceptual similarities, I will maintain that there are separate kin selection predictions for societies and organisms at all four levels of organizational complexity as specified in Figure 3.2.

The idea that any form of mutualistic cooperation builds on division of labor has a long history, as Huxley (1912) and the opening quote of this chapter illustrate. However, from a Darwinian point of view, mutualisms are expected to suffer from inter-symbiont conflict for the acquisition of host resources (Frank, 1996a) whenever a host commits to a redundant set of symbionts, typically multiple strains from the same gene pool. Modern reviews of mutualism differ greatly in how explicitly they emphasize these potential conflicts and the need for hosts to avoid or suppress them (e.g., West et al., 2002a; Herre et al., 2008; Douglas, 2010; Leigh, 2010a; Bronstein, 2015). Neglecting symbiont conflict is unproductive, for it is impossible to evaluate the stability of mutualisms over evolutionary time without having stared their potential and realized conflicts in the face (e.g., Sachs and Simms, 2006). As we will see, the only way for mutualistic cooperation to be harmonious by default is to have partners commit for life in non-redundant single genotypes (see Table 3.1). This should generally hold true when partners do not belong to the same co-replicon (see Box 2.2), that is, when they have potential resource allocation conflicts in later joint reproduction. As with intraspecific studies of social evolution (see Chapter 4), understanding mutualistic cooperation and conflict between representatives of different gene pools also benefits from using a Tinbergen framework of four complementary types of questions (see Figure 1.1; Box 1.3). Only such an approach will guarantee that ultimate aspects of adaptation and phylogenetic contingency are separated from proximate questions of joint mutualistic ontogeny (e.g., asking whether it matters when in a life cycle symbiotic partners join) and mechanisms of gene expression, protein synthesis, and physiology across partner species. Likewise, mutualism research should not only consider descriptive patterns but also include experimental manipulations that challenge the depth of interspecific harmony.

Distinguishing between hosts that maintain societies of symbionts and hosts that commit to a single partner genotype is important—as in cooperation within species, only the latter arrangement has the potential to realize an irreversible MTE towards higher organizational complexity. Mutualisms based on association with societies of symbionts are asymmetric and may even function as biological markets if neither hosts nor symbionts practice closure (Noë and Hammerstein, 1995; Werner et al., 2014; Wyatt et al., 2014; Kiers et al., 2016; Noë and Kiers, 2018). However, such dynamics are absent in non-redundant and symmetrical LTC mutualisms, with the possible exception of a brief period in which lifetime partners find each other. Because the focus of my analyses will be on the necessary and sufficient conditions for MTEs, I will not cover classic (e.g. Hutson et al., 1985) or modern (e.g. Thompson, 2005; Bronstein, 2015) aspects of ecological dynamics in symbioses—they are of interest but are nested within the overarching Darwinian questions of adaptive design. In the next section, I will summarize how inclusive fitness approaches have come a long way in explaining the maintenance of symbiotic cooperation for mutual benefit. In the subsequent section, I focus on the LECA to see whether non-redundant social closure can explain egalitarian MTEs as well as it could explain the fraternal MTEs in Chapter 4. Finding that ultimate LTC causation applies in both cases implies that intraspecific altruism and interspecific cooperation are two sides of the same coin in the limit case of pairwise non-redundant commitment that makes r_x/r_o cancel out of Hamilton's rule. In the final section, I will argue that emergent evolution by gradual conflict reduction does not provide a viable alternative for understanding the origin of LECA and other egalitarian MTEs, just as such arguments did not appear credible for explaining the origins of the fraternal MTEs in Chapter 4.

The gene's eye view of cooperation between nonrelatives

Inclusive fitness theory is fundamentally about relatedness, costs, and benefits, and relatedness is traditionally about similarity between individuals owing to recent common descent. However, restricting attention to related individuals is not absolutely necessary for social evolution theory as first noted by Nee (1989). Frank (1998, p. 4) drew attention to the possibility of capturing interspecific social interactions with statistics of association like those that we use for intraspecific interactions. He argued that relatedness coefficients are partial covariances between partner phenotypes and actor genotypes without mathematically requiring that these partners have common ancestry. Traditional regression relatedness is thus merely a convenient scalar of transmission fidelity, that is, the efficiency by which genes are passed on to future generations via relatives who sacrifice their own reproduction. Such a coefficient represents conditional information, that is, it is a statistical estimate of an expected fitness response to a (set of) predictor variable(s) (Frank, 1998, p. 39). This logic was also embodied in Queller's (1992a, 1992b) models that partitioned the selection components affecting a social trait to fit a multiple regression scheme. His re-derived version of Hamilton's rule then emerged as the overall regression of phenotypic fitness on the breeding values of traits that could be alleles, but also group characteristics, environmental variables, or cultural beliefs (Frank, 1998, pp. 53–54). This universally applicable result also emerged less explicitly from earlier purely phenotypic Price equation models (Seger, 1981; Michod, 1982). The overall conclusion was that the outcome of selection is always determined by the statistical association between social interactants. Co-ancestry is just one factor imposing association, albeit a common and very important one (Frank, 1998, p. 56).

Frank (1998) concluded that social payoff depends on the regression of partner behavior on the breeding value of a focal individual for a relevant trait (see also Frank, 2013a), because that relation captures the necessary information about behavioral genes that partners will transmit when co-breeding with

the focal individual. In open systems with repeated encounters and specific probabilities of promiscuous partner change, such focal individuals rely on Bayesian inference to combine prior and context-dependent information. Depending on perceived statistical information about social partners, interacting agents will then be under selection to maximize utility in inclusive fitness terms (Frank, 1998, p. 113). This measure is fractional, so deviations from maximal information can be used to predict the extent to which individuals are selected to exploit their group, that is, to express mutualistic virulence (Frank, 1996a). Frank used Hardin's (1968) tragedy of the commons as illustrative example of group success being negatively affected in proportion to resource acquisition competition within groups (z). This quantity, already briefly introduced towards the end of Chapter 4, appeared to be inversely related to relatedness (r), such that equilibrium group fitness equals $1 - z^*$, where $z^* = 1 - r$ (Frank, 1998, pp. 130–133). No group can therefore maximize its resource acquisition fitness unless individual and group interests are perfectly aligned ($z^* = 0; r = 1$), consistent with the invariably maximal relatedness argument that I presented in Chapter 4 as a necessary but not sufficient condition for the origin of fraternal MTEs. Frank's analysis suggested that interspecific mutualisms might be explainable in Hamiltonian terms provided each partner maximizes its own inclusive fitness while participating in an interspecific mutualism. If that claim holds, it should imply that the origins of fraternal and egalitarian MTEs (Queller, 2000) can both be explained with inclusive fitness theory.

Frank's argument (see also Frank, 1994, 1995a, 1997a) was framed as an analysis of interspecific altruism, but I will refrain from using that term because a Darwinian definition should separate subject and object and thus focus on "others only" effects rather than group effects (see Chapter 4).

Because deciding agents exchanging services with nonrelatives cannot obtain indirect fitness benefits instantaneously, the Frank model required a neighbor-modulated notation (see Box 4.2) where benefits are returned from the social environment, rather than having to be "reaped" immediately by an altruistic actor herself. Models of this kind therefore use phenotypic versions of Hamilton's rule, which makes them conceptually less general than the genetic theory of kin selection (e.g., Gardner et al., 2011). Frank's model also had to assume that "willing cooperators" already exist—they could not evolve de novo because synergistic b/c benefits needed to exceed some minimal threshold value—so his approach remained silent about the origin of mutualistic cooperation, that is, the stage where b/c ratios would barely exceed 1. A decade later, two studies elaborated on Frank's pioneering work. Foster and Wenseleers (2006) used a general differential equation approach based on Frank (1998) to focus on the relatedness structure within each of the cooperating partners, which produced partial regression slopes equivalent to Hamilton's rule with cost and benefit terms scaled by relatedness. The authors concluded that stable interspecific mutualism requires both high relatedness in own gene-copy currency and a high association fidelity between partners. As in Frank's model, the dynamics of change were captured, but the origins of interspecific mutualisms remained opaque. However, the basic idea that unrelated partners could evolve stable mutualistic cooperation provided each of them was able to increase its own inclusive fitness had gained further support.

Around the same time, Fletcher and Zwick (2006) extended a modeling approach initiated by Queller (1985) and Michod and Sanderson (1985) that did not use breeding value predictors but concentrated on interactions between behavioral phenotypes of cooperating species. This allowed costly investments in cooperation to be randomly distributed as long as the expression of these phenotypes is conditional on the responses of partners. The outcome of this model was consistent with the approaches of Frank and Queller, but it required a specific coefficient to account for nonadditivity because the interactions modeled represented forms of conditional cooperation where an actor only invests when the partner responds in kind. We should remember, however, that this model was framed explicitly in terms of behavioral rather than genetic population structure (Frank, 2013a). In this behavioral approach, very similar mathematical expressions apply to questions of kin selection, reciprocity, assortment, and deme-structured group selection because what causes the genetic similarity between donor and recipient remains unspecified (Grafen, 1985b). Fletcher and Zwick (2006) confirmed that Queller's (1985) version of Hamilton's rule always correctly predicts the direction of selection for conditional cooperation between species, but their claim to have proven "altruism between species" is unwarranted. They only showed that the dynamics of reciprocity are predictable when both species already have evolved the relevant cooperative traits—their behavioral conceptualization precluded interpretation in terms of invasion by cooperative genetic mutants. One could also argue that the nonadditivity coefficient that Fletcher and Zwick (2006) required for their model is a mutual cooperation parameter, not an altruism one, bringing us back to the ambiguities that emerge as soon as one abandons "others only" altruism. The Fletcher and Zwick contribution represents a somewhat broader spectrum of models that allowed similar conclusions (e.g., Fletcher and Doebeli, 2006, 2009; Zink, 2015) but without considering genetic factors—similar, in fact, to Frank (1998).

The most informative analysis so far appears to be the one by Wyatt et al. (2013), which confirmed that it is crucial that benefits of helping nonrelatives come back to own off-spring. These authors followed Foster and Wenseleers (2006) in making genetics explicit and used a dynamically sufficient model with an infinite number of patches, each containing one haploid individual of two partner species. Costs and benefits were made explicit, and the model also included the necessity for agents to compete for patches becoming vacant by random mortality. This approach is attractive for several reasons. First, it is spatially and temporary explicit and allowed validation of analytical results by simulation, as Fletcher and Zwick (2006) had done without considering genetics. Second, it illustrates that deep causation can only be captured by considering intraspecific altruism within each partner species, confirming the result of Foster and Wenseleers (2006) but now for lifetime pairs—the central element of the LTC and MTE logic developed in this book (see Table 3.1). Third, the model allows an alternative group selection interpretation, but the kin selection interpretation does not have to assume that interspecific helping already exists, a handicap that the models by Frank (1998) and Fletcher and Zwick (2006) also encountered. That left intraspecific altruism combined with interspecific reciprocity as the most parsimonious explanation, a partitioning that did not emerge with the same clarity when using trait-group approaches.

Wyatt et al. (2013) opened their paper with a quote from Darwin's *Origin* (1859) that "[i]f it could be proved that any part of the structure of any one species had been formed for the exclusive good of another species, it would annihilate my theory, for such could not have been produced through natural selection." This was possibly Darwin's most daring statement in the later sense of Popper (1962), illustrating his unwavering confidence in natural selection as the exclusive driver of adaptive design. The starting G.C. Williams quote of Chapter 1 echoes this

trust, as does the W.D. Hamilton quote at the beginning of Chapter 2. The adjective *exclusive* is crucial because it implies that Dawkins' (1982) well-known example of zombie ants as extended phenotypes of a fungal parasite does not challenge Darwinism because the workers' adaptive somatic design had originally evolved to serve ant inclusive fitness. Wyatt et al. (2013) concluded that natural selection can indeed favor the evolution of indiscriminate helping between species, but before one calls this interspecific "altruism," one should ask whether that result can more fully and parsimoniously be accounted for by intraspecific altruism. Gardner and Grafen (2009) obtained an analogous result when they mathematically "captured" the (bottom-up aggregative) superorganism of Wilson and Sober (1989), only to conclude that this could not be done unless all individuals were clone mates (see Chapter 4). Wyatt et al. (2013) thus showed that benefits and costs continue to be expressed in species-specific gene-copy currency, but that effects need to be conceptualized across generations when within-pair relatedness is zero in each specific generation.

Darwin's interspecific *exclusive good* statement applies with the same force when taking an intraspecific gene's eye perspective because it can be generalized to read "for the exclusive good of a non-relative." This underlines the specific meaning of invariably zero relatedness (see Figure 4.2b) where altruism cannot evolve and where mutualistic cooperation always needs to incur direct reciprocal fitness gains if partnerships are to remain evolutionary stable. Extensions to negative relatedness appear valid as well, because such interactions invert the statement into "for the exclusive harm" of a partner, implying that social harming cannot evolve unless it benefits the harmer's gene copies. Darwin's *exclusive good* statement also explains why mathematically valid group selection models with implicit individual agency can be drawn up (see Figure 4.3). As it turns out, the equivalent neighbor-modulated notation that can be flipped to inclusive fitness will always show that inclusive fitness remains the maximand. The all-inclusive group selection definition of altruism (e.g. Pepper, 2000) thus ensures direct benefits to *self* in a group of finite size, so Darwin's *exclusive good* statement is not violated, but the key difference between cooperation for mutual and altruistic benefits has become muddled (see Figure 4.1). Finally, invariably maximal relatedness (see Figure 4.2b) also has a deeper meaning in light of Darwin's statement, for it represents the only trajectory where lower- and higher-level fitness interests remain perfectly aligned (so selection at the higher level is unconstrained) when joint partner efforts yields consistent Hamiltonian $b > c$ benefits in resource acquisition. After I made the LTC condition for fraternal MTEs explicit in Chapter 4, Wyatt et al. (2013) thus offered compelling evidence that the same logic must apply to the origin of egalitarian MTEs such as LECA, as I will elaborate in the next section.

Before embarking on the origin of stepwise egalitarian transitions to higher-level organismality, it is fitting to better understand the continuous society dynamics of cooperation between nonrelatives. Two recent papers by David Queller (Box 5.1) elegantly illustrate the conceptual merit of modeling approaches that focus on phenotypic causality as a complement to the genetic models summarized in Chapter 4 (e.g. Gardner et al., 2011).

The latter are so general in laying out the overall concept of inclusive fitness that they can be accurate about how natural selection affects correlations between genes and fitness without caring about direct causation (Gardner et al., 2007). Queller's approach alleviates some of this abstraction by framing analyses in terms of breeding values while also using path analysis as developed in his own and Steve Frank's previous work. In the first of these papers (Queller, 2011), he showed that relatedness-like coefficients exist both for deviations from additivity and for reciprocal feedback between interacting partners. This produces an extended dynamical version of Hamilton's rule consisting of a cost term and three partial regression slopes that capture both classic kin selection based on genetic relatedness, and interactions with unrelated neighbors of any type (what he termed kith selection) and between isolated copies of genes or non-recombining supergenes that interact irrespective of the interests of the genomes they belong to (what he termed kind selection). Kind selection thus encompasses green-beard dynamics and non-additive games between two unrelated players, but it is kith selection that is most relevant for the questions addressed here. This is because kith selection can either be positive to produce reciprocal cooperation or negative when selection favors mutual exploitation, complementing the insight that kin selection can both mediate helping (when relatedness is positive) and harming (spite, when relatedness is negative).

Box 5.1 Kin, kith, and kind selection on separate or joint phenotypes

In the mid 1980s, David Queller introduced two conceptual tools to explain why Hamilton's original inclusive fitness predictions may deviate from simple single-locus population genetics models. He first used game theory to show that fitness-relevant deviations in behavior that depend on who agents interact with can be captured with a synergistic selection coefficient of positive or negative sign that complements

Box 5.1 *Continued*

the inclusive fitness effect (Queller, 1984). He further noted that this synergistic effect does not depend on interactants actually being related, suggesting that synergistic selection and kin selection are comparable albeit also very different. In a following paper (Queller, 1985), he then sacrificed the complete recursion characteristic of single-locus population genetics to obtain an alternative version of Hamilton's rule that is essentially qualitative, but gains in general applicability. In a *News & Views* commentary, Alan Grafen (1985) endorsed this approach because it can handle interactions in pedigrees, in viscous populations, and even those between mere lookalike genotypes. However, he also noted that there are more general conceptualizations of Hamiltonian benefits and costs that would make the synergy term superfluous, as Hamilton (1970) in fact proposed himself. In retrospect, it seems fair to conclude that two different approaches to inclusive fitness theory came to light at this point in time: an exact genetic one that would develop into the partial regression approach based on abstract benefits, costs, and breeding values (Gardner et al., 2011), and an approximate one using phenotypic predictors for analyzing the benefits and costs of cooperation.

Spelling this history out is relevant, because Grafen (1985) stipulated that the nonadditivity problem (and thus the need for a synergy term in Hamilton's rule) disappears when we make the "plausible assumption that the gene effects in question are small." This is R.A. Fisher's "microscope argument" that adaptations close to their focal optimum in terms of evolutionary design can only be improved by genes of small effect, that is, larger-effect mutations will be much more likely to be deleterious. This argument also featured in Grafen's (1984) phenotypic gambit that we encountered in Chapter 2 and it is at the core of the recently suggested strategic reference gene by Fromhage and Jennions (2019) that we met in Chapter 4 (see also Garcia-Costoya and Fromhage, 2021), and of a recent analysis by Levin and Grafen (2021). However,

although the phenotypic gambit implies that complex phenotypic traits are better captured with quantitative genetics than with single-locus models, it was not until the early 1990s that the alternative notation, based on the additive genetic variance or breeding value of social traits, was formally developed by Queller (1992a, 1992b). That approach produced what can be referred to as a general marginal value version of Hamilton's rule (Frank, 2013b) in which different social phenotypes can be evaluated in a multiple regression design based on the Price equation. This way of conceptualizing social evolution is explicitly causal in a proximate sense because predictor variables can be concrete measurements of fecundity, season, or habitat quality, which can then be compared for their relative predictor importance via path analysis (Lande and Arnold, 1983).

Frank (1998) reviewed these developments to unify different approaches for partitioning phenotypes and fitness. As he wrote more recently about developments in the 1990s (Frank, 2013b), it had become necessary to "unite Queller's causal approach through regression and path analysis with the analytical power of the maximization and marginal value techniques," because "causal analysis provides the foundations for reasoning about complex problems, and marginal value analysis provides the techniques for applying that reasoning to particular cases." In Chapter 4, I summarized how contributions focusing on genetic abstraction (e.g., Gardner et al., 2011) have generalized inclusive fitness theory in spite of occasional tensions with concepts focusing on causal (i.e., measurable) variables (see Levin and Grafen, 2021). For this chapter, it is most relevant to start with phenotypic models of correlated selection, even though we will inevitably return to underlying forces of genetic kin selection. In the main text, I introduced Frank's (1998) phenotypic kin selection model (see also Frank, 1994) and summarized how Fletcher and Zwick (2006) elaborated on Frank's and Queller's logic of correlated

Box 5.1 *Continued*

phenotypic selection, whereas Foster and Wenseleers (2006) developed an approach that remained close to genetic kin selection theory and that would later be elaborated by Wyatt et al. (2013). Here, I explain in more detail the connections between these models and their convergence on a more general understanding of inclusive fitness theory across multiple forms of association between unrelated partners, which resulted in the so far most encompassing causal phenotypic approach by Queller (2011, 2014).

While Gardner et al. (2007) generalized how kin selection works on correlations between genes and fitness, Queller (2011) operationalized causal factor analysis by expanding neighbor-modulated fitness theory to make two additional kinds of social selection explicit: kind selection and kith selection. Kind selection applies when an agent's social trait has different effects depending on whether recipients share the trait, with green beards and nonadditive games as prime examples. Kith selection refers to situations where interactants are neither kin nor kind but unspecified neighbors that may engage in mutually beneficial cooperation or in manipulative mutual exploitation. Assuming that social phenotypes have genetic variation for agency to maximize inclusive fitness under the usual assumptions of Grafen's (1984) phenotypic gambit, Queller pursued a compromise between mathematical generality on one hand and accessibility for empirical biologists, who measure real variables when they test hypotheses, on the other hand. He proceeded to derive an expanded Hamilton's rule that has separate terms for each of the possible kin, kind, and kith interactions, which yielded the inequality $-c + \Sigma br + \Sigma ds + \Sigma mf > 0$. The first two terms are the classic version of Hamilton's rule. Then follow the summed products for deviation from additivity (d) and synergism (s), and for feedback (f) and mutualism or manipulation (m). The summation terms are thus always products of a selection term relating a social action to fitness (b, d, m) and an association coefficient that captures the relative heritability of these factors (r, s, f).

Queller's approach has parallels with generalizations of inclusive fitness theory that Steve Frank developed around the same time (Frank, 2012c, 2013a). In his approach, the focus is primarily on the population-wide accumulation of adaptive information by selection versus the decay of that same adaptive information by other evolutionary processes. Like Queller, Frank always recovers a Price equation where change has a first component comparable with a slope that quantifies the strength of selection and a second component that adjusts for the extent to which other effects on phenotypic fitness preclude selection from being maximally effective in forging directional change. As both Frank's and Queller's arguments are based on the breeding values of social traits, it is illuminating to see that they end up with logic reminiscent of standard quantitative genetics where the response to selection R is equal to the product of the selection differential S and the narrow sense heritability h^2 (see Box 2.3). In this equation, adaptation also accumulates by selection and decays by incomplete transmission of the adaptive information gained. Queller's (2011) Price equation approach is straightforward because the only assumption he had to make is that there is no correlation between the breeding value of interest and the unexplained residual variation in fitness that remains after evaluating a series of causal predictor variables (see Lande and Arnold, 1983), which is a general validity condition for any regression analysis.

In his section on causality, Queller (2011) explains that it is technically correct to only consider an agent's own breeding value and to ignore a partner's breeding value (as all earlier models did), but that crucial insights may be missed when considering partners in isolation if they are in fact interdependent, socially or ecologically. In Queller's view, the inclusive fitness approach of keeping at least two regression terms in an expanded Hamilton's rule is more than just a useful heuristic for understanding the causes and consequences of social evolution. It is in fact necessary to maintain the phenotypic gambit's conceptual validity because the

Box 5.1 *Continued*

assumption of uncorrelated breeding value and residual variance can only be upheld if the breeding values of one's relatives (which are correlated by definition) are partialed out. Queller also explains that the theoretician is obliged to consider only predictor variables that are likely to maintain selection pressures in the social contexts of the cells or organisms considered. An important consequence of this notion is that any byproduct effects of altruistic or mutualistic interactions need to be excluded from formal analysis. This insight goes back to Williams (1966a), whose understanding of adaptation as an "onerous" concept (see Chapter 1) was fundamentally based on the search for adaptive functional design, subject to natural selection but ignoring any correlated traits that can be dismissed as side effects. It is also consistent with byproduct mutualisms not being considered in a recent review by Leigh (2010a). Against this background, Queller set out to develop a general inclusive fitness approach based on covariance ratios of gene–phenotype associations.

Starting with kith selection for mutualistic cooperation across species, Queller shows that the Σmf term that now replaces the relatedness term Σbr becomes positive when the actor's phenotype predicts the partner's phenotype, even when these phenotypes can no longer be expressed in the same currency (e.g., carbon and nitrogen production by the algal and fungal partners in a lichen). Synergy is thus conditional on cooperators preferentially associating with other cooperators or, if association is random, cooperators giving higher benefits to partners that reciprocate in kind. The key to kith selection is that both partners are required to actively invest in their interaction, consistent with the definition of mutualism, which is different from kin-selected altruism where the recipient can be passive, as, for example, in parental care. This is why we need the structural feedback coefficient f instead of the relatedness coefficient r that suffices in kin selection. However, while causation in kin-selected

altruism is obvious because costs and benefits are expressed in the same gene-copy currency, causation remains opaque in the positive Σmf term. This can be resolved by writing the adjusted version of Hamilton's rule in neighbor-modulated notation, but that also shows that the acting agent only values the cooperating partner in proportion to the extent to which the interaction results in direct fitness returns to herself, both when the partner's change in fitness is incidental (i.e., a byproduct) and in a lichen example where partners have a number of coadapted traits. This conclusion is similar to the results obtained by Foster and Wenseleers (2006) and Wyatt et al. (2013).

Kith selection works with benefits across species that can both be positive (mutual benefits) and negative (mutual manipulative exploitation) as long as benefits are scaled by a heritability-like feedback coefficient that downgrades these benefits in proportion to an actor's phenotypic inability to perfectly predict the partner's phenotype. In the final section, Queller (2011) shows that similar logic applies in what he coined kind selection, but here the selection benefit emanates from a nonadditivity coefficient d, and a downscaling heritability-like factor s—a synergy coefficient. Queller applies kind selection to green-beard dynamics and simple 2×2 game-theory payoff matrices. In both cases, the interactions can be helpful or harming, giving d a positive or negative sign, and they are either facultative (condition dependent) or obligate. Phenotypic similarity is key here in contrast to kin selection where benefits often go to unlike phenotypes such as dependent and passive immature offspring. A kind selection approach can also be used when there are nonadditive fitness effects in a kin selection model and one aims to separately understand these deviations from simple phenotypic gambit logic. Overall, Queller (2011) adopts a middle position between fully general genetic inclusive fitness theory and highly specific population genetic models to elucidate similarities across most categories of social evolution models.

Box 5.1 *Continued*

In a sequel paper, Queller (2014) further developed general inclusive fitness theory for mutualistic interactions focusing on joint phenotypes across species. Once more taking an explicitly Fisherian route, he shows that coadapted joint phenotypes can evolve when selection targets the sum of the additive genetic variances of the relevant species-specific traits, weighted by the strength of selection on each of them. This provides simple mathematical formalism for mutualisms where the mean population-wide fitness is affected by heritable traits in partner species as well. The results therefore extend Fisher's fundamental theorem of natural selection to also include coevolution. Because this more encompassing theory focuses on phenotypes and their breeding values, it is also applicable within species for evaluating mutualistic cooperation and conflict between the sexes or between parents and offspring. The models thus provide a general framework for understanding evolutionary conflicts because the signs of selection on joint phenotypic change can be opposite across partners. Technically, Queller's (2014) approach is reminiscent of an extended phenotype concept (Dawkins, 1982a) but it is more general, similar to indirect genetic-effect analyses (e.g., McGlothlin et al., 2014), except that combined ownership of joint phenotypes is now causal rather than consequential. The Price equation thus allows changes in relative fitness to be captured in covariance terms for each party, after which nonadditive components end up in the final term of the equation which is assumed to be independent of the sum of the breeding values.

In his final analyses, Queller (2014) obtained very similar differential equations for expected evolutionary change in joint phenotypes when interacting individuals are of the same or different species. Variances in breeding values are always adjusted for the number of interactions experienced by average individuals of both partners, while joint phenotypes and mean breeding values are defined for focal partnerships. Queller works out an example of cheetahs and gazelles interacting in an antagonistic arms race with selection forces being of opposite sign despite each having some "co-ownership" in the other's phenotype. The joint phenotype approach thus identifies breeding values for gazelle and cheetah genes that

affect both gazelle and cheetah fitness. As expected, these equations reduce to Fisher's fundamental (non-interactive) theorem if partner species have no heritable effects on each other's fitness. However, the mean fitness of gazelles declines if cheetah genes for gazelle fitness are either more variable or more strongly selected and vice versa. In the end, any such "other species" effects (there may be multiple ones if cheetahs prey on several gazelle species) are in large part what Fisher (1930) identified as deterioration of the environment and what the Red Queen hypothesis captured with the Alice in Wonderland phrase that it takes all the running you can do to stay in the same place (Van Valen, 1973). Applying the same logic at the within-species level only required formulating Fisher's fundamental theorem in inclusive fitness terms when analyzing antagonisms between age groups (e.g., cannibalism) or between sexes with different roles in joint reproduction.

Queller (2014) made several further points of general interest. First, just as Fisher's fundamental theorem is a deep principle of design by natural selection, so does the joint phenotype approach formalize the general analysis of conflicting design criteria. Second, the gazelle–cheetah interaction represents an example of realized conflict because potential conflict between parties is always expressed, but in many social situations one party may lack the power to respond (Beekman and Ratnieks, 2003). Somatic domestication (Table 3.1) will then be a logical consequence in a closed organismal setting (e.g. gene loss in mitochondria), while open social interactions can be expected to lose stability when response power is asymmetrical. For example, contemporary cheetahs may not have much power of agency because their genetic variation is seriously depleted, which may not bode well for their long-term adaptive potential. Third, and most relevant here, the joint phenotype approach clearly demonstrates that inclusive fitness theories of mutualism and antagonism are two sides of the same coin, with the former being mutual exploitation and the latter being unilateral exploitation (see also Frank, 1998; Herre et al., 1999). This is also true for cases where selection forces on joint phenotypes are both positive, a scenario that Queller (2014) elaborated using the concept of kith selection (Queller, 2011). In such

---

Box 5.1 *Continued*

cases, mutualisms only remain free of potential conflict when the positive selection forces on both parties are identical, which is valid only for stringent forms of LTC (see main text). If the plusses are unequal, selection will only maintain unconditional mutualism until cooperation starts failing to be beneficial for the partner with the lowest gain rate, typically because chimeric redundancy creates a tragedy of the commons among symbionts, as first recognized by Frank (1996a).

In a sequel paper, also summarized in Box 5.1, Queller (2014) extended Fisher's fundamental theorem to apply to co-adapting phenotypes when selection targets the sum of the relevant heritable traits in each of the partner species. This implies that genes in one species take partial co-ownership in the phenotypes of the partner species. Such mutual effects can be asymmetrical so that intermediate situations between mutualism and exploitation can be conceptualized via reciprocal kith selection coefficients that have the same sign but a different magnitude. This insight is consistent with Frank's earlier (1996a) idea that hosts may experience virulence by mutualists when symbionts do not perform at their maximal rate, typically because they compete for host resources with other symbiont lineages of the same gene pool that are redundant for the host (see also Kiers and Denison, 2008). Queller (2014) illustrated the mutual exploitation version of kith selection with the interaction between cheetah-predators and gazelle-prey, which produced another interesting analogy with previous understanding because asymmetric dynamics of this kind are reminiscent of the life–dinner principle suggested by Dawkins and Krebs (1979). The kith selection approach thus allows a very broad understanding of antagonistic arms races that are normally referred to as Red Queen processes (e.g. Van Valen, 1973; Decaestecker et al., 2007). It also appeared that interspecific antagonistic effects can in large part be captured by what Fisher (1930) identified as the environmental deterioration that always tends to compromise naturally selected improvements in adaptive design. Hosts that are forced to work with multiple strains of symbiont in the sense of Frank 1996a, 1996b may thus be analogous to predators depending on multiple prey populations—both situations preclude reaching the maximal level of co-adaptation across the different partner species.

Queller's contributions (Box 5.1) conceptualize many coadapted, beneficial or harmful, social interactions in a single framework. His modelling allows firmer conclusions about adaptation than Fletcher and Doebeli's (2009) locus-specific inferences of relatedness as a coefficient of statistical information about social partners (West and Gardner, 2013). Queller's approach also has the advantage of framing novel testable predictions. For example, he acknowledged that interacting partners, both mutualists and antagonists, are likely to have asymmetric power of agency because their standing genetic variation for coevolving traits will differ, and he attempted for the first time to jointly capture the conflicting design criteria shaped by natural selection across almost all cases of social conflict. His analyses reinforced the idea that mutualisms are essentially always examples of mutual exploitation (Huxley, 1912;

Herre et al., 1999; Kiers and Denison, 2008) so that mutualistic and antagonistic coevolution are two sides of the same coin. Around the same time, Frank (2013a, 2013b) obtained several similar insights. Both Frank and Queller used the breeder's equation (see Box 2.3) to show that adaptive genetic variation always accumulates in population gene pools via beneficial effects, scaled by a relatedness-like coefficient between 1 and −1, and is always lost at rates inversely proportional to heritability-like coefficients between 0 and 1, which resolves limitations of earlier nongenetic models. As in the explanation of altruism (see Chapter 4), the openness of a partnership also matters greatly in mutualistic cooperation. Queller's (2014) analyses assume that interactions are asymmetrical by definition, a generality consistent with mutualisms having retained varying degrees of symbiont redundancy, that is, with hosts maintaining societies of symbionts. In contrast, the model by Wyatt et al. (2013) captured symmetrical LTC scenarios with egalitarian MTE potential, the topic to be explored in the next section.

Capturing LECA and other mutualisms in terms of closure and symmetry

The LECA is special because it was a dual fraternal (meiotic sex) and egalitarian (nuclear-mitochondrial) MTE. The origin of meiotic sex endowed the archaeal precursor cells, empowered by a clonal mitochondrial power plant (Lane and Martin, 2016), with the capacity for highly structured recombination that must have been instrumental for LECA's lineage to be favored by natural selection. First recognized as chromosomal crossing over by Weismann, it was Dobzhansky (1958) who made the tremendous diversity-generating power of meiotic sex explicit. As he wrote, in a world with "only 1,000 kinds of genes, each represented by only ten different alleles [...]

[s]exual reproduction would [...] be potentially capable of engendering 10^{1000} gene combinations [...] an almost unimaginably large number [given that] [p]hysicists estimate the number of electrons and protons in the visible universe to be a mere 10^{73}." When de novo mutation rates tend to hover around 10^{-10}, the importance of mutation for generating genetic diversity must therefore be minor compared to the reshuffling of genes by recombination. Dobzhansky (1958) may also have been among the first to note that "sex is just as efficient in combining adaptations as it is in breaking them up" and that "[u]nlimited interbreeding of populations adapted to different ways of life would result in disintegration of adaptive gene complexes set up by natural selection." To this he added that "the gaps between species correspond to gene combinations most of which would yield low fitness, or which would make their possessors wholly unfit to survive," so that biological species are inevitable facts of nature, separated by "gaps [that] correspond, in Wright's metaphor, to the adaptive valleys." The validity of this Fisherian (Frank, 2012d) species-level adaptive landscape appears to be obvious in contrast to the within-species version of Wright's shifting balance theory (Coyne et al., 2000).

It is no surprise, therefore, that questions about the maintenance of recombination have been intensively studied, particularly in the 1970s and 1980s (e.g., Williams, 1975; Maynard Smith, 1976b, 1984; Stearns, 1987) (see Chapter 2), but it is striking how little, in comparison, the origin of sex has been contemplated. Questions about evolutionary origins and secondary maintenance or elaboration are very different. Distinguishing between them is fundamental in evolutionary studies (Griffiths, 2009; Bourke, 2011), because the latter refer to extant lineages while the former are inferred reconstructions handicapped by extinction often having eroded early diversification (e.g., Wilkins and Holliday, 2009; Bernstein and Bernstein, 2010). Three inferences seem worth highlighting in this context: (1) in contrast to previous suggestions (e.g., Szathmáry and Maynard Smith, 1995), it is no longer tenable to consider meiotic sex as an MTE in its own right (Szathmáry, 2015)—sex was just part of LECA's adaptive syndrome. (2) While meiotic recombination has been secondarily lost in variable frequencies, both in unicellular eukaryote clades and in some multicellular and superorganismal lineages, all MTEs that came after LECA evolved from sexual ancestors. Meiotic sex thus remained a defining trait of all major lineages of multicellular and superorganismal eukaryotes (see Figure 3.2a). (3) Many prominent evolutionists have pursued the idea that sex might have originated by higher-level selection. This included George Williams, who noted, in an introductory section of an edited compilation of early papers pro and contra the original idea of group selection, that "[t]he nearly universal existence of the sexual cycle of meiosis and fertilization [...] is perhaps the most crucial evidence of the importance of group selection" (Williams, 1971, p. 161) (see also Maynard Smith, 1976b, 1976c; Bell, 1982).

But can a protist exconjugate (zygotic) cell, consisting of two lifetime committed gametes and multiplying by mitotic division, be conceptualized as an emergent property of group selection, or is that genetic information commitment itself the causal factor of enhanced organizational complexity, similar to the equally strict LTC between a nuclear and mitochondrial genome that characterized LECA? In the following paragraphs, I will use analogy arguments from multicellular symbioses to shed light on the strictness of the LTC principle in partnerships of very different identities. To start this evaluation it is relevant to note that division of labor, not sex, is the logical a priori starting point of any mutual benefit symbiosis, as Huxley (1912) established

and Leigh (2010a) confirmed, while division of labor is an a posteriori consequence in the fraternal MTEs, which are based on altruistic germline–soma differentiation realized by a single gene pool. LECA is therefore defined by its closed, non-redundant (1:1) nuclear–mitochondrial symbiosis, with meiotic sex as a clever add-on. However, comparing the single-celled LECA with the types of macroscopic mutualisms lined up towards the left of Figure 4.1 indicates that details of commitment have often remained implicit. In what follows, I will therefore emphasize aspects of interaction symmetry and commitment closure in terms of genetic information, rather than the classic (e.g., Herre et al., 1999) dichotomy between voluntary (horizontal) and enforced (vertical) partner association. No matter how one classifies mutualistic symbioses in ultimate terms, three, largely mutually exclusive, proximate mechanisms for establishing stable mutualism emerge: (1) entry screening for symbiont quality; (2) host sanctions to remove or punish underperforming, that is, free-riding symbionts; and (3) reciprocity by default when resource acquisition interests are fully aligned (see Figure 4.1).

Symbiont choice has earlier been proposed as a unilateral mechanism in voluntary partner associations (Bull and Rice, 1991; Sachs et al., 2004). However, active choice is hard to imagine unless it is mediated by screening (Archetti et al., 2011; Scheuring and Yu, 2012), a selection process that is less challenging than evaluating quality signals for each candidate symbiont separately (Oldroyd, 2013). In contrast to choice, screening can be conceptualized as a host-enforced self-evaluation process among symbionts that makes candidate strains run a gauntlet so that only the best in terms of benefit–cost performance remain by the time hosts confirm their commitment. Several interactive systems appear to rely on screening; I use two here to illustrate the principles. The first are the luminescent *Vibrio* bacteria of

Euprymna bobtail squid (McFall-Ngai, 2014a). In this symbiosis, a newly hatched squid needs only a few hours to reduce a huge number of possible bacterial symbionts to just one or two cells that are ultimately admitted to each of six crypts in a highly complex light organ. After settling there, the rapidly dividing *Vibrio* cells offer their hosts lifetime predation protection by providing countershading services in exchange for housing, nourishment, and reproductive success—ca. 90% of the *Vibrio* cell copies are released in the seawater every day (McFall-Ngai, 2014b; Aschtgen et al., 2016). Secondary colonization of light organs is impossible, so the outcome of screening is a form of LTC. However, several *Vibrio* strains may coexist within and across host crypts, so relatedness among *Vibrio* cells may end up not being maximal (Wollenberg and Ruby, 2009). Whether diversity ultimately becomes reduced to complete non-redundancy (Table 3.1) is unclear, but symbiotic *Vibrio* strains are known to have multiple behavioral phenotypes that either pursue competitive dominance or sharing of host resources (Koehler et al., 2019).

The fungus-farming termites and their *Termitomyces* cultivars (Aanen et al., 2002) are a second example of mutualistic symbiont choice by screening. An incipient colony founded by a future queen and king initially associates with a substantial environmental selection of *Termitomyces* haplospores introduced with detritus collected by the first foraging workers. These spores survive gut passage to be deposited on a spongy labyrinth of primary feces (the fungus comb), where they hatch and grow a haploid fungal thread that merges with a compatible neighboring filament to become a dikaryotic mycelium. The actively growing dikaryons then compete for re-inoculating new fungus comb by producing nodules with asexual spores that are eaten by the termites whose inoculated feces form the new comb substrate. This results in positive frequency-dependent selection favoring the best performing strain to become the only dikaryon that remains. This process also ensures that no rare alternative cultivar can ever become established later on, and that the prevailing cultivar is the best possible match to the abiotic, biotic, and social characteristics of a particular termite colony (Aanen et al., 2009). The gauntlet that candidate dikaryons run is created by the joint efforts of termite hosts and the competing symbionts, unlike the squid–*Vibrio* symbiosis that appears to be exclusively host-controlled (Aschtgen et al., 2016; Koehler et al., 2019), although more work is needed to establish this. Both screening symbioses achieve stable partnerships, but the termite one is known to result in non-redundant 1:1 interaction symmetry while the squid symbiosis at least initially retains several symbiont strains that passed the screening process. In Figure 5.1 I have subdivided symbioses into categories based on closure, redundancy/symmetry, and symbiont acquisition mode, such that the horizontally acquired *Vibrio* symbiosis ends up at the bottom left and the equally horizontally acquired *Termitomyces* symbiosis at the top right, because it quickly and stably becomes symmetrical and remains a non-redundant partnership in spite of being an ectosymbiosis.

Neither the squid–*Vibrio* symbiosis (Figure 5.1, bottom left) nor the termite–*Termitomyces* symbiosis (Figure 5.1, top right) qualifies as a LECA-like non-redundant endosymbiosis (Figure 5.1, top left), but their respective physical and functional closure sets them apart from the more open symbioses that maintain variably redundant societies of not previously screened ectosymbionts (Figure 5.1, bottom right). The *Vibrio* and *Termitomyces* examples indicate therefore that horizontal symbiont acquisition can produce harmonious mutualisms provided they are closed, and suggests that closure may be a better predictor for the stability of symbioses than vertical versus

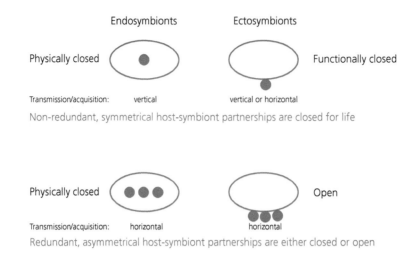

Figure 5.1 Closure, redundancy, and symmetry in symbiotic interactions for mutual benefit. A conceptualization of symbiotic mutualisms complementing the left hand side of Figure 4.1. Instead of emphasizing whether hosts and symbionts associate freely and voluntarily (i.e., horizontally), or by coercion (i.e., vertically under host control), this diagram highlights whether hosts internalize mutualists as endosymbionts or whether partners remain ectosymbionts (columns), and whether partners join for life in a non-redundant, symmetric liaison or exploit each other in an asymmetric, redundant relationship (rows). Redundant interactions can become a lifetime commitment when hosts can endosymbiotically close the partnership, but not when the symbiosis remains open. When hosts maintain multiple, redundant strains (bottom row), these symbionts will compete for host resources— tendencies that can be suppressed by selection on hosts (Frank, 1996a, 1996b). Host interference or policing may then reduce the impact of lower-level selection but without enabling its elimination, unless symbiont diversity becomes secondarily reduced to non-redundant symmetry, in which case the interaction would move to the upper row (e.g., when *Vibrio* strains in bobtail squid would invariable become reduced to one per light organ). The bottom row thus distinguishes between permanently redundant open societies of symbionts (right) and redundantly closed societies of symbionts (left) and conjectures that these forms of egalitarian association lack the potential for realizing MTEs to higher levels of organismality, analogous to the society argument for single gene pools in Chapter 4. In contrast, permanently symmetrical non-redundant symbioses (top row) should have MTE potential, particularly when they are physically closed, as LECA was. As drawn, the ellipses in the diagram represent unitary hosts, but they could be subdivided to represent hosts with modular growth in order to illustrate that modular hosts (e.g., plants) are more likely to have stable mutualisms with redundant symbionts when host compartments house only one or a few strains each (e.g., figs and pollinating wasps and root nodules housing *Rhizobia*). This diagram does not consider facultative symbioses, cases where mutualistic opportunities are largely unilateral, or partnerships that trade complementary services in brief market exchanges (Leigh, 2010a; Werner et al., 2014). Neither does it consider multispecies communities such as biofilms or gut microbiomes (Foster et al., 2017), although any form of even partial compartmentalization is likely to reduce conflict also in these open systems (see Preussger et al., 2020, Figueiredo and Kümmerli, 2020 for an interesting exchange on this topic).

horizontal transmission, the explanation most often used since Herre et al. (1999). When it is closure that aligns host and symbiont interests, it also becomes legitimate to ask whether and to what extent more open, usually modular, symbioses in fact practice forms of partial closure to limit strain diversity (and thus competition) within host compartments, and whether they can better control symbiont performance if they do (Kiers and Denison,

2008; Leeks et al., 2019). In all cases with closed non-redundant partnerships (top row in Figure 5.1), we should expect that obligate interaction symmetry selects symbionts to lose any costly competitive agency towards alternative strains when they never coexist with competing strains in the same host compartment. This appears to be the case in *Termitomyces* (Aanen et al., 2009), but not in *Vibrio* or in the fungal cultivars of leaf-cutting ants. The latter symbiosis has many analogies with the farming termite mutualism, but lacks positive frequency-dependent cultivar propagation so that secondary forms of ectosymbiont chimerism remain possible. This appears to have maintained selection for joint ant-cultivar sanctions to defend and preserve the original 1:1 partnership for the lifetime of a colony (Poulsen and Boomsma, 2005).

While symbiont screening is an a priori discrimination measure in voluntarily partnerships, host sanctions are an a posteriori control mechanism (see Figure 4.1). Screening can evolve when hosts (with or without the help of a resident symbiont) can impose forms of primary or rapid secondary closure, as in the *Vibrio* (physical closure) and *Termitomyces* (functional closure) examples. In contrast, evolving a posteriori sanctions appears to be the only way to protect host interests in the ectosymbiotic interaction types positioned at the bottom right of Figure 5.1. Sanctions in such symbioses have been well studied, in fig trees that are selected to discourage cheating pollinating wasps (Herre et al., 2008; Jandér and Herre, 2010; Jandér et al., 2012) and in legume host plants selected to constrain underperforming *Rhizobia* and mycorrhizal root symbionts (Kiers et al., 2003; Kiers et al., 2011; see also Werner and Kiers, 2015). While these ancient mutualisms appear at first sight to be stable and harmonious, the gene's eye theory of cooperation and conflict clearly suggests that wasteful tragedy of the commons situations (for hosts) should threaten mutualism stability. Hence such conditions are expected to have selected for the evolution of sanctioning mechanisms, which should imply that potential conflict can be provoked to become actual conflict in manipulation experiments (Frank, 1996a; West et al., 2002a, 2002c; Kiers and Denison, 2008; Steidinger and Bever, 2016). Such tests indeed confirmed that cheating symbionts are common and that

conditionally expressed sanctioning mechanisms against them are adaptively designed to increase host-control efficiency for reasonable costs (Ågren et al., 2019; Chomicki et al., 2020; Zhang et al., 2021). These findings are consistent with ongoing selection on hosts to protect their interests when they are strongly if not obligatorily dependent on the assistance of a specific class of symbiont (Fisher et al., 2017) whose services have structural elements of redundancy and can therefore not be expected to have fully aligned interests with the host (see previous section).

Apart from screening and sanctioning, there is a third proximate category of mutualistic reciprocity by default at the left hand side of Figure 4.1 that remains to be interpreted in the terms of Figure 5.1. This is a tricky category for we cannot just assume that any symbiosis where screening or sanctions are unknown will have harmonious reciprocity—this needs to be actively demonstrated. Proving that cheating and sanctions are operational requires an elaborate combination of molecular and experimental approaches, similar to proving screening (see references cited above), so many cases of host interference may remain to be discovered. The key question is whether reciprocity by default always requires both interaction symmetry (non-redundancy; see Table 3.1) and structural or functional closure, that is, a position in the top row of Figure 5.1, independent of interactants belonging to the same or different gene pools. I have lined up illustrative examples in Figure 5.2, starting with the strict physical membrane closure of the last universal common ancestor (LUCA) and the eukaryote founder LECA (Figure 5.2a). I have added an image of a eukaryote zygote that represents the ultimate form of meiotic sexual closure that remained the default starting point of individuals in essentially all more complex lineages of eukaryote life (Figure 5.2b). Immediately below (Figure 5.2c)

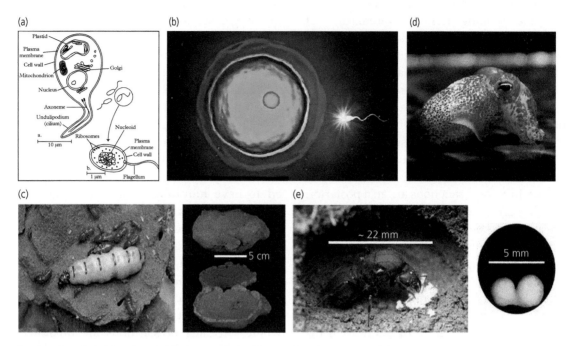

Figure 5.2 Functional similarity in lifetime commitment between founding partners of the same species or different taxonomic lineages. (a) Simplified diagram of the physically (membrane) closed prokaryote cell (bottom) and the complex endosymbiotic eukaryote cell (top). The prokaryote cell has a circular chromosome, a simple cytoskeleton, and a simple flagellum, while the symbiotic eukaryote cell has a membrane-surrounded nucleus with multiple linear chromosomes, mitochondria, a complex cytoskeleton with endomembranes, several other organelles (e.g. plastids), and a cilium. Republished with permission of The University of Chicago Press, from *In Search of Cell History*, F.M. Harold (2014); permission conveyed through Copyright Clearance Center, Inc. (b) LECA's ancestor evolved meiotic sex, so its symmetrical non-redundant symbiosis between a single nuclear and mitochondrial genome became accompanied by another, equally strict, lifetime commitment between two gametes that usually become merged into a zygote—a cellular arrangement that would remain the dominant reproductive mode throughout all more complex eukaryote life (Image credit: Karl-Ludwig Poggemann, Flicker.com, CC BY 2.0). (c) Non-redundant lifetime monogamy in *Macrotermes* fungus-growing termites, with a single queen and male (just above her) per colony surrounded by lifetime altruistic soldiers, large workers, and small workers (image by the author). The royal couple is surrounded by a royal cell of hard clay, which they cannot leave because the small entrance holes allow only workers to enter and leave to feed their parents and carry away eggs. This is a clear case of functional closure that becomes established as a colony grows, that is, some time after the dispersal flight that brought the pair together—it closes the colony in terms of genetic information for its several decades-long life. Images of royal cell David Nash; collection Michael Poulsen. (d) A bobtail squid housing luminescent *Vibrio* bacteria in a closed but asymmetric (i.e., redundant as defined in Table 3.1) symbiosis that involves more than a single *Vibrio* strain per squid, obtained after an elaborate screening process to only acquire cooperative bacteria well adapted to symbiotic life (Kremer et al., 2013; Wernegreen, 2013). After admission, unrelated bacteria may still compete for host resources (Speare et al., 2018), which might increase the cost of the symbiosis for hosts because more surplus cells might need to be released into the environment than if the symbiosis had been non-redundant. Whether hosts manage to ultimately reduce symbiont diversity to a single strain per light organ crypt is unknown. If that would happen, the interaction might become functionally similar to the fungus-growing termite symbiosis of panel (c). Photo credit Margaret McFall-Ngai. (e) A colony-founding queen of *Atta sexdens* with its incipient garden and brood. Shortly before digging her nest burrow, she was inseminated by two or three males (Fjerdingstad and Boomsma, 2000) whose ejaculates were stored in a large (5 mm) spermatheca to last her lifetime of up to ca. 20 years. As she will never re-mate later in life, her spermatheca is a physically closed compartment containing redundant sperm symbionts (which are haploid clones

(*Continued*)

is an image of a strictly monogamous couple of *Macrotermes* fungus-growing termites and the royal cell in which they live their functionally closed decades-long lives in permanent darkness. All three examples have non-redundant, strictly 1:1 partnerships of, respectively, a nuclear and mitochondrial genome, a female and male gamete genome, and a female and male breeder genome. This strict physical or functional closure makes them all belong in the top row of Figure 5.1—whether partners practice physical or functional closure and belong to the same gene pool or to different taxonomic lineages does not appear to matter.

The final two panels have the bobtail squid previously discussed for its screening practice (Figure 5.2d) and a newly inseminated queen of *Atta* leaf-cutting ants in her nest-founding burrow (Figure 5.2e). The *Atta* queen ant is as committed for life to the limited genetic sample of sperm that will fertilize her eggs as the fungus-growing queen termite is to her king, but she has stored the ejaculates of several males in her physically closed sperm storage organ, expressing aspects of male symbiont redundancy similar to the *Vibrio* symbionts of bobtail squid. Both symbioses therefore represent physically closed partnerships (bottom left in Figure 5.1) that are stable only because hosts actively interfere, by a priori screening in squid and by a remarkable form of a posteriori elimination of male–male conflict in the *Atta* spermatheca. While the ejaculates stored are mutually hostile and incapacitate each other's sperm on microscope slides in vitro, this male–male strife is almost immediately eliminated in vivo by protein secretions from the female sperm storage organ (Den Boer et al., 2010; Dosselli et al., 2019). This has likely been driven by consistent selection on queens, because every sperm cell lost is one too many when she is long-lived, needs to be extremely fertile, and cannot re-mate later in life. Both the aggressive ejaculate traits and the superior female defense traits evolved only ca. 15 million years ago from ancestors with strictly singly inseminated queens (Villesen et al., 2002; Nygaard et al., 2016) whose male ejaculates peacefully coexist on microscope slides because competitive agency was never selected for (Den Boer et al., 2010). All cases of harmonious non-redundant reciprocity by default (Figure 5.2 a,b,c and the ancestors of e) derive from hosts that acquire their symbionts horizontally, so it seems appropriate to ask how special physically closed, vertically transmitted endosymbiosis—the kind that produced LECA—really is.

Symmetric symbiosis of the non-redundant type (top row in Figure 5.1) should be equally stable no matter whether partners associate vertically or horizontally (Figure 5.2). However, vertically transmitted symbionts such as mitochondria have the handicap of no recombination with the maternal nuclear genome, a problem that does not apply for partners that find each other horizontally. This is unlikely

---

Figure 5.2 (*Continued*)
because of hymenopteran haplodiploidy). This redundant symbiosis is evolutionarily derived because the ancestral attine ants were obligatorily monogamous (Villesen et al., 2002) with only a single non-redundant ejaculate stored by each queen—most likely, paternity redundancy was secondarily selected for to amplify the genetic diversity of superorganismal worker offspring (Boomsma et al., 2005b; Kooij et al., 2015). *Atta* queens suppress redundancy conflicts shortly after insemination (see main text) but, as in the bobtail squid, suppression of sperm competition might impose a cost on host queens although selection is likely to have gradually minimized that cost (Baer-Imhoof et al., 2022). Images by Boris Baer, reproduced from Sperm storage induces an immunity cost in ants, B. Baer, S.A.O. Armitage, J.J. Boomsma, *Nature*, Volume 441 (2006), with permission of Springer Nature and Boris Baer.

to have been a problem for LECA and the unicellular lineages that she spawned, because viability selection among unicellular zygotic exconjugants could efficiently purge mutationally compromised mitochondria. However, mitochondrial disease is known from multicellular animals and has been well documented in humans (Adashi et al., 2021). Incomplete elimination of male mitochondria leading to chimeric heteroplasmy is also associated with disease (Cummins, 2000), but clean nuclear–mitochondrial transplants that maintain 1:1 non-redundancy have so far seemed to be unproblematic (Tachibana et al., 2009; Wallace, 2018), provided 100% replacement can indeed be secured which is far from certain (Adashi et al., 2021). This makes one wonder whether a lower level of selection might have reappeared after obligate multicellularity evolved, reactivating a level of competitive conflict that had disappeared when LECA domesticated the proto-mitochondrion in a 1:1 non-redundant manner. This would seem reasonable to expect because mitochondria accumulate considerable mutational load in long-lived bodies (Aanen et al., 2014), so that discriminatory removal by mitochondria-specific autophagy (Lionaki et al., 2015) might also target genetically different, artificially introduced replacement symbionts. All this implies that vertical symbiont transmission secures having a non-redundant symbiont adapted to the maternal lineage, but not necessarily a top-quality one.

Considering the regularities of increasing foundational "ploidy" of closure across the subsequent MTEs (Figure 5.3), it appears that strict physically closed LTC can be hypothesized to have removed all lower-level selection when LECA appeared as doubly non-redundant unicellular MTE (Figure 5.2a). This principle prevailed throughout the subsequent fraternal MTEs, but it is less clear whether the origin of LUCA, which established organized

cellular life, has an analogous explanation. Physical cell membrane closure must have secured all-in-the-same-boat harmony (see next section), but competitive redundancy challenges between enclosed replicators competing for the same function might have compromised LUCA's stability. Such challenges may also have plagued the early ancestors of bobtail squid just after associating with *Vibrio* symbionts, but this interaction still managed to evolve an extant mutualism of high stability and sophistication. However, if MTE potential would generally depend on exclusive non-redundancy up front, a LUCA origin based on single-replicator-single-vesicle closure might seem more parsimonious—a scenario that I will briefly discuss in Chapter 7. While LUCA's origin is still shrouded in mystery, it is clear that LECA's principle of non-redundant genetic information closure at conception has been maintained in the ancestral founding units of the multicellular organisms and colonial superorganisms (see Chapter 4). In sum, there appears to be, at least in theory, a single general and parsimonious principle for explaining and interpreting the origins of both the fraternal and the egalitarian MTEs, from superorganisms down to LECA. The empirical Chapters 6 and 7 will evaluate how well currently available data match these conjectures, and the final section of Chapter 8 will tentatively explore the extent to which the human MTE, uniquely characterized by cumulative culture, is conceivable as a non-redundant egalitarian MTE, and what the implications might be in case that idea would prove to have merit.

Putting emergence, conflict, and levels of selection in their rightful places

Despite our conviction that gradual change has predominated the history of life, we are

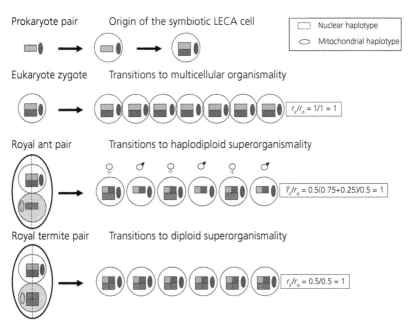

Figure 5.3 The expansion of non-redundant closure (Table 3.1) across the nested MTE origins. The origins of all non-human MTEs in organizational complexity can be parsimoniously conceptualized as having been singular events of strict lifetime commitment (LTC) between non-redundant carriers of genetic information at the foundation of a new cell, multicellular organism or colonial superorganism. The start of organized life (LUCA) required physical closure in the form of a prokaryote cell membrane to maintain aligned interests among a string of replicating genes on a circular chromosome (Harold, 2014). Two of these closed prokaryote cells merged in a higher-order structural closure event that established the complex (LECA) cell as egalitarian MTE, after which subsequent fraternal MTEs could emerge via higher-level functional closure, expanding the nested hierarchy of organizational complexity until its maximal possible (core family) level. Foundational closure (left column) thus stepwise increased the organizational complexity of organisms (see Figure 3.2b) while always requiring strict non-redundancy in partnership, in LECA to preclude costs of symbiont chimerism and in the fraternal MTEs to secure that relatedness to cell copies or sibling offspring remains maximal (i.e., $r_x/r_o = 1$; see Figure 4.2b). None of these MTEs were predictable in absolute time but their hierarchical order was. The stepwise increase in organizational complexity is reflected in the functional ploidy of the haploid genomes in founding units, which increased from one in free-living prokaryotes (not counting possible plasmids), via three in LECA and five in a haplodiploid royal pair of ants, to six in a royal pair of termites. Ancient plastid symbionts whose domestication allowed hosts to obtain autotrophic metabolism added yet another haploid genome (Keeling, 2004), but plastids never induced a new enhanced level of organizational complexity in the sense of Figure 3.2b—they functionally joined the mitochondrial co-replicon (Box 2.2) as a distinctly different non-redundant symbiont not competing for offering the same service to a host cell. When drawing the diagram, I have assumed that the archaeal ancestor that captured the bacterial proto-mitochondrion preceded the evolution of diploidy and meiosis. A reverse order of these two symmetrical commitment events before LECA appeared would not affect the argument, but seems unlikely because division of labor is normally the prime driver of mutual benefit symbiosis. A yin and yang inspired imaginary version of the left column of this diagram appears on the cover of this book. Republished after minor modification from Lifetime monogamy and the evolution of eusociality, J.J. Boomsma, *Philosophical Transactions of the Royal Society* B, 364, 3191–3207 (2009), with permission of The Royal Society conveyed through Copyright Clearance Center, Inc.

aware that the tree of life is punctuated by a few fundamental discontinuities, referred to as grades by Huxley (1912), as levels of complexity by Bonner (1974), and as MTEs by Maynard Smith and Szathmáry (1995) (see Figure 3.2). However, we have been living without understanding how natural selection mediated the origins of these transitions, which is puzzling because MTEs must have represented adaptive syndromes, both to have originated at all and to have been able to persist and radiate. Neither Maynard Smith and Száthmary nor later edited volumes (Calcott and Sterelny, 2011; Ruiz-Trillo and Nedelcu, 2015) have had more to offer than conceptualizing MTEs as elusive emergent properties that are somehow out of deeper ultimate causality reach. At the end of Chapter 4 I quoted Maynard Smith (1988) for erroneously connecting the resolution of multiple conflicting levels of selection with the origins of MTEs, and concluded that conflict reduction is only relevant within specific domains of social evolution and for understanding the expression of facultative (i.e., condition-dependent) altruism in society-like arrangements. Also the present chapter on the egalitarian, i.e. mutualistic, MTEs has offered no evidence for MTEs having originated by gradual conflict reduction. In contrast, we obtain a parsimonious general framework of nested conjectures for the origins of MTEs by strictly defining life's organizational complexity as a series of discontinuous pairwise closure events around fixed quantities of genetic information (Figure 5.3), securing that external soma could arise by directional kin selection. In this and the previous chapter, I have implied that these conjectures can be considered as the contours of a kin selection theory for the levels of cellular, multicellular and colonial organismality (see also Chapter 7).

Both the generalization of inclusive fitness theory via the Price equation (see Chapter 4) and Maynard Smith's (1988) focus on MTEs

as levels-of-selection phenomena have contributed to a general narrative in evolutionary biology that downgraded *family* to just being a special case of *group*. This semantic shift has contributed positively to the general conceptual transparency of inclusive fitness theory, but it clouded understanding of MTE origins by implicitly suggesting that these crucial events could not be explained by natural selection, let alone by inclusive fitness theory. Rather, as we saw in Chapter 3, the progressionist aspects of MTEs were denied in an ultimate sense, so we ended up with "contingent irreversibility" as a dismissive proximate placeholder. The alternative non-redundant closure conjecture that I have developed and placed in a deeper historical context reverses the causation of MTEs, and explains their origins as first-principle extensions of Fisherian thinking in terms of genetic information combined with Hamiltonian inclusive fitness theory. No matter whether partners are organelles, gametes, or parents, MTEs can be conjectured to have originated by cooperative pairs simultaneously maximizing their inclusive fitness until the ensuing division of labor had become an irreversible advance in efficiency, externally in the fraternal MTEs and internally in the egalitarian MTEs. LTC and invariably maximal relatedness implied that developmental somatization of organelles, cells, or multicellular individuals could accumulate gradually without conflict over resource acquisition until reproductive isolation from a less organizationally complex sister lineage was achieved. As I explained in earlier chapters, Weismann (1893), Wheeler (1910, 1911), Huxley (1912), Sturtevant (1938), Williams (1966a), and Lack (1966, appendix 3) understood that selection works on the potentialities of founding zygotes and higher-level organismal equivalents, so why was this understanding lost?

The terminology transition that accompanied sociobiology redefined the zygote as a genetic bottleneck, language normally used for a genetic diversity constriction in populations, which removed emphasis from the obvious fact that a zygote represents absolute, obligate, and irreversible commitment by a pair of gametes. Against this flow of semantic inclusiveness, the idea of unavoidable cooperation when parties are "in the same boat" for life had to be reinvented, first for stored sperm by queens of social insects (Boomsma et al., 2005a) and later by Leigh (2010a) who wrote that absolute partner fidelity (i.e., LTC) "links the fates of interacting individuals of two species involved in mutualism, just as the fates of two people in a rowboat on a stormy sea are linked if both must row for either to survive." This is because "injuring one's partner injures oneself," so that "the common interests among such mutualists can be as absolute as if transmission were vertical." In such a setting, two always makes up a non-redundant pair while three is a redundant crowd that doubles paternal or symbiont genetic diversity. The ultimate question is whether redundant closure as in *Vibrio* and stored *Atta* ejaculates (Figure 5.2 d,e) will still make all 'same boat passengers' do all the rowing they can muster, or whether aspects of simmering unilateral exploitation remain or could reappear. It seems reasonable to conjecture that redundant closure precludes reaching the organismal harmony of non-redundant symbioses such as LECA—why else would multiple strains of mitochondria in the same cell never have become a stable arrangement? Also in the unicellular domains, it may thus be parsimonious to assume that MTE origins required 'cellular boats' to be populated by pairs of complementary coreplicons, which seems to make moot any claim that these metaphorical boats needed to be big or fancy. This is consistent with MTEs having originated as "unnoticeable" ratchet clicks initiated by higher-level

foundational commitment as a closed social exaptation relative to prevailing custom in the ancestral domain. Higher-level selection could then take it from there when the new incipient organizational complexity paid off, i.e. by Hamiltonian b/c ratios continuing to increase.

The nested ratchet view of MTEs can be hypothesized to apply to all small origin circles in Figure 3.2a that symbolize the domain transitions in organizational complexity illustrated in Figure 3.2b, with the human domain as only exception. Taking this view changes how we think about emergent properties, about the role of conflict, and about what multilevel selection can achieve. If the origins of MTEs were always products of novel adaptive design, non-adaptive emergent properties have no role to play in the explanations of MTE origins. This does not imply that emergent phenomena are always irrelevant—they may affect processes of maintenance and elaboration during adaptive radiation of MTE lineages (Bourke, 2011), but not origins. The same is true for conflict reduction. Coerced harmony is important for achieving and stabilizing higher complexity in societies but is never 100% effective (El Mouden et al., 2010). Policing can therefore not induce a stepwise transition from lower-level direct fitness agency into domesticated indirect fitness agency as happened in the fraternal MTE origins. Fruiting bodies of cellular slime molds are societies of cells in the unicellular domain, and colonies of paper wasps or naked mole rats are societies of individuals in the multicellular domain. The slime mold lineage did not achieve an MTE towards permanent multicellularity, and neither have paper wasps or naked mole rats become superorganismal. Redundant, society-like developments also originated as later modifications of superorganismality, for example, when lineages evolved multiple insemination of queens. This induced sophisticated forms of conflict reduction, as with worker policing in the honeybee (Ratnieks et al., 2006), but this conflict reduction did not make honeybees transition to a new level of organizational complexity, nor could this new habit of paternity redundancy reverse the earlier MTE to permanent somatization of the worker caste.

Implicit arguments of emergence and explicit ones about conflict reduction as necessary conditions for MTEs are common in the literature of the last 25 years (e.g., Calcott and Sterelny, 2011; Ruiz-Trillo and Nedelcu, 2015). However, I maintain that all one can say is that these aspects are relevant for societies and that complex societies of cells or multicellular individuals appear to have never been ancestral to, respectively, multicellular organisms or colonial superorganisms, similar to LECA not having arisen from a redundant assembly of prokaryote partners. The same MTE irrelevance must apply to organismal size, which has often been highlighted as a correlate of organizational complexity, both for multicellular organisms and for social insect colonies (Bonner, 1955, 1988; Bourke, 2011). Such arguments have pooled domains of organizational complexity that should have been kept apart. This was clear to early naturalists, but inclusive definitions of multicellularity and eusociality that applied to all levels in Figure 4.7 came to hide that there are almost certainly different domain-specific correlations between body/colony size and organizational complexity across the (black) society branches versus the (red) (super)organismal branches. Scatter plots of complexity versus colony size may superficially look like a single gradient because societies are on average smaller and less complex than superorganisms. However, if origins of MTEs only required a novel pairwise LTC condition for the foundation of new levels or higher organismality (Figure 5.3), they might well

have originated with the smallest possible body or family size, so increases in collective mass were domain-specific markers of later elaboration, not of origins. The same applies to distinguishing between facultative and obligate multicellularity (see Figure 4.7a). It seems puzzling that keeping domains apart has been obvious for prokaryote cells versus much larger protist cells, and for permanently multicellular bodies relative to protists, but that a similar distinction was abandoned when sociobiology merged all multicellular social evolution in an all-inclusive complexity gradients that muddled the concepts of caste- and germline-soma differentiation (Boomsma and Gawne, 2018).

I have used the present and previous Chapter to review how successful and truly integrated inclusive fitness theory has become to offer an encompassing view of the social dynamics of cooperation and conflict. The key omission that I have criticized and attempted to rectify with the contours of a static kin selection theory for the origin of MTEs may seem so beguilingly simple that it is puzzling that this extreme state argument has been overlooked. I maintain that the fact that this set of conjectures has not been considered, in spite of incisive understanding in the early 20th century, has obstructed progress in conceptualizing some of the most fundamental aspects of the evolution of organized life. As I will argue in Chapter 8, a main reason for this neglect may have been anthropomorphic bias, that is, an intuitive tendency to impose our own efforts to make societies more harmonious upon the rest of nature, rather than realizing that our societies are exceptional and unsuitable as templates for understanding the deeper principles and implications of inclusive fitness theory. The Hamiltonian explanation for the origins of MTEs proposed here is more parsimonious than any previous framework, and it includes explicit acknowledgment of the key hierarchical layers of biological organization (see Figure 3.2b)—what it denies is that levels of selection had anything to do with MTE origins. As I indicated before, all

this remains conjecture, so it seems timely to leave theorizing behind and focus on empirical evidence in Chapters 6 and 7, before addressing some wider implications of the theory in Chapter 8.

CHAPTER 6

The multicellular organisms and colonial superorganisms

Summary

Comparative data indicate that differentiated multicellular organisms and colonial superorganisms always originated by somatic adherence to a diploid (zygotic) cell and by comparable loyalty of a worker caste to monogamous parents. However, the functional analogies between metazoan organisms and colonial superorganisms remain ambiguous because multicellular animals ultimately die from somatic failure while single-queen colonies die from germline failure. This difference relates to the forms of corruption that plague the two levels of organizational complexity, metazoan cancers and inquiline social parasites, both initiated by mosaicism. Parental monogamy also shaped condition-dependent reproductive altruism in societies of vertebrate cooperative breeders, but without permitting the evolution of superorganismal colonies. In addition to evidence for ultimate analogies I explore three proximate parallels between multicellular animals and colonial superorganisms. First, the ways in which germline and soma differentiate in bodies and colonies; second, the putative principles by which superorganismal (but not society) immune defenses reached impressive efficiencies, particularly in ants and termites that defend non-overlapping territories; third, the extent of developmental similarity between cell differentiation in metazoan bodies and caste differentiation in superorganismal colonies. Early organismal biologists appreciated these parallels in ontogenetic development more than many modern scientists, and even pre-Darwinian naturalists deserve credit for being remarkably competent observers of life's social organization. Overall, the empirical data appear consistent with expressions of condition-dependent altruism by cells or multicellular individuals not being ancestral to obligate and unconditional reproductive altruism. This challenges the reproductive groundplan concept for the origin of castes and suggests that the clarification of unique gene regulatory networks for obligate somatic altruism needs to replace the reductionist identification of toolkit genes.

> *As in the case of the individual animal, no further purpose of the colony can be detected than that of maintaining itself in the face of a constantly changing environment till it is able to reproduce other colonies of a like constitution. The queen mother of the ant colony displays the generalized potentialities of all the individuals, just as the Metazoan egg contains in potentia all the other cells of the body.*
>
> **Wheeler, *Ants*, (1910, second (unchanged) edition 1925, p. 7)**

In the preceding chapters, I conceptualized the significance of these few lines that William Morton Wheeler wrote more than a century ago. As late as the 1960s this view was uncontroversial. For example, in Sturtevant's (1965, p. 2) *History of Genetics*, we find that "[Aristotle's] general conclusion was that what is inherited is not characters themselves in any sense but only the potentiality of producing them. Today this sounds self-evident,

Domains and Major Transitions of Social Evolution. Jacobus J. Boomsma, Oxford University Press. © Jacobus J. Boomsma (2022).
DOI: 10.1093/oso/9780198746171.003.0006

but at that time it was an important conclusion, which was not always fully understood, even by the early Mendelians." This statement reflects a deep, albeit implicit interest in ontogenetic development not unlike the prose that Huxley (1912) used in his early neo-Darwinian appreciation of the overarching patterns of nature's organizational complexity (see Chapter 3). Sturtevant (1965, pp. 17–18) also wrote that, according to Weismann's scheme, "the germ line is the continuous element, and the successive bodies of higher animals and plants are side branches budded off from it, generation after generation. [...] Weismann recognized that in the higher plants and in many animals the visible distinctness of the germ line only appears late in development and, in fact, that many cells that will not normally give rise to germ cells still retain the potentiality of doing so." This suggests that the purportedly metazoan bias in applying Weismann's principles (Buss, 1987) has grown historically, not because of Weismann's original thinking. In this chapter, I will evaluate the empirical evidence for Wheeler's statement today in terms of germline–soma differentiation of cell lines and caste phenotypes. I will also test Wheeler's implicit appreciation of LTC between gametes or parents that I used in Chapter 4 to theorize about parallel kin selection explanations of the fraternal MTEs (Queller, 2000) towards multicellular organismality and colonial superorganismality (see Table 3.1 for terminology).

Although often phrased in gene's eye language, my overview will be explicitly organismal and non-reductionist. While molecular genetics and genomics have made tremendous advances in recent decades, I will leave them largely aside because the purpose of this book is to develop a framework of general conceptual understanding in the neo-Darwinian synthesis tradition. I will thus take a top-down view of the discontinuous

organizational complexity steps illustrated in Figure 3.2b to see how the fraternal MTEs make sense in terms of adaptation and ontogenetic development. In this review, I will assume that the nested layers of irreducible complexity mostly represent self-organized processes that are monitored by natural selection (Johnson and Lam, 2010; Newman, 2022) (see also Chapter 8). Summarizing the empirical understanding of fraternal MTEs in a single chapter is a challenge because all of modern organismal biology of plants, algae, fungi, and animals is involved, and so is the social biology of insects, vertebrates, and hydrozoan Cnidaria (see Figure 3.2.a). Reviews have for a long time refrained from casting their net so widely (but see Bourke, 2011), so my account had to omit many studies that were not of immediate relevance for understanding the origins of the fraternal MTEs. As the organismal biology of animals and plants is generally better known than the colonial lives of ants, bees, wasps, and termites, my focus will mostly be on the latter to make analogies with the former transparent. I will, however, start with a summary of insights in social life that inspired Darwin's thinking before I refocus on modern comparative phylogenetic data and proximate aspects of development. In the final section, I will argue that recent evidence for permanent castes having arisen directly from parental care challenges the predictive value of purely proximate approaches that emphasize a generalized reproductive groundplan and toolkit genes.

Phylogenic evidence for a single adaptive explanation of the fraternal major transition origins

Ultimate questions are about the Darwinian principles of adaptation through natural selection and historical descent with modification—two distinct quadrats of evolutionary causation in Tinbergen's scheme (see Figure 1.1; Box 1.3). While phylogenetic trees were pure conjecture in Darwin's youth, adaptations of animals and plants were explicitly appreciated, albeit often from an incorrect creationist perspective (Gardner, 2009). Yet, this body of knowledge was crucially important for Darwin to resolve how adaptive characteristics of what he called the neuters of social insects could have evolved, as Pittendrigh (1958) noted explicitly. Tracing back this long history is of interest because the careful observations by Kirby and Spence (1818), both natural theologians, identified many examples

of colony-level adaptation in ants and bees, and also some expressions of within-colony conflict. These facts helped Darwin to understand how natural selection at the family level settled his initial despair about explaining lifetime sterile castes and their specific somatic adaptations. Box 6.1 pays homage to these British natural historians of the early 19th century and illustrates what Williams (1966a) and Mayr (1982) meant when they noted that scientific facts may be known in considerable detail before they are understood (see Chapter 1).

Almost two centuries after Darwin drew his first notebook tree, we now map putatively

---

Box 6.1 Darwin's social evolution inspiration: Kirby and Spence (1818)

Darwin realized that his theory of adaptation by natural selection would not be credible if it had major exceptions. He therefore struggled to explain the sterile "neuters" of ants, bees, wasps, and termites, because it seemed incomprehensible that natural selection could produce their special morphology and behavior without most of them being able to reproduce (Darwin, 1859; Cronin, 1991; Ratnieks et al., 2011). In Darwin's formative years, the go-to survey of everything worth knowing about insects was the four-volume *Introduction to Entomology* by Kirby (1759–1850) and Spence (1783–1859), which appeared in the years 1815–1826 and travelled with Darwin during his five-year voyage on the *Beagle* (Egerton, 2013). However, Darwin did not read Kirby & Spence until 1843 (Richards, 1987, p. 155), and when he did he realized that his inability to explain social insect neuters was a potentially fatal problem for his whole theory. After various intermediate and partial solutions (Richards, 1987), he finally arrived at the correct answer that the difficulty "disappears, when it is remembered that selection may be applied to the family, as well as to the individual, and may thus gain the desired end" (Darwin, 1859). To appreciate the state of social insect knowledge that Darwin

encountered when contemplating the evolution of neuter castes, I have summarized some of the fascinating observations by Kirby and Spence (1818, volume II) below. It is revealing that their descriptions of intra-colony conflict situations were correct but would not be understood until the late 20th century. They exemplify the superb quality of early British naturalists that some are too eager to depreciate or dismiss entirely (e.g., Godfrey-Smith, 2009).

Volume II of Kirby and Spence (1818) is subdivided in letters with Roman numerals, which only cover the social insects in the later sections. In **Letter XVI**, the authors show an acute awareness that social insect colonies are different in principle from mere aggregations because they are families acting for a common cause. It is interesting to note that 19th-century British natural theology, particularly the later version of Lord Brougham, who acknowledged rational behavior in animals, used precise language for bottom-up groups and top-down families and that this appears to have helped Darwin develop his materialistic theory of adaptive design by natural selection (Richards, 1987). Kirby and Spence distinguished between "*perfect* societies [. . .] that are associated in all their states, live in a common habitation, and unite their labours to

Box 6.1 *Continued*

promote a common object; and [. . .] *imperfect* societies, those that are either associated during part of their existence only, or else do not dwell in a common habitation, nor unite their labours to promote a common object." In their words, the imperfect societies included five categories: "associations for the sake of company only – associations of males during the season for pairing – associations formed for the purpose of travelling or emigrating together – associations for feeding together – and associations that undertake some common work." They then note that only the last of these five categories approach, but do not reach, perfect society status because they only "unite in some common work for the benefit of the community," mentioning examples of beetles, gregarious Lepidopteran larvae, and sawflies (see Costa, 2006, for a recent review).

In these descriptions, Kirby and Spence are aware that the first four categories have nothing to do with family life, while the fifth category always has, albeit only for part of the life cycle. Clear distinctions of this kind are important because even now one can find papers containing vague and pooled notions of "group living" that fail to acknowledge this crucial distinction—for example, presenting ecological complexity arguments in populations, communities, ecosystems, and even planet Earth in terms similar to ontogenetic development of multicellular organisms or family-based superorganisms (e.g., Brooks, 2017). In their next **Letter XVII**, Kirby and Spence turn to the perfect societies of the ants and termites, with occasional remarks on bees and wasps, and write that "their nests contain a numerous family of helpless brood." They further outline that pairs of termites associate for life and start copulating only after colony founding, whereas colonies of ants, bumblebees, and vespine wasps are founded by a single inseminated female without any further role in social colony life for the males. They are aware of exceptions to the default of lone dispersal in ants that practice colony budding and provide detailed descriptions of earlier naturalists' observations on how workers of wood ants recruit newly inseminated queens back

into existing colonies. Finally, they write about slave-making ants and their "predatory expeditions, for the singular purpose of procuring *slaves* to employ in their domestic business," and about the obligate mutualism between *Lasius flavus* ants and their aphid "*milch cattle*" that are "residing in hemispherical formicaries, which are sometimes of considerable diameter."

In **Letter XVIII,** Kirby and Spence focus on the perfect societies of vespine wasps and bumblebees (then called humble-bees) and explain that they have both completely sterile "neuters" and "small females," which only lay male eggs and are in fact assumed (erroneously as we now know) to lay all male eggs in bumblebees. These were later referred to as egg-laying workers, but it aptly illustrates that 19th-century naturalists including Darwin were very precise when they defined the functions of social insect family members in terms that would make it feasible for Darwin to see the woods for the trees when he proposed his natural selection explanation of perfect societies. Kirby and Spence's perfect society definitions overlap completely—as far as example cases are given— with what Wheeler (1911) coined (super)organisms and what Huxley (1930) referred to as caste societies (see Chapter 3). Thus, when Darwin offered his functional natural selection explanation for the origin of the neuter castes, his understanding was based on correct terminology that had been established for decades and that remained in use until Williams' (1966a) monograph *Adaptation and Natural Selection*. Most surprisingly, however, Kirby and Spence also report the first evidence ever of the expression of reproductive conflict. They offer detailed observations of a queen and "small female" workers of bumblebees competing for laying eggs in specific brood cells with various interfering roles of "neuter" workers. They also note that, in contrast to honeybee drones, males of vespine wasps and bumblebees partake in hygienic, undertaking, and nest repair tasks, but they never mention males to be involved in brood nursing tasks—an assessment that remains correct today.

Box 6.1 *Continued*

In **Letter XIX,** Kirby and Spence provide an even more detailed account of early 19th-century understanding of the perfect societies of honeybees, which they refer to as the hive-bee. They are aware of honeybee hives never tolerating more than a single queen and mention that this contrasts with ants where multiple queens, when they occur, peacefully coexist in the same colony. They also note that ant colonies can bud, but that this type of multiplication is impossible in cavity nesting honeybees. They give a detailed description of lethal fighting between virgin queens hatching around the same time without interference by the workers, and they explain why it is therefore prudent (i.e., adaptive *avant la lettre*) that the old queen leaves with the swarm. In the same Letter XIX, we also find that honeybee queens normally have their mating flight just a day after an old queen left with a swarm and that workers whose antennae are removed cannot recognize their queen. They further refer to Huber who discovered that honeybees need to embark on a mating flight to be inseminated and who claimed that a single drone provides a queen with sufficient sperm to remain fertile for life. This explains that all 19th-century and early 20th-century naturalists implicitly considered honeybee hives to be inhabited by a single full-sibling family. This inference was not corrected until World War II when Roberts (1944) reported that honeybee queens mate with many drones on (usually) a single mating flight, so that most of their worker bees are half-sisters. It is therefore reasonable to assume that Darwin had full-sibling families in mind when he solved his social insect neuter problem, as did Weismann, Wheeler, and young Huxley (see Chapter 3). While the documentation of multiple insemination in some social insect lineages came surprisingly late, the hymenopteran idiosyncrasy of haplodiploidy was already known in the mid 1800s, but nobody realized that this peculiar sex determination system creates asymmetric relatedness between full sisters (0.75) on the one hand and sisters and brothers (0.25) on the other until the inception of inclusive fitness theory by Hamilton (1963, 1964a). This illustrates Mayr's (1982, pp. 29–30) conclusion that "facts acquire significance only when an inquiring

mind asks an important question."

In **Letter XX,** Kirby and Spence describe in detail that honeybee workers always kill any remaining drone brothers towards the end of the summer, unless there is no hive queen. This context-dependent aggressive behavior makes perfect sense in light of modern inclusive fitness theory, because queenright colonies can save the maintenance costs of drones during inevitable winter food stress to rear new ones of the same kin value in spring, while a doomed colony that can no longer produce new workers clings to even the most marginal chance for reproductive success via the drones already raised. Kirby and Spence further extensively report on the nastiness of African honeybees and on the perfection of thermoregulation in honeybee hives, concluding that "lessons may be learned of patriotism and self-devotion to the public good; of loyalty; of prudence, temperance, diligence, and self-denial," a biblical moralistic perspective that lingered on even though major scholars such as Herbert Spencer, Julian Huxley, and George G. Simpson explicitly refuted the organizational analogies between human societies and caste differentiated insect colonies (see Chapter 3).

In a final **Letter XXVII** with general considerations about the instincts of insects, Kirby and Spence return to the honeybees with remarkable inferences of the prudence of these social insects in the divine design tradition that had been disseminated by the Reverend Paley (Gardner, 2009). Without a materialistic theory to explain design of complex instincts they had to reluctantly infer that honeybees have obtained some form of rational thinking via an otherwise impeccable chain of arguments. As they write, "[w]e cannot reasonably suppose insects to be gifted with instincts adapted for occasions that are never likely to happen. If therefore we find them [. . .], still availing themselves of the means apparently best calculated for ensuring their object; and if in addition they seem in some cases to gain knowledge by experience; if they can communicate information to each other; and if they are endowed with memory – it appears impossible to deny that they are possessed of reason." I spell this out here to illustrate what able scholars these

Box 6.1 *Continued*

Victorian clergymen were and how wrong it would be not to acknowledge their pioneering observations (as e.g., Godfrey-Smith, 2009 tends to do), just because they could not confront their observations with a scientific conceptual frame of reference. Without Kirby and Spence's detailed accounts, Darwin would not have been able to offer a natural selection explanation for "neuters" that made the idea of social insects being endowed with reason obsolete—a good example of a Paleyan perfection idea being overturned by parsimonious teleonomic logic without qualms in admitting that adaptations are not expected to be perfect in an absolute sense (Pittendrigh, 1958).

Kirby and Spence's principled way of categorizing social interactions and the distinctions among them remained essentially unchanged for about one and a half centuries, all the way through to Williams (1966a), as documented with direct quotations by Boomsma and Gawne (2018). Towards the end of his life, Wheeler (1934) offered essentially the same categories of "togetherness behavior of whole organisms from Bacteria to Anthropoid apes" as Kirby and Spence (1818) had done. He carefully kept apart the ecological communities (biocoenoses) from the societies of mammals (including humans), which he considered distinct from those of insects, and he separated these three from the mere cellular or individual aggregations and mating congregations, the non-differentiated nutritive consociations (e.g., bryozoans and tunicates), the allospecific nutritional food associations (predation, parasitism, mutualism), and the flocks and herds. By that time, Wheeler no longer used his ontogenetic superorganism concept, which may be related to the Chicago School of ecology having hijacked it for describing supposedly homeostatic ecological communities (see Chapter 3). He sufficed with a strong objection against that kind of opaque thinking, writing that "[s]ome authors have referred all the heterogeneous consociations I have enumerated to a single 'cause,' variously designated

as the 'social,' 'gregarious,' or 'herd instinct,' but this is mere animistic verbalism." Yet, that is what Wilson's sociobiology tried to formalize a few decades later with a shallow gene's eye narrative and an all-inclusive eusociality definition that turned out to discourage in-depth Darwinian understanding of social evolution.

It is sobering to realize that many of the key issues of social evolution were captured in beautiful simplicity by Huxley, Wheeler, Weismann, Darwin, and even Kirby and Spence not because they were necessarily understood but because the patiently documented behavioral and functional facts were not in doubt. After World War II, this comparative thoroughness diminished because organismal biology had to increasingly compete with molecular biology for the public's and funder's attention. Where reductionist molecular biology can consider a mission of discovery completed when all loose ends have been tied up in a single model organism, an organismal biologist cannot even begin to understand adaptive meaning unless she will have amassed substantial comparative data—the most crucial requirement for understanding adaptation as Williams (1966a, p. 271) noted explicitly. The study of social insect mating systems offers an instructive example of how organismal understanding can be led astray by generalizing from isolated new discoveries. Even though early (allozyme) genetic marker studies found evidence for both single and multiple insemination of social insect queens, the discovery of honeybee queens mating with many drones set the tone of reviews and even suggested that this finding challenged kin selection theory (e.g., Page and Metcalf, 1982). This in spite of Hamilton (1964b) having cautioned there would not be such a problem if multiple queen-insemination would always be an evolutionarily derived trait, as subsequent work started to show (Boomsma and Ratnieks, 1996; Strassmann, 2001) and was finally proved by Hughes et al. (2008) ca. 25 years later.

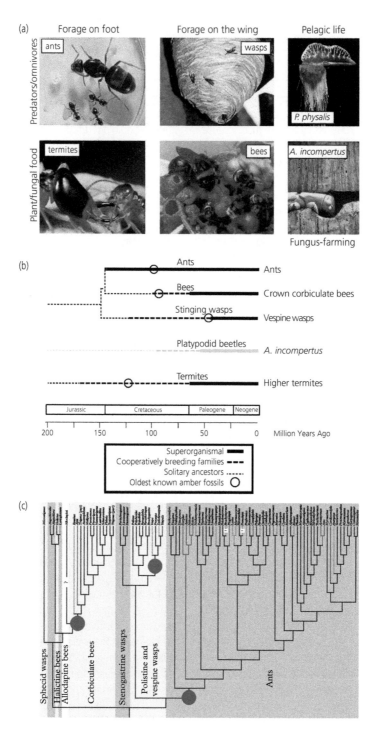

Figure 6.1   The superorganismal lineages with permanent reproductive division of labor. (a) Representative images of ants (*Lasius niger*), higher termites (*Macrotermes* species), corbiculate bees (*Bombus terrestris*)—also the perennial stingless bees and honeybees belong to this superorganismal clade—and the vespine (yellowjacket) wasps (*Dolichovespula saxonica*). These four

(*Continued*)

adaptive traits on phylogenetic trees of exquisitely resolved detail, using the comparative method that Williams (1966a) singled out as the most powerful instrument for understanding adaptive evolution (see Chapter 2). As I will summarize below, such comparative analyses have provided considerable support for the conjecture that outbred LTC, either between two gametes or between two monogamous parents, was indeed the universal ancestral condition that allowed the fraternal MTEs to emerge.

The colonial superorganisms consist primarily of four monophyletic social insect lineages that independently realized global adaptive radiations from lifetime monogamous ancestry (Hughes et al., 2008; Boomsma, 2009). An ambrosia beetle genus also fulfilled the necessary conditions but failed to radiate, and a derived hydrozoan lineage differentiated in a modular way by obligate adherence of complete polyps. Representative images are given in Figure 6.1a, lined up in a 3 × 2 matrix that specifies the approximate nutritional niches of the respective lineages and their modes of resource acquisition. All six of them are heterotrophs and retained the same unitary (the five terrestrial cases) or modular (the single pelagic case) ontogeny as their putative ancestors in the previous domain of organizational complexity (see Figure 3.2a). This suggests that modular growth, as practiced by basal Metazoa and autotrophic, photosynthesizing multicellular lineages (plants and algae) has rarely offered suitable pre-adaptations for transitions to superorganismal life. All origins of the MTEs towards superorganismality are relatively recent, dating back ca. 50–150 million years (Figure 6.1b). Except for the ants, the superorganismal clades coexist with extant sister lineages that evolved more open forms of family-based society life (precise sister clade relations are unknown for the Siphonophora). Looser forms of social breeding have evolved

---

Figure 6.1 (*Continued*)

lineages (left and central column) have considerable ecological footprints (Wheeler, 1928b; Wilson, 1971, 1990), feeding either on plant/fungal material or as predators/omnivores, and foraging either on foot or on the wing. The Siphonophora (*Physalia physalis*) are modular pelagic superorganisms, as already established by Huxley (1912) (see also (Bourke, 2011). The platypodid ambrosia beetle *Austroplatypus incompertus* is added here for proof of concept because it represents an independent but incipient origin of superorganismality that never adaptively radiated. Images by the author, except for *A. incompertus* (electron micrograph Geoff Avern) and *P. physalis* (photo credit Casey Dunn). (b) An approximate timeline for the origins of successful (black lines) and unsuccessful (gray line) superorganismal clades based on phylogenetic reconstructions by Branstetter et al. (2017), Peters et al. (2017), and Smith et al. (2018). Before the Cretaceous, colonial social life was restricted to some families of lower termites that largely lived inside logs that served both as nest-site and food (Engel et al., 2016; Korb and Heinze, 2016; Bucek et al., 2019; Evangelista et al., 2019). In general, cooperative breeding in families appeared or was elaborated during the Cretaceous but, outside the ants, superorganismality did not evolve until ca. the Paleogene (see Boudinot et al., 2022, LaPolla et al., 2013; Rust and Wappler, 2016, for reviews of fossil data). Sister-lineages that evolved cooperative breeding continued to radiate in parallel, and are thus extant in considerable diversity except for the ants where no such sister lineage has been identified. See Box 6.2 for details and references. (c) The first comparative evidence for the monogamy hypothesis (Boomsma, 2007) by Hughes et al. (2008), illustrating that single insemination of nest-founding females is ancestral across the ants, bees, and vespoid wasps (black branches) while a number of independent origins of multiple insemination of queens (red branches; dotted when facultative and solid when obligate) arose only after superorganismal caste differentiation had been irreversibly established. Reproduced with permission of AAAS, from Ancestral monogamy shows kin selection is key to the evolution of eusociality, W.O.H. Hughes, B. Oldroyd, M. Beekman, F.L.W. Ratnieks, *Science* 320, 1213-1216 (2008); permission conveyed through Copyright Clearance Center, Inc. The respective origins of superorganismality from what likely were ancestors with simple (subsocial) parental care (see Figure 4.7b) have been marked with blue dots (added with permission of W.O.H. Hughes). A comparable blue dot could be placed on any termite phylogeny at the base of the higher termite crown group (e.g. Revely et al., 2021).

in many more animal lineages than those that share direct common ancestry with the superorganismal clades, both among invertebrates (aphids, thrips, spiders, shrimps, beetles) and primates, other mammals, birds, reptiles, and fish (Costa, 2006; Rubenstein and Abbot, 2017). However, for the purpose of this book, I will focus primarily on sister clade comparisons between superorganisms and societies, because these distinctions have been most obviously blurred by eusociality gradient thinking (e.g., Sherman et al., 1995).

In Chapter 4, I claimed that all superorganismal lineages must have arisen directly from subsocial, maternal or biparental, care and not as paraphyletic lineages within clades of cooperative breeders with advanced societies (see Figure 4.7b). The evidence for this conjecture relies on a large body of work summarized in Box 6.2 (reviewed while focusing mostly on the lineages of Figure 6.1). The match between prediction and empirical support is quite reasonable, although more studies are clearly needed. Recent molecular phylogenies have contributed greatly to increasing the resolution of comparative analyses. They have established that the Apoidea (specid wasps and bees) are the extant sister group of the ants (Johnson et al., 2013), which implies that solitary maternal care must have been the deeper ancestral condition, both for the ants and the social bees. The larger clade encompassing the ants, sphecid wasps and bees, combined with the pompiloid, scoliid, and typhoid wasps, is the sister group of the rhopalosomatid and vespid wasps, suggesting that subsocial breeding was the ancestral condition for all Hymenoptera that evolved advanced societies and superorganisms (Johnson et al., 2013). We do not have fossils indicating that the ants ever had a direct sister clade of family-based cooperative breeders (see Boudinot et al., 2022

for a recent review), in contrast to the termites where only the crown-group Termitinae are unambiguously superorganismal with reproductives, somatic workers, and soldiers all having irreversible developmental pathways. Some lower termite lineages also evolved true workers, but they retained the option to develop into replacement reproductives, indicating that germline and soma do not segregate for all colony members early in development (Revely et al., 2021). The bee and vespoid wasp phylogenies have the highest diversity in social systems and support the conjectures that I presented in Figure 4.7b most consistently (see Box 6.2 for details).

While the first comparative analysis by Hughes et al. (2008) indicated that lifetime parental monogamy was indeed likely to have been the necessary condition for the three hymenopteran MTEs to superorganismality (Figure 6.1c) (see Hartke and Baer (2011) and Boomsma (2013) for evidence in termites), a similar analysis by Fisher et al. (2013) showed that clonal group formation (emanating from a single zygote or dikaryon in eukaryotes) must have been the necessary condition for obligate multicellularity (Figure 6.2a). The phylogeny by Fisher et al. (2013) elaborated earlier versions by Grosberg and Strathmann (2007) and Baldauf (2008) while omitting the clades that universally retained a unicellular lifestyle. Further analysis showed that a number of lineages fulfilling the necessary clonal relatedness condition remained facultative multicellular (inset histogram), suggesting that the Hamiltonian $b/c > 1$ conditions were not consistently met here. Another tree by Lang and Rensing (2015) highlighted the photosynthesizing protist clades and the multicellular lineages that underwent bouts of whole genome duplication (animals, plants, fungi, ciliates, but none of the algae), indicating that

Box 6.2 The origins and adaptive radiations of superorganismal lineages

The superorganismal lineages (Figure 6.1a) evolved their convergent permanent forms of caste differentiation in separated niches (Misof et al., 2014) and relatively late (Figure 6.1b), but that does not imply that these MTEs therefore had to emerge from already fairly complex societies that might have evolved earlier. Below I investigate in more detail whether and how superorganismality can be explained as having evolved from simple subsocial ancestors along the outer edge of the Figure 4.7b tree where simple parental care systems remained conserved while society-forming lineages that would not end up being ancestral to superorganismal MTEs branched off. Before presenting the details below, I want to stress that the hypothetical society branches of Figure 4.7b are heuristic as every detail may not necessarily apply in precisely the same way across all lineages considered. The following text also offers a brief survey of the studies that underpin Figure 6.1b.

The ants (>13,000 extant species) originated up to 145 MYA as full-fledged superorganisms with permanent reproductive division of labor between a queen and worker caste (Brady et al., 2006; Schmidt, 2013; Ward et al., 2015; Barden and Grimaldi, 2016). Single insemination of females appears to be the rule, consistent with the ancestral state reconstruction by Hughes et al. (2008). Compared to the vespoid wasps and bees, the ants evolved a huge array of secondary social elaborations, primarily because winglessness allowed workers to become miniaturized in body size (Peeters and Ito, 2015). This includes many forms of polygyny, analogous to germline mosaicism combined with secondary chimerism for the unrelated males that inseminate daughter queens. Polygyny yields several short-term ecological benefits owing to nest-budding dispersal on foot but also made possible various forms of social parasitism in the longer run (Bourke and Franks, 1991; Boomsma et al., 2014). In several ant lineages, polygyny selected in turn for cheaper reproductives in the form of permanently wingless (ergatoid) queens or inseminated (gamergate) workers as replacement reproductives. The latter could evolve only in ant subfamilies where the ancestrally unmated workers had not yet lost the sperm storage organ so they could start reusing it despite their caste morphology having become unspecialized for reproduction (Ito and Ohkawara, 1994; Gobin et al., 2008; Peeters and Ito, 2015). These derived social arrangements could even result in the "re-invention" of dominance hierarchies (Peeters, 2012), but with characteristics (e.g., Heinze, 1990; Heinze, 2016) incomparable to dominance hierarchies in non-superorganismal societies. The fact that society-forming sister lineages are unknown suggests that superorganismality evolved rapidly from subsocial maternal care along the outer edge of Figure 4.7b once the necessary and sufficient Hamiltonian conditions (see Figure 4.2b) were met. Wheeler's superorganism concept implies that colonies should not only vary in their family characteristics, but that this variation is heritable, a contention that was recently confirmed (Walsh et al., 2020).

The termites (ca. 3000 extant species) evolved up to 170 MYA from wood roaches as log-dwelling conifer decomposers. They became unambiguously superorganismal around the Cretaceous–Paleocene boundary ca. 70 MYA when the monophyletic higher termites (Termitinae) with true, central-place foraging workers and irreversible developmental pathways for all castes emerged as a phylogenetic crown group (Inward et al., 2007; Bourguignon et al., 2014; Engel et al., 2016; Evangelista et al., 2019). All termite (sub)families initiate new colonies with lifetime monogamous pairs of reproductives—secondary deviations towards multiple parentage are rare. The subsocial family life of the *Cryptocercus* cockroach sister lineage has been amply documented (Nalepa and Bell, 1997; Bourguignon et al., 2014; Evangelista et al., 2019), including recent evidence that *Cryptocercus* females are facultatively promiscuous and therefore lack the lifetime monogamy of the termites (Yaguchi et al., 2021), as predicted by the monogamy hypothesis (see Figure 4.6). A *Cryptocercus*-like lineage becoming strictly lifetime monogamous for exaptive reasons independent of existing family life may thus have been the most decisive factor for colonial termite sociality to evolve. The basal termite subfamilies form

Box 6.2 *Continued*

a grade rather than a clade, rendering the term lower termites paraphyletic. Some of these lineages have true foraging workers while the wood-dwelling lineages never forage outside their log. Almost all these taxa have one or a few differentiated soldier individuals (Lüscher, 1953), but their worker altruism appears to have remained conditional otherwise as most workers can still become replacement reproductives (Korb et al., 2012). A likely reason for the wood-dwelling lower termites having failed to realize superorganismal MTEs is that they have a non-zero likelihood of parental promiscuity. This is because neighboring log-sharing colonies are forced to merge when most of the wood has been consumed, something that is not feasible for termites with central-place foraging workers (Boomsma, 2013). This violation of the lifetime monogamy condition is proximately due to female and male reproductives both surviving—in ants, bees, and wasps, fathers only survive as stored sperm, which cannot be transferred to another queen (Boomsma et al., 2014). In terms of Figure 4.7b, the lower termites appear to match the characteristics of the closest sister group of the superorganismal higher termites, with the foraging lower termites being somewhat more advanced in partial caste differentiation. Whether the latter's workers are as altruistic as those of the higher termites remains to be seen, but it is quite striking that none of these lower termite branches produced an adaptive radiation that is even remotely comparable to what the crown-group higher termites produced (Bucek et al., 2019). Because the termites lack the asymmetries in relatedness that characterize the family-based social Hymenoptera, their biology has primarily been reviewed from an ecological perspective (e.g. Korb, 2010; Korb and Heinze, 2016), which notes the importance of kinship but did not develop testable Darwinian predictions. Although basal lineages with facultative or fully promiscuous society life are lacking, progression in termite social evolution appears to match the Figure 4.7b tree for there is no doubt that parental LTC (absent in *Cryptocercus*) was the ancestral state that guaranteed maximal full-sibling relatedness early in colony development when the lower termites evolved specialized

colony life and altruistic soldier castes. This transition coincided with a distinct change in sperm morphology that greatly reduced and soon totally eliminated sperm motility (Dallai et al., 2014), consistent with a complete absence of sperm competition (Hartke and Baer, 2011). However, what kept lower termite lineages with true foraging workers from evolving their own superorganismal MTEs remains unresolved.

The vespoid wasps (ca. 4200 extant species) evolved at the Jurassic–Cretaceous boundary ca. 145 MYA and are phylogenetically more distant to ants and bees than ants and bees are to each other (Johnson et al., 2013). They have basal lineages with mostly solitary or cooperative breeding species, but evolved obligate colonial breeding in the ancestor of the polistine and vespine wasps around the Cretaceous–Paleocene boundary ca. 70 MYA (Figure 6.1b). Of these two subfamilies, only the Vespinae (yellowjackets and hornets) made the transition to annual superorganismality of modest complexity (Archer, 2012) ca. 45 MYA (Branstetter et al., 2017; Peters et al., 2017). The polistine wasp sister lineage is more speciose, has greater diversity in social systems, and can reach much larger colony sizes in the polybiine branch than do the vespine wasps (Turillazzi and West-Eberhard, 1996). The swarms in polybiine species can also be much larger than those of superorganismal *Provespa*, which contain fewer than 100 workers (Jeanne, 1991; Zucchi et al., 1995). In contrast to the vespines, the polistines do not have clear differentiation between reproductive and helper roles, and they only occasionally show signatures of pre-imaginal caste differentiation for some, but far from all, colony members (Noll and Wenzel, 2008; Noll et al., 2021). Size and longevity of mature colonies are thus improper predictors of hierarchical organizational complexity across lineages. Phylogenetic analyses (Hines et al., 2007; Piekarski et al., 2018) revealed that the closest relatives of the combined polistine and vespine clade are the solitary or facultatively social potter wasps with maternal nest provisioning, of which the subsocial Zethinae appear to be the direct sister lineage. These phylogenetic reconstructions are consistent with a direct transition from simple

Box 6.2 *Continued*

subsocial maternal care to superorganismality (vespines), with lineages of society forming wasps branching off independently (see Figure 4.7b). The stenogastrine wasps, earlier believed to be the sister lineage of all other social vespoids, form the most basal lineage (Piekarski et al., 2018) and merely evolved facultative and variably promiscuous cooperative breeding (the basal lineages in the hypothetical Figure 4.7b tree) without any later elaboration to more complex societies (Turillazzi, 1991; Field, 2008). Overall, the conjecture that all vespoid lineages arose directly from monogamous parental care appears well supported (see also Perrard et al., 2016), but it is interesting that the sperm storage organs of vespine wasp workers have remained morphologically indistinguishable from those of queens even though they remain un-inseminated for life (Gotoh et al., 2008).

The bees (>20,000 extant species) evolved ca. 100 MYA and initially encompassed a number of basal ground-nesting and stem-boring lineages with various forms of cooperative breeding. Superorganismality evolved once (Romiguier et al., 2015) in the late Cretaceous ca. 65–70 MYA when the crown group of the corbiculate bees (bumblebees, stingless bees, and honeybees) arose as sister lineage of the orchid bees (Branstetter et al., 2017; Peters et al., 2017; Murray et al., 2018), which are largely solitary. This superorganismal crown group realized substantial adaptive radiation even though the bumblebees retained annual life cycles similar to the yellowjacket wasps. The superorganismal honeybees and stingless bees became ecologically dominant, but their species diversity remained modest compared to the more basal bee lineages (Michener, 2000). The bee phylogeny offers the clearest evidence for annual superorganismality, as exemplified by the bumblebees, having evolved directly from a univoltine subsocial ancestor, a possibility that was in fact recognized more than 35 years ago (Michener, 1985), well before molecular phylogenies existed. The allodapine bees appear to have socially variable colonies similar to some of the orchid bees and paper wasps, consistent with basal hypothesized branches in Figure 4.7b (Schwarz et al., 1998). The long-tongued bees, to which both the corbiculate

and the allodapine and xylocopine bees belong, also have strictly solitary lineages such as the megachilid bees (Cardinal and Danforth, 2013), similar to what has been documented in the vespoid wasps (Piekarski et al., 2018). This is once more consistent with social clades of varying complexity arising from simple subsocial ancestry and with non-family communal breeding remaining taxonomically isolated and failing to produce adaptive radiations. In the short-tongued bee sister clade, social behavior evolved almost only in the halictid bees, which appear to have representatives across all the hypothetical black branches of Figure 4.7b, also including facultative cooperative breeding (Batra, 1966; Michener, 2000; Danforth et al., 2013; Quiñones and Pen, 2017). The Ammoplanidae wasps are now considered to be sister taxon to the entire bee clade (Murray et al., 2018; Sann et al., 2018)—these are tiny wasps that hunt flower-visiting and pollen-consuming thrips, making subsequent nutritional specialization on a pollen diet a parsimonious scenario for the ancestral bees. In comparison to the yellowjacket wasps, there are clearer indications of structural and functional reductions in the sperm storage organs of bumblebee workers relative to queens (Schoeters and Billen, 2000). An interesting parallel between superorganismal bees and wasps is that colony founding by swarming evolved in the stingless bees and the honeybees, just as the yellowjacket wasps evolved a swarm-founding *Provespa* lineage while the remaining vespine wasps retained independent colony founding by single queens and annual lifecycles similar to bumblebees.

The ambrosia beetles are a highly speciose assembly of haplodiploid or diploid, wood-boring, and fungus-farming lineages (Kirkendall et al., 2015). Solitary breeding with biparental or maternal care is the rule, but a number of genera, particularly in the haplodiploid Scolitidae, have species where older siblings can help their mother to raise younger siblings before they disperse to found their own colony, particularly in the sib-mating genus *Xyleborinus* (Biedermann and Taborsky, 2011; Biedermann et al., 2011). A focal species of interest here is the platypodid ambrosia beetle *A. incompertus*, which

---

Box 6.2 *Continued*

tunnels in live *Eucalyptus* trees and has colonies with a single, long-lived foundress mother and maximally ca. ten somatic (lifetime unmated and similarly long-lived) daughter helpers that maintain the burrow's fungus garden but never forage outside. *Austroplatypus* originated ca. 55 MYA as an incipient superorganismal lineage, but remained an evolutionary dead end because it failed to adaptively radiate (Smith et al., 2018). Discovered as a putative case of "eusociality" (Kent and Simpson, 1992), recent work revealed that *A. incompertus* has exclusively full-sibling colonies, which implies that both the necessary superorganismality condition of relatedness equivalence between offspring and siblings, and the sufficiency condition of sib-rearing being more efficient than dispersal and personal reproduction, are fulfilled (Smith et al., 2018). While shorter-lived ancestors practiced biparental colony founding, the females of *A. incompertus* store a single ejaculate of sperm before starting their burrow and apparently never re-mate later in life (Smith et al., 2018). The fact that worker sperm storage organs are invariably empty and their ovaries vestigial is therefore an unambiguous signature of obligate caste differentiation by reproductive division of labor. This beetle thus appears to be a rare example of obligate family life with functionally specialized reproductive division of labor being trapped in its incipient superorganismal potential (red branch in

Figure 4.7b) because of extreme niche specialization. Lifetime storage of sperm evolving from biparental care strongly suggests that hymenopteran lifetime sperm storage must have been an enabling pre-adaptation for both family-based cooperative breeding and superorganismality to emerge. Overall, this single ambrosia beetle species supports the hypothesis that lifetime caste differentiation originated from simple subsocial ancestry, while underlining that reaching an irreversible MTE origin is no guarantee that a lineage with superorganismal organizational complexity will become ecologically successful.

The Siphonophora (ca. 175 described species) are all superorganismal in a modular, non-sessile, and pelagic fashion. Clonal multicellular polyps remain attached to each other while expressing extensive morphological differentiation and division of labor (Dunn, 2009). Phylogenetic reconstructions including hydrozoan outgroups have only recently started (Dunn et al., 2005, Munro et al., 2018), offering considerable insight in branching patterns, basal and derived superorganismal traits, convergent evolution of separate sexes, and occasional secondary reductions of organizational complexity, but no evidence for secondary reversals to an undifferentiated cnidarian lifestyle. However, no dated phylogeny is yet available, fossils appear to be lacking, and the closest non-superorganismal sister lineage remains unknown.

---

such proximate developments were important covariates during adaptive radiation. The diversity of cell types is another key aspect of permanent multicellularity. Comparing patterns across the different lineages is complex (Bell and Mooers, 1997), but produced an overall log–log correlation between cell type diversity and body size (Figure 6.2b). A recent analysis showed that the cell type slope is steeper in small compared to large multicellular organisms (Fisher et al., 2020), with ca. five cell types and 100,000 cells as

inflection point. It is intriguing that plotting caste diversity versus the number of individuals per social insect colony produced a parallel line with an intercept that is about an order of magnitude smaller (Strassmann and Queller, 2007).

The three algal lineages that independently evolved permanent multicellularity (Figure 6.2) all belong to the Archaeplastida, the eukaryote lineage that obtained their photosynthesizing chloroplasts from a single cyanobacteria-domesticating event

(Bengtson et al., 2017). This seems relevant because no forms of more redundant secondary and tertiary endosymbiosis (i.e., heterotrophic protists engulfing another primary photosynthesizing protist) appear to have produced permanent multicellular organismality (Yoon et al., 2006). Primary chloroplast domestication was the second most important innovation after LECA was defined by its acquisition of mitochondria and meiotic sex and recombination. Recent estimates place this key event more than 2 billion years ago, while noting that an independent cyanobacterial endosymbiosis evolved ca. 500 million years ago (MYA) in the unicellular *Paulinella* lineage (Macorano and Nowack, 2021). New fossils are now pushing the first appearance of permanent and somewhat differentiated multicellularity in green and red algae back to ages of a billion years or more (Bengtson et al., 2017; Tang et al., 2020), but most MTEs towards permanent algal and fungal multicellularity are of similar age as the metazoan MTE (Nagy et al., 2018) (see Figure 6.2 legend), which likewise produced modular lineages until the Bilateria evolved. Only the obligatorily multicellular brown algae evolved more recently (Silberfeld et al., 2010) and are about as old as the termites (Figure 6.1b). None of the phylogenies obtained in recent years have provided evidence for secondary losses of permanent multicellularity once the number of differentiated cell types had surpassed the ca. five that can also be found in lineages with facultative multicellularity (see Chapter 7). This is consistent with the fraternal MTEs to multicellularity being irreversible advances to higher organizational complexity, as were the superorganismal lineages.

The volvocine green algae evolved three-dimensional multicellularity independently (Rokas, 2008; Umen, 2014), albeit with minimal organizational complexity in the form of cellular division of labor (Herron and Nedelcu, 2015). Weismann (1889) and Huxley (1912) were among the first to give detailed accounts of the putative forces of natural selection that have shaped the *Volvox* lineage from basal unicellular *Chlamydomonas* to undifferentiated multicellular *Gonium* and a crown group of *Volvox* species (Umen, 2014). In a functional way the volvocines resemble the ambrosia beetles, which often evolved simple cooperatively breeding societies (of parent–offspring individuals rather than cells) with only a single genus barely qualifying as having evolved incipient superorganismality (Figure 6.1c). Substantial recent volvocine research has focused on explaining the evolution of multicellular complexity (Herron, 2016), working largely with Buss's (1987) idea of putative conflict between cell lineages (Michod and Roze, 2000; Michod et al., 2003; Herron and Michod, 2008). However, when MTEs originate from single zygote commitment and the benefits of joint resource acquisition, it seems doubtful that levels of selection considerations would be relevant in the wild even when interaction dynamics between clonal cells may leave the impression that competition can occur (Box 6.3). Unconditional cooperation with the appearance of competition has recently been documented at the multicellular level for two species of colonial Bryozoa, which have a clear higher-level developmental biology in spite of zoids remaining undifferentiated, but they lack heritable variation among zoids. This precludes lower-level selection for differential propagation of traits within a colony but, as the authors show, heritable variation at the whole colony level is significant (Simpson et al., 2020).

Recent studies have shown that permanent multicellularity with differentiated cell types has arisen multiple times within a number of lineages previously considered to be single

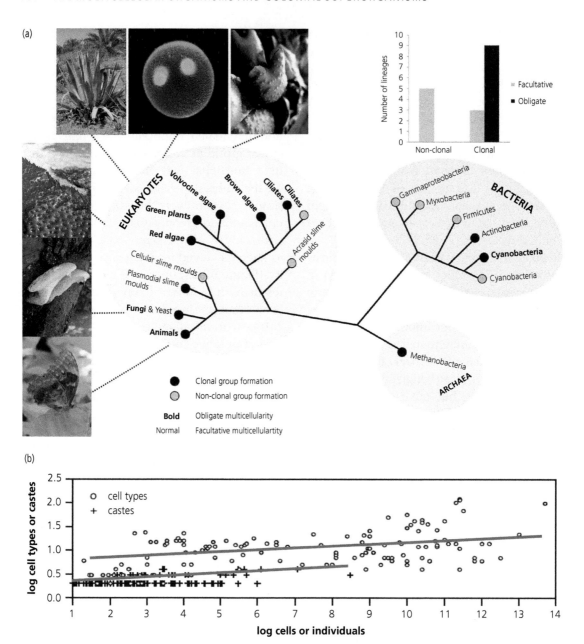

Figure 6.2 The multicellular lineages with permanent cellular division of labor. There are ca. six such lineages with three-dimensional bodies (all eukaryotes), of which two are heterotrophic (animals and fungi) and four are autotrophic (plants and three major algal lineages). The animals, comprising more than a million extant species, evolved permanent multicellularity > 750 MYA (Briggs, 2015), which implies that the oldest colonial superorganisms (Figure 6.1b) were around for only the last 20% of that period and that the average superorganismal MTE originated in the last ca. 10% of that total time span. The fungi appear to have evolved more than a billion years ago and the Basidiomycota and Ascomycota clades that independently produced permanent multicellularity date back 500–700 million years. They have a deep aquatic origin (Lutzoni et al., 2018), but are now mostly terrestrial living in water-saturated environments. A recent review (Nagy et al., 2018) indicates that a number of other fungal lineages also evolved multicellular fruiting bodies later in evolutionary time. The emergence of the multicellular Basidiomycota and Ascomycota likely coincided with the colonization of the continental land masses together with lineages of the Viridiplantae (Stajich et al., 2009; Choi and Kim, 2017; Lutzoni et al., 2018; Naranjo-Ortiz and Gabaldón, 2019). One of these lineages (the

(Continued)

MTEs (Leliaert et al., 2012; Umen, 2014; Sebé-Pedrós et al., 2017; Lutzoni et al., 2018; Nagy et al., 2018). This implies that the comparative analysis of Fisher et al. (2013) could now be repeated with higher statistical power, although nothing in the biology of these lineages suggests that the LTC principle for explaining MTE origins will become compromised. Fisher et al. (2013) also documented cases of multicellularity among the prokaryotes, but here only one lineage (Cyanobacteria) became truly filamentously multicellular, and its cell type differentiation remained very low. The comparative analysis of social insects by Hughes et al. (2008) also deserves to be updated, both because phylogenies have improved in resolution and because their analysis did not discriminate between cooperatively breeding societies and superorganismal colonies of hymenopteran social insects (blue dots in Figure 6.1c). A key objective of such updates should be to challenge the prediction from Chapter 4 that aggregative group formation has never produced an obligatorily

multicellular or superorganismal lineage, a conjecture that has so far been upheld.

Finally, while there are a handful each of fraternal MTEs to permanent multicellularity and colonial superorganismality, it is worth noting that the richness of their adaptive radiations is very different because the multicellular domain originated much earlier than the superorganismal domain. The superorganismal clades of Figure 6.1c may have a total number of ca. 50,000 species, while the descendants of MTEs to permanent multicellularity (Figure 6.2) likely have an overall species diversity surpassing that number by at least two to three orders of magnitude.

There is no way to understand the origin and subsequent stability of social systems without staring potential conflict and corruption in the face. However, societies and (super)organisms are very different, and the types of conflict that constrain them differ as well. Society members always compete for access to resources, by scramble competition when participants are cells, but often by contest competition when

---

Figure 6.2 (*Continued*)

Streptophytes) produced the embryophyte land plants (>10^5 species) and the other the multicellular Chlorophytes (ca. 4500 species) (Umen, 2014). It is now also well documented that the red algae (Rhodophytes) are phylogenetically more related to the green algae than to the brown algae (Glaucophytes) (Cock et al., 2010; Finet et al., 2010; Brawley et al., 2017). Extant species numbers of green, brown, and red algae may hover around ten thousand each, but many of these remained unicellular or facultatively multicellular. (a) A comparative study by Fisher et al. (2013) showed that individuals in all lineages with obligate two- and three-dimensional multicellularity are clonal because they start from a single cell, usually a single zygote in eukaryotes, whereas non-clonal initiation never produced obligate multicellularity. However, not all protist lineages with clonal cell adhesion became permanently multicellular (inset histogram), consistent with predictions for MTE origins (main text). A recent study by Fisher et al. (2020) further showed that the lineages with clonal group formation almost invariably originated in aquatic rather than in terrestrial environments, and that this difference is even more pronounced when considering MTE origins. However, facultative multicellularity originated more commonly in terrestrial habitats as Bonner (1998) already noted. Republished with permission of Elsevier, from Group formation, relatedness and the evolution of multicellularity, R.M. Fisher, C.K. Cornwallis, S.A. West, *Current Biology* Volume 23 (2013); permission conveyed through Copyright Clearance Center, Inc. Images by the author, except for red and brown algae (Patrick Keeling) and *Volvox* (Casey Dunn). (b) Comparison between somatic differentiation of cells in multicellular organisms and caste differentiation in social insects as functions of, respectively, the number of cells in a body and the number of individuals in a colony (the latter data originally from Bonner (1993, 2004)). Two approximately parallel trend lines (fitted by hand) are about an order of magnitude apart. Reproduced with permission from Insect societies as divided organisms: The complexities of purpose and cross purpose, J.E. Strassmann, D.C. Queller, *PNAS*, Volume 104 (2007); Copyright (2007) National Academy of Sciences, U.S.A.

Box 6.3 Redundancy due to chimerism, promiscuity, and mosaicism before and after major transitions

---

Offering a functional definition of redundancy is challenging (see Table 3.1). If we assume that genetic heterogeneity within (super)organisms is always proportional to non-alignment of ultimate reproductive interests, one could ask how little redundancy suffices as a proxy for apparent competition. For example, could very slight variation among clonal cells in a body induce competition for access to resources and could similar minute heterogeneity among superorganismal siblings have an analogous effect? This point was raised by Buss (1987), and recent evidence appears to support this claim (e.g., Meyer et al., 2014; Shakiba et al., 2019), but is it really competition? The interests of clonal cells should be 100% aligned (Maynard Smith and Szathmáry, 1995, p. 244), because all have the same genome-wide inclusive fitness. Might it then be that epigenetic marks make clonal cells "compete" for division of labor assignments just to make sure that joint tasks proceed with maximal efficiency for the common good? Gawne et al. (2020) suggest that this interpretation is likely for metazoan bodies. This conceptualization also appears reasonable for superorganisms, because worker ants can be observed to "compete" for tasks that all could do but some can likely do better (e.g., Anderson et al., 2002). Could such dynamics therefore just be proximate mechanisms that mediate self-organization (Bonabeau et al., 1997; Camazine et al., 2001; Duarte et al., 2011)? This conjecture seems compelling because directional kin selection and epigenetic pseudo-competition could then have jointly self-organized the irreducibly complex *b/c* synergies (white zones in Figure 3.2b) that mediated irreversible MTE origins, irrespective of the level of organismality.

A more general question is when closure is non-redundant enough to forge an MTE (see Figures 5.1 and 5.2) or, the other way around, when internal genetic heterogeneity starts to induce division rather than division of labor? That MTE origins can be explained by non-redundant closure did not preclude that new forms of higher-level chimerism such as multiple queen insemination could arise later on (Figure 6.1c). In that process, closure was maintained

and promiscuity in the sense of serial monogamy did not return, that is, the evolutionarily derived mating systems of superorganisms with multiply inseminated queens never came to resemble anything like *Drosophila* (Boomsma et al., 2005a) where females remain receptive and mate throughout adult life (e.g., Chapman et al., 1995). Promiscuity increases genetic variation of family members over time while chimerism refers to spatial genetic heterogeneity within bodies. These aspects appear to be comparable within the domains of social evolution but not necessarily between them. For example, we expect intriguing similarities between promiscuity in plants, fungi, and solitary animals because the reproductive target is always differential success in future zygote or dikaryon participation—the egg, pollen, nucleus, and sperm agents that create these commitments are convergently analogous (Dickinson and Grant-Downton, 2009). Also, the convergent secondary evolution of multiple queen insemination across the ants, corbiculate bees, and vespine wasps is amenable to comparative analysis (Figure 6.1c). Multiple insemination of superorganismal queens is always about creating closed, paternally polyploid zygote analogies adding redundant green rectangles to the founding royal ant pair in Figure 5.3. However, analogous elaborations are impossible for organismal zygotes because spindle formation during cell division proximately precludes polyspermy to be viable (Hemmings and Birkhead, 2015; Yu et al., 2021).

Regarding effects on social cohesion, forms of primary chimerism are fundamentally different from manifestations of mosaicism that emerge later in a life cycle (Santelices, 2004). In permanently multicellular organisms, the most pervasive form of mosaicism is cancer, a normally non-contagious form of somatic homeostasis failure where cell lineages revert to ancestral unbridled division after losing their higher-level somatic commitments (Frank, 2010; Aktipis et al., 2015). Multicellular organisms have evolved extensive DNA-repair and immune-monitoring defenses against the somatic mutations and other instabilities that typically initiate cancer

Box 6.3 *Continued*

(e.g., Crespi and Summers, 2005; Frank, 2007; Sonnenschein and Soto, 2020). The efficiency of these defenses is related to average life span because selection for them is largely synonymous to selection for somatic maintenance as captured by Medawar, Williams, and Hamilton when they developed the key evolutionary principles for understanding aging (see Chapter 2). This explains that Peto's paradox—the incidence of cancer being independent of the number of cells and cell divisions in a metazoan body—is not a real paradox because somatic repair and cancer risk tend to become positively correlated during life-history evolution (Tollis et al., 2017; Sun et al., 2022). Shallow cancer analogies can be identified at the superorganismal level, but they are normally about invasive germline individuals of a secondary chimeric kind, that is, exploitative unrelated queens or reproductive workers with a priori reduced loyalty to a colony, and not about somatic individuals wreaking mosaic havoc after having become mutationally compromised.

As long as superorganismal colonies have a single founding queen, either singly or multiply inseminated, her functionality is analogous to a metazoan germline (see Figure 5.3). Just like a zygote, such a newly mated queen is initially on her own but becomes sequestered as protected germline as soon as her first somatic workers hatch. When the colony develops, survives, and reproduces, all somatic workers will ultimately die when the queen germline dies, not the other way around as in metazoan bodies where aging is primarily somatic. That situation changed, however, when lineages with superorganismal colonies such as the ants secondarily evolved adoption of newly inseminated gyne daughters to become additional queens. Social insect researchers often use the term *polygyny* for the coexistence of multiple breeders in a colony, but polygyny means something very different on either side of an MTE separating societies from superorganisms, and both are different again from polygyny specifying the extent of harem formation in otherwise solitary vertebrates. Colonies of *Polistes* paper

wasps are often referred to as polygynous when they have dominance hierarchies of totipotent breeders, many of whom can realize both direct and indirect fitness (West-Eberhard, 1978), but when superorganismal ant colonies become polygynous, their germline becomes chimeric because each of these "mosaically" related daughter queens is inseminated by an unrelated male. Even if the lifetime unmated original worker caste remains committed to its somatic tasks, the less related somatic workers that new queen germlines add to the colony will have much more destabilizing potential than primary redundancy due to a single queen being multiply inseminated—this is because a spermatheca storing multiple ejaculates remains closed (see Figure 5.2) while queen readoption polygyny is largely an open market.

Additional superorganismal queen germlines are largely restricted to the ants (De la Mora et al., 2020), similar to inquiline social parasites (Buschinger, 2009; Cini et al., 2019), so it is logical to ask whether there is a connection (Bourke and Franks, 1991). Secondary polygyny may have resource acquisition benefits because germline rejuvenation extends colony life span and allows monopolization of habitat patches for future kin (Boomsma et al., 2014). However, the long-term downside is that these chimeric germlines are likely to evolve exploitative tragedy-of-the-commons traits, initially disadvantaging kin in favor of direct reproduction, but ultimately becoming reproductively isolated species that specialize on behaving as horizontally transmitted parasitic germlines. In that later phase, inquiline social parasite species tend to partly or entirely lose their somatic castes when workers of host colonies are manipulated into offering these services to alien queens (Boomsma and Nash, 2014). Ant inquilines insinuate themselves into mature host colonies by stealth, as expected when they evolved gradually from milder intraspecific social parasitism via daughter queen adoption, where the female redundancy element is mosaic and only the sperm tucked away in storage organs is a chimeric element. This is very different from aggressive intra- or interspecific

---

Box 6.3 *Continued*

nest usurpers that replace lone colony-foundresses—these also evolved in bees and wasps (Cini et al., 2019). Higher termites appear to lack sexual (i.e. outbred) germline rejuvenation, but can do this asexually (Matsuura et al., 2009), which has led to intriguingly cryptic forms of intraspecific inquilinism (Tamaki et al., 2021). The closest known cancer analogue of superorganismal soma is the parthenogenetic Cape honeybee whose workers infest colonies of the sexual African honeybee (Oldroyd, 2002; Yagound et al., 2020).

Finally, it is worth noting that organismality in fungi is special. Their hyphae remain assemblies of totipotent haploid nuclei, which evolved many mechanisms for avoiding or regulating chimerism (Roper et al., 2011). The best known of these are the clamp connections of most Basidiomycota, which during cell division ensure faithful inheritance of the two haploid nuclei that formed the zygote-like dikaryon, a pseudodiploid state without nuclear fusion. Dikaryons can donate nuclei to haploid monokaryons, turning them into dikaryons, and with one of the two haploid nuclear lineages often being a more successful donor than the other (Nieuwenhuis et al., 2013). Fungal nuclei are thus only committed as long as they remain cellularly enclosed (Vreeburg et al., 2016), and violation

of this rule increases the likelihood of reproductive chimerism conflicts in ways unknown in other multicellular lineages (Bastiaans et al., 2016; Goncalves et al., 2019; Grum-Grzhimaylo et al., 2021). A striking superorganismal analogy is that males of ants, bees, wasps, and *A. incompertus* ambrosia beetles survive as stored sperm (analogous to fused gametes in a diploid zygote), whereas the male reproductives of termites survive as separate reproductive agents while continuing to inseminate the same queen (analogous to haploid fungal nuclei). As we saw, independent male agency is constrained by lifetime functional closure in a royal cell (see Figure 5.2c) in the higher termites, but not in the wood-dwelling lower termites where occasional colony mergers followed by parental re-mating promiscuity occur regularly. In this sense, basidiomycete fungi seem functionally analogous to lower, not higher termites (Vreeburg et al., 2016). This may be perceived as a violation of the overall analogy between superorganismality and permanent multicellularity and underlines that both fungi and termites need further study to better understand the details of their trajectories towards ontogenetically stable multicellularity and superorganismality.

---

members have sufficient individual agency. When such stand-offs become ritualized dominance hierarchies, that does not mean they have become mutualistic. As Williams (1966a, pp. 95, 243) phrased it, "[m]any animals form dominance hierarchies, but these are not sufficient to produce an evolutionary advantage in mutual aid. [...] If an animal is continuously exposed to aggressive competition it may be adaptive, even if the animal is currently healthy and well fed, to curtail or delay its efforts to reproduce." This captured, *avant la lettre*, what later became known as reproductive skew (Vehrencamp,

1983), a concept that is dynamically meaningful in animal societies, but hardly or not in superorganismal colonies (Boomsma et al., 2014). Within (super)organisms, competition is highly constrained, because opting out to try one's luck elsewhere is rarely an option. Somaticized individuals can rebel only in subtle ways by responding to different co-replicon interests (see Box 2.2) or to forms of secondary partnership redundancy that allowed lower-level selection to return (see Woyciechowski and Kuszewska, 2012, for an illustrative example in honeybees). To capture the Darwinian logic of potential conflict in societies and superorganisms, it is important to carefully distinguish between forms of primary chimerism that preclude MTEs in societies and forms of mosaic heterogeneity or secondary chimerism that arose in lineages with ancestrally homogeneous full sibling colonies. (Box 6.3).

The first explorations of the coherence and empirical validity of the lifetime monogamy hypothesis (see Figure 4.6) indicated that vertebrates have never been monogamous enough to make transitions to caste-differentiated superorganismality, but that their monogamous tendencies should be a more powerful predictor of family-based cooperative breeding than ecological factors. In other words, if necessary relatedness conditions come before sufficiency conditions of ergonomic benefit in the evolution of superorganismality, the same principle is expected to also apply across the family-based cooperative breeders. This prediction (Boomsma, 2007) went against the prevailing thought that environmental constraints are the prime drivers of vertebrate cooperative breeding (see Shen et al., 2017, for a recent review), but it has held up well in subsequent comparative analyses across birds and mammals (Figure 6.3). Initial studies confirmed that sibling relatedness drives transitions to cooperative breeding (Cornwallis et al., 2010; Lukas and Clutton-Brock, 2012), but the ecological uncertainty explanation also gained new support (Jetz and Rubenstein, 2011). This inconsistency was resolved by the insight that environmental uncertainty does not select for cooperative breeding per se, but facilitates colonization by family-based cooperative breeders that evolved sib-rearing in more benign habitats (Cornwallis et al., 2017). Lukas and Clutton-Brock (2018) and Downing et al. (2020a) confirmed that distinguishing between family and non-family groups is essential when testing whether indirect fitness

benefits drive sib-rearing in vertebrate cooperative breeders (Figure 6.3). However, it also became clear that delayed dispersal can be just extended parental investment, and that core family life is necessary but not sufficient for selecting older offspring to actually become helpers at the nest (Griesser et al., 2017).

Overall, the comparative evidence (Figures 6.1–6.3; Box 6.2) appears to be reasonably consistent with the terminology proposed in Figure 4.7, suggesting several analogies between social evolution in the multicellular organismal and the colonial superorganismal domains, the two hands full of clades that represent all known fraternal MTEs. In Table 6.1, I revisit the hypothetical social evolution trees of Figure 4.7, filling in details that seem reasonable generalizations for carving out a neo-Darwinian, trait-based understanding of how natural selection has shaped societies and (super)organisms. The bottom (solitary) row and the top (post MTE) row have only distinct (binary) specification terms, while the three middle rows have mostly ambiguous terms, illustrating the gradual variation in social complexity (relative to solitary states) and the associated deficiencies that preclude MTEs because they imply that the necessary (relatedness equivalence) and sufficient ($b > c$) conditions are not being universally met. It is revealing that the sociobiological definition of *eusociality* as presently used in the literature applies jointly to the four upper rows (numbered 2–5) in Table 6.1. This illustrates that *eusociality*'s all-inclusive defining traits of cooperative brood care, reproductive division of labor, and generation overlap, apply to all forms of social breeding jointly. To avoid such confusion, it would be better to just use the term *social insects* (as was done before sociobiology), which would encourage precise qualification of the kind of sociality that a particular study focuses on. Such precision is always opportune when asking *why*-questions *sensu* Tinbergen (see Figure 1.1). In Darwinian analyses of the greatest-common-divider type (see Box 1.3), unspecified *eusociality* and unspecified *multicellularity* are meaningless terms, for questions should focus on traits that can, in fact, respond to selection (see Table 6.1). Reified broad-brush *eusociality* and

multicellularity are clearly not traits (Boomsma and Gawne, 2018), which means that composite terms like eusocial behavior, eusocial nesting, eusocial communication, and eusocial evolution are similarly imprecise and bound to cause more confusion than clarity (see also Box 1.3 and earlier remarks in Chapters 2 and 3).

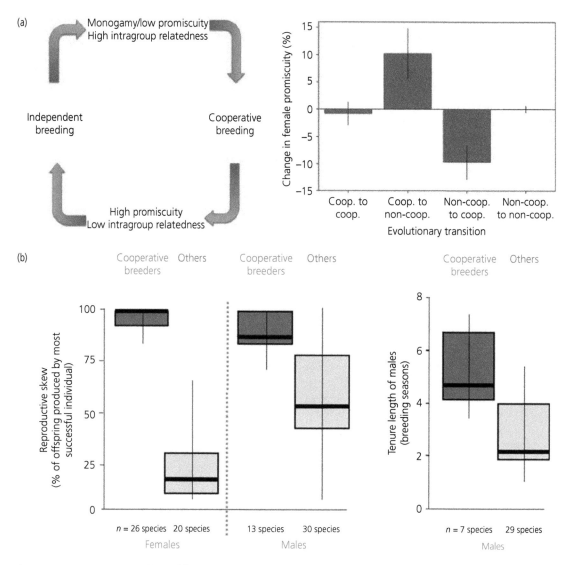

Figure 6.3 Monogamy as predictor of family-based cooperative breeding in vertebrates. Several comparative studies have investigated the extent to which parental monogamy predicts cooperative breeding in birds and mammals. The analyses were restricted to cooperative breeding in family groups, asking whether parental promiscuity discourages older offspring to take up helper roles when their prospect is to help raising half-siblings rather than full-siblings. The comparisons thus excluded the

(Continued)

Germline integrity, closed immune defenses, and levels of organismal development

After addressing ultimate questions of social adaptation and phylogeny, it is appropriate to look at proximate explanations of mechanism and development. Following Tinbergen's organismal biology approach (see Figure 1.1; Box 1.3) and the phenotypic gambit (see Box 2.3), I will continue to black-box the specifics of molecular mechanisms. When doing that, *three topics* seem particularly suitable to illustrate analogies between metazoan bodies and caste-differentiated colonies. These may vary in solidity but have in common that they cannot apply to any form of society, as the opening quote of this chapter illustrates. *The first of these* is the *differentiation and integrity of organismal and superorganismal germlines*. Societies of cellular or multicellular individuals lack differentiated germlines by definition. In contrast, after Weismann (1892) established the metazoan organismal germline concept, he almost immediately extended it to ant colonies in an exchange with Herbert Spencer (Weismann, 1893), which shows that the organismality of caste-differentiated colonies was appreciated well before Wheeler formalized that analogy (Boomsma and Gawne, 2018). When Buss (1987) challenged the validity of Weismann's concept for multicellular organisms, he argued that germlines are only sequestered early in the Bilateria while they appear somatically produced in the basal Metazoa and other multicellular lineages. Such soma-derived, germlines have compelling analogies with workers that may reproduce in haplodiploid superorganisms, as Wheeler (1910, pp. 115–116) explained and later briefly reviewed (Wheeler, 1928b) (see Bourke, 1988 for a more recent review).

Figure 6.3 (*Continued*)
relatively rare cases of non-family cooperative breeding (e.g., Riehl, 2013) where indirect fitness gains must be unimportant so that including these data points would merely introduce noise. (a) Cornwallis et al. (2010) amassed a comparative data set of 267 bird species and used independent phylogenetic contrasts to reconstruct ancestral states to assess the likelihood that each of the four arrows was an important driver in gaining or losing cooperative breeding. They found that parents in cooperatively breeding species were generally less promiscuous than parents that were solitary breeders, both on average and when plotting the proportion of nests with helpers against parental promiscuity. They further showed that the adoption of cooperative breeding was significantly associated with reductions in parental promiscuity, whereas reversals to solitary breeding were significantly related to increases in parental promiscuity. The controls, that is, lineages that retained solitary or cooperative breeding continued the same level of parental promiscuity that they already had. More recent comparative studies (Downing et al., 2020a, 2020b) confirmed these findings by even more explicitly distinguishing between family groups and non-family groups. These studies documented that helpers have no effect on breeder fecundity in non-family groups, while they enhance breeder reproductive success in family groups, as predicted by kin selection theory, a Hamiltonian *b/c* benefit that became visible after the primary relatedness effect was partialed out. Republished with permission of Springer Nature, from Promiscuity and the evolutionary transition to complex societies, C.K. Cornwallis, S.A. West, K.E. Davis, A.S. Griffin, *Nature*, Volume 466 (2010); permission conveyed through Copyright Clearance Center, Inc. (b) Lukas and Clutton-Brock (2012) showed that a similar pattern applies in cooperatively breeding mammals, using a data set of ca. 45 species. Here the best measure of monogamy was reproductive skew (Vehrencamp, 1983), with high skew indicating that one breeder produced almost all offspring (which were thus siblings), whereas low skew implied that breeders raised offspring with multiple partners so that offspring is less related (see also Lukas and Clutton-Brock, 2018). Also here, the prediction of high relatedness inducing helper behavior was met, although some species did not practice cooperative breeding in spite of males having high skew. This is consistent with the benefits and costs of pursuing indirect fitness being important in addition to relatedness, as envisaged by Hamilton's rule (Chapter 4). The reproductive skew result was confirmed by male-tenure data, that is, the number of subsequent breeding seasons that a male remained the exclusive father of a female's offspring. An even stronger effect was found when the authors focused on the females, which are often the dominant individuals in cooperatively breeding mammals. Republished with permission of the Royal Society, from Cooperative breeding and monogamy in mammalian societies, *Proceedings of the Royal Society* B, Volume 279 (2012); permission conveyed through Copyright Clearance Center, Inc.

Table 6.1 The social complexity parallels between obligate multicellularity and colonial superorganismality

Multicellular organisms with developmentally hardwired germline–soma division of labor	(5) Obligate	Obligate	Developmentally canalized	(5)	Colonial superorganisms with developmentally hardwired germline–soma division of labor
		Irreversible major transition			
	Multicellular individual colony life	Multicellular individual behavioral differentiation	Phenotypic plasticity for multicellular reproductive or altruistic soma roles		
Obligate clonal multi-cellularity with low division of labor	(4) Always	High	Socially regulated	(4)	Obligate family life with low division of labor
Variably chimeric cell clusters	(3) Always	Medium	Environmentally and competitively regulated	(3)	Variably promiscuous cooperative breeders
Facultative multicellularity	(2) Variable	Low	Present	(2)	Facultative cooperative breeding
	Cellular colony life	Cellular differentiation	Phenotypic plasticity for germline or soma roles		
Solitary cells	(1) Absent	Absent	Undeveloped	(1)	Solitary animals

(left vertical label) Societies of cooperatively reproducing cells

(right vertical label) Societies of cooperative breeding multicellular individuals

This table avoids unspecified multicellularity and eusociality as convenience terms that would apply to all levels except the bottom one. Instead, it focuses on three specific phenotypic traits that can respond to natural selection (central columns with blue headers and green footers) to develop a hypothetical unifying framework for understanding types of sociality between cells (left green text) and multicellular individuals (right blue text). The hypothetical trees of Figure 4.7 are repeated in simplified form in two of the corners, starting with solitary life (1) to reiterate the earlier conjecture that whatever aggregative or promiscuous social system (3) that evolved after facultative social life (2) had become obligate, it was not ancestral to MTEs based on obligate clonal cell adherence (Figure 6.2) or obligate full-sibling altruism (Figure 6.1) that required the strict relatedness equivalence condition (4) and only therefore had the potential to realize an MTE (5) towards irreversible differentiation into germline and soma for all cells or individuals involved. In societies (levels 2, 3, and 4) the three key social traits that can be independently shaped by natural selection (central columns and rows) have to do with colony life having to become obligate, with social roles having to become differentiated, and with plasticity for these roles having to become regulated. However, in permanently multicellular organisms and superorganismal colonies (red text) germline–soma differentiation is developmentally hardwired and can be assumed to have evolved directly via selection for unconditional altruism of entire offspring cohorts rather than by selection for conditional altruism by individuals, the prevailing behavior in society branches. Phenotypic plasticity for staying at the nest versus breeding independently is assumed to be externally (competitively or environmentally) regulated in promiscuous societies (2 and 3) and internally (socially) in clonal or full-sibling societies (4). That superorganismal caste development is canalized similar to germline-soma differentiation during metazoan development (Figure 6.6c; see also Wagner et al., 2018) has recently been confirmed for two ant species (Qiu et al. 2022).

His key point was that worker reproduction questions Weismann's immutable germline theory because worker sons may pass on somatically selected traits to future germlines.

In fact, only very few evolutionarily derived genera of ants, and fewer if any in corbiculate bees and vespine wasps, have workers that lost their ovaries completely (Peeters and Ito, 2015), so most superorganisms practice some combination of unitary reproduction via their foundational germline and modular reproduction via somatic worker individuals. Various forms of structured germline replacement in swarming corbiculate bees and vespine wasps (Boxes 6.1 and 6.2) and of mosaic or chimeric germline amplification in ants (Box 6.3) have evolved, so the question is how analogous these derived reproductive schedules are with reproductive division of labor in the multicellular bodies of basal Metazoa, plants, or fungi. To conceptualize the organization of germline integrity and somatic altruism, Lanfear (2018) emphasized that germline segregation is very different from later germline differentiation when pluripotent stem cells divide into specialized gametes (see also Haig, 2016). There is also an important spatial factor, clarified by Aanen (2014) who showed that undifferentiated stem cells remain at the tip of roots, shoots, and hyphae in modular plants and fungi that grow by cell division in the subapical regions while they remain at the (inside) base of differentiating cell lineages in animals. Using the same symbolism as in Figure 5.3, I summarized the haplodiploid superorganismal version of this argument in Figure 6.4, both with an abstract diagram and two concrete examples of leaf-cutting ant genera. Worker-male production rarely happens by default. It is often curtailed by policing mechanisms, as predicted when workers are differently related to queen-sons and worker-sons (Ratnieks et al., 2006; Wenseleers and Ratnieks, 2006). However, selection might also prioritize collective reproductive efficiency (Hammond et al., 2003; Hartmann et al., 2003; Hammond and Keller, 2004; Helanterä and Sundström, 2007), as somatic worker contributions to future germlines often occur only after queen death.

Apart from superorganismal germlines being potentially affected by purely somatic selection, senescence of these germlines is independent from aging in the worker caste, which seems inconsistent with somatically-driven aging in the bilateran Metazoa. However, in recent years, Weismann's idea of an immutable germline has become more of a typological Platonic ideal than a viable scientific concept. Monaghan and Metcalfe (2019) have argued that evidence for the so-called Weismann barrier being leaky has become quite overwhelming even in bilaterian Metazoa. As they show, rather than retaining constant quality across generations, animal germlines deteriorate when offspring are produced late in life, due to mutation accumulation (sperm) and more complex forms of nuclear and cytoplasmatic damage from long-term storage (eggs). These phenomena, often referred to as the Lansing effect, suggest that germline quality is an optimized life-history trait that could, for example, be correlated with Fisher's (1930) well-known reproductive value curve, but many confounding factors need to be considered (Monaghan et al., 2020). These explorations suggest that germline aging is largely, but not completely, independent of somatic aging in Metazoa, but there is no apparent superorganismal analogy of the Lansing effect. Honeybee queens clearly suffer from reproductive aging (Al-Lawati and Bienefeld, 2009), but cause and effect are hard to disentangle because the genus *Apis* has evolved queen-replacement procedures. A study on *Atta* leaf-cutting ants, which are highly unlikely to replace queens in existing colonies, showed that queen germlines age in the redundancy of their sperm use (Den

Boer et al., 2009), but whether queen aging is affected by the age of her parents is unknown.

Like Buss (1988), Monaghan and Metcalfe (2019) also consider how primordial germ cells give rise to gametes when the male and female germlines differentiate. Some metazoan germlines originate by induction, which implies that the primordial germ cells are zygote produced after fertilization and thus sequestered late during gastrulation. Other metazoan germlines are inherited via the female line and set aside already in the oocyte, so they sequester earlier in the blastula stage, just a few cell divisions after the paternal genome becomes expressed.

There is an interesting analogy here with the ways in which superorganismal germlines are established. Termites follow an induction model with symmetrical roles for the sexes and delayed caste determination among their immature neuter offspring. In contrast, the ants, bees, and wasps follow a germline inheritance model because virgin gynes (oocyte analogues) are asymmetrically destined for the germline role, which just needs to be confirmed by insemination and sperm storage (Nagel et al., 2020).

Reproductive division of labor in cooperatively breeding bird families selects for both higher fecundity and greater longevity of

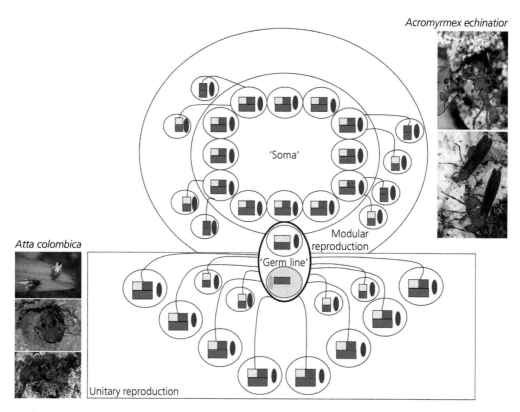

Figure 6.4 The unitary and modular components of superorganismal germlines. The diagram (using the same colors and symbols as in Figure 5.3) illustrates that hymenopteran superorganismal germlines have somatic elements as long as workers retain ovaries that can produce viable unfertilized eggs. The central ellipse represents the germline of a colonial superorganism, consisting of a queen with a diploid nuclear genome (blue and red) and a haploid mitochondrial genome (brown), inseminated by a single male

(Continued)

breeders (Downing et al., 2020a), while super-organismal caste differentiation can induce extreme differences in lifespan between queens (in termites also kings) and workers (Keller and Genoud, 1997). This suggests that the classic solitary trade-off between reproduction and survival is moot or applies to separate castes at different parallel levels (e.g. Korb and Heinze, 2021). As it appears, the challenge of explaining the phenotypic plasticity mechanisms that mediate these differences between queens and workers in the same colonies is far from resolved (e.g. Korb and Heinze, 2021, Kramer et al., 2021).

There is no doubt that superorganismal germlines are costly to maintain, an idea that was recently also proposed for Metazoa (Maklakov and Immler, 2016; Maklakov and Chapman, 2019), consistent with evidence that stressed soma in zebrafish recovers better when germlines are knocked down (Chen et al., 2020). Metazoan germlines are indispensable by definition, similar to inseminated queen germlines that cannot be replaced, but ant colonies have often evolved cheaper secondary germlines in the form of gamergates (workers that store sperm to reproduce sexually) or smaller ergatoid (permanently wingless)

Figure 6.4 (*Continued*)

with a haploid genome (green) that is passed on in its entirety to the queen's daughters (workers and gynes) but not to the queen's haploid sons who have no father. As is almost universally true in eukaryotes, the father's mitochondrial genome (yellow) is not passed on to any offspring. The bottom rectangle represents a colony's unitary reproduction in which only colony germline genes from the inseminated queen are passed on to daughters (larger ellipses) and sons (smaller ellipses), while the top double ellipse represents somatic reproduction (large outer ellipse) by some of the colony's worker daughters (smaller inner ellipse) who are never inseminated (so they cannot produce diploid female offspring) but can produce sons from unfertilized eggs to be reared by sibling workers. During this form of modular somatic reproduction of a colony, the nuclear father genes are also passed on to male offspring, because haploid grandsons have a maternal grandfather. The two sets of images illustrate superorganismal germlines in two sister genera of leaf-cutting ants. In *Atta*, the workers have vestigial ovaries that can no longer produce viable eggs (Dijkstra and Boomsma, 2006). The dispersing (winged) gynes (virgin queens) and males (left, top image) are fully analogous to egg and sperm as gametes of multicellular organisms (Boomsma et al., 2005a), except that polyspermy is impossible while multiple insemination of queens is feasible (Box 6.3). An *Atta* queen inseminated by two or three males sheds her wings and digs her colony founding burrow (left, middle image); she is endowed with substantial body fat reserves and thus an apt analogy of a zygote deposited as an egg with yolk or a seed with endosperm (Baer et al., 2006). Ca. six weeks later she has reared her vertically transmitted fungus garden to macroscopic size (1–2 mm) and added eggs that are by now worker-destined larvae in various stages of development (Figure 5.2e), analogous to a germinating acorn or crustacean nauplius larva that have produced their first soma. Finally, a queen as the functional germline of a mature colony lays thousands of eggs per day, almost all of which develop into small, medium, and large workers, or soldiers to attend to the colony's somatic needs of resource acquisition, defense, and protection of the queen germline (left, bottom image). At this stage the colony will also produce gynes and males in the optimal season for mating and dispersal (top image) analogous to an adult crustacean shedding eggs or sperm in the water column. All this is similar in the *Acromyrmex* sister genus except that inseminated queens have lower fat reserves, so they forage while rearing their first brood, and that most large workers (right, top image) have retained ovaries that can produce viable male eggs to be reared into functional adult males (Dijkstra and Boomsma, 2007). This usually but not always happens after a colony has lost its queen and implies that somatic worker individuals can deliver a non-negligible share of the spermatozoa that fertilize gynes in a mating swarm, not unlike a dioecious plant producing only pollen. Winged dispersing sexuals (right, bottom image) are thus 100% germline produced when female (bottom individual) but less than 100% when male (top individual). Note that none of this applies in diploid termites where lifetime unmatedness automatically implies complete sterility. Republished with permission of the Royal Society, from Lifetime monogamy and the evolution of eusociality, J.J. Boomsma, Philosophical Transactions of the Royal Society, Volume 364 (2009); permission conveyed through Copyright Clearance Center, Inc. Photo credits by the author except for the bottom *Atta* image (Boris Baer) and the top and bottom *Acromyrmex* images (Klaus Lechner with permission from the AAAS, and David Nash, respectively).

queens when resource acquisition constraints are severe (Peeters and Ito, 2015). The general impression is that there are many intriguing germline-soma similarities, but few solid analogies in the sense of Box 1.4, which may be due to (1) superorganisms having no equivalent of dormant and sequestered stem cells—their "stem individuals" need to be continuously produced as eggs; and (2) colony death in monogamous superorganisms being induced primarily by germline senescence and not by failure of the somatic workers (Boomsma, 2013). Superorganismal germlines thus appear to be "farmed" by the worker soma until they drop dead and they can even be actively discarded, both in honeybees (Box 6.1) and in annual bumblebees and yellowjacket wasps, when it serves the worker fitness interests (Bourke, 1994; Loope, 2015). Whether menopause is an analogous germicide syndrome continues to deserve active research—see Stearns and Medzhitov (2016) and Cant and Croft (2019) for recent reviews.

A second topic that allows illustrative comparisons between permanently multicellular organisms and colonial superorganisms *sensu* Wheeler (1911) is *immune defense*—a crucial activity under continuous natural selection and obviously dependent on the level of organismal closure, be it superorganismal, multicellular, or cellular. At each level immune defenses are mediated by self/nonself discrimination, detection of potential danger, and specificity, based on relative rather than absolute criteria (Guerrieri et al., 2009; Pradeu et al., 2013; Tranter et al., 2014; Eberl and Pradeu, 2018). The question is, therefore, whether higher-level closure of superorganismal colonies made social immune systems evolve that are distinctly more efficient than those in open animal societies of comparable size. To make such comparisons, it is useful to distinguish between short-lived versus long-lived social systems, and between

those that operate in two-dimensional and three-dimensional space. Selection for efficient social immunity in superorganisms with annual life cycles (bumblebees and yellowjacket wasps) will be weak particularly in the months preceding predictable seasonal colony death, so comparisons should focus on perennial societies and superorganisms. However, only some swarm-founding polistine wasps might qualify as large perennial insect societies, and they do not appear to have been surveyed to quantify disease pressure. Among the convergently evolved superorganisms, infection pressure is inevitably higher in the bees that forage on the wing without maintaining territories than in the ants and termites that forage on foot and usually have non-overlapping foraging ranges (Boomsma et al., 2005b). As expected, perennial bee colonies suffer from a number of specialized chronic pathogens while the ants and termites hardly have such diseases (Schmid-Hempel, 1998; Boomsma et al., 2005b; De Bekker et al., 2018). However, the complexity of these cross-lineage comparisons is huge, and we still have few well-studied model systems two decades after these conjectures were made (but see Milutinović et al., 2020; Alciatore et al., 2021).

To understand why natural selection may have discouraged potentially virulent pathogens to specialize on perennial ant and termite superorganisms, it is instructive to realize that their worker soma can confront omnipresent disease propagules far away from the colony germline, while individual metazoans that each maintain their personal germline, even when living in societies, cannot. Behind this simple truth are some staggering numerical differences. It is useful to qualitatively compare organismal bodies and superorganismal colonies by drawing them on the same scale (Figure 6.5a), but it is also important to realize that there are ca. 10^{10} obligatorily altruistic somatic cells in a mouse body of 30 g, while an ant or termite colony of the same mass would have ca. 10^4 workers and would occupy a nest volume three orders of magnitude larger than the mass of a mouse body. In addition, each disease-fighting ant or termite worker has a brain and six legs and the ability to freely move around in interstitial nest space (Figure 6.5b) guided by self-organized naturally selected instincts, rather than being fixed in some organ. It thus seems obvious that colonial superorganisms might have evolved better abilities to eliminate infections than societies where cooperation is condition dependent (Figure 6.5c), or single animal bodies where altruistic cells are fixed in tissues or defenses are constrained by what circulation systems can reach (Figure 6.5c).

Apart from internal disease transmission, also the infection rates between colonies matter for the likelihood of achieving cross-colony herd immunity. Foraging on the wing in three-dimensional landscapes must preclude reaching that result for a substantial number of pathogens, possibly explaining that honeybees have many viral diseases (McMahon et al., 2015) that ants and termites seem to lack unless they secondarily evolved unusual levels of colony connectedness (Brahma et al., 2022). Non-overlapping two-dimensional territories in ants and termites may thus explain that these lineages appear to be almost free of virulent infectious diseases.

It is also relevant to compare the principles of single-body immune defense and collective social immune defense (Cremer et al., 2007) beyond the analogies presented in Figure 6.5a,b. Animal immune defenses are both intracellular and intercellular (humoral) and have, particularly in vertebrate bodies, specificity and long-term memory (Eberl and Pradeu, 2018). These characteristics have been confirmed to have analogies in the social immune defenses mounted by ant colonies (Rosengaus et al., 2013; Tranter et al., 2014;

Cremer et al., 2017; Konrad et al., 2018)) and even to some extent in plants (Jones et al., 2016). It is also increasingly acknowledged that animal immune systems encompass immune tolerance (Miller et al., 2006) and that immune defenses appear to react to discontinuous changes in organismal homeostasis rather than being informed only by the self/non-self identities of cellular or molecular fragments (Pradeu and Vivier, 2016). Asking whether there are superorganismal analogies of these immune-defense priorities is feasible in colonies of ants (e.g., Stroeymeyt et al., 2018) and would likely be equally rewarding in termites. However, for practical reasons, termite research has largely focused on non-superorganismal lower termites that lack foraging territories (Rosengaus et al., 2011) and have immune defenses that go rather little beyond mutual grooming and nest sanitation (e.g., Korb et al., 2012; Cremer et al., 2017; Cremer, 2019). Comparing disease pressure in cooperatively breeding lower termites and superorganismal higher termites seems a timely agenda because transcriptomic analyses indicate that superorganismal immune defenses in fungus-growing termites are so advanced that these colonies are essentially free of any disease (Otani et al., 2019).

Like individual immunity in a multicellular animal, superorganismal social immunity requires a set of highly interactive and communicative somatic interactions. In solitary animal bodies, extensive cell–cell signaling detects infections and initiates defensive responses, which include advanced discrimination of aberrant cells by killer cells and the expression of adaptive cellular suicide (apoptosis). Superorganismal colonies maintained these individual defenses, but they added an entire higher-level layer of integrated social defenses involving disposable individuals (Aanen, 2018). Optimal operation of any set

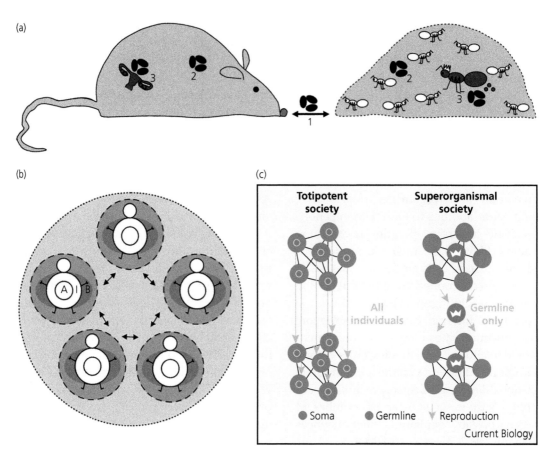

Figure 6.5 Immune defense analogies between metazoan organisms and colonial superorganisms. (a) Cremer and Sixt (2009) visualized the conceptual analogy between organismal and superorganismal immunity, comparing the body of a mouse with a colony of ants. Both levels of organizational complexity need to have multilayer defenses against pathogens (small black ellipses), preventing them from (1) entering the (super)organism (border defense), (2) damaging or annihilating the peripheral cellular or worker-individual units (soma defense), and (3) damaging or killing the cellular or multicellular germline units that produce the gametes or the female and male reproductives. (b) A cross section of pathogen defenses at the superorganismal level, showing that social immunity resides in a nested set of modules each with its own brain and six legs. The outer dotted circle is the outline of the colony in panel (a), while the nested colony members have their own hierarchy of defenses starting with avoidance behaviors (B) and then supplementing with innate immune defenses (I) and acquired immune defenses (A), which all combine to interact with similar defenses mounted by other colony members (see Cremer et al., 2007, for details). The extra level of functionally closed superorganismal organization can therefore make combined immune defenses extraordinary efficient. (a+b) Republished with permission of the Royal Society, from Analogies in the evolution of individual and social immunity, S. Cremer, M Sixt, Philosophical Transactions of the Royal Society B, Volume 364 (2009); permission conveyed through Copyright Clearance Center, Inc. (c) The efficiency of social immunity is not merely a function of relatedness between colony members but primarily depends on informational closure, that is, on the iteroparous superorganismal commitment of specialized colony members acting as soma to the colony germline (Cremer, 2019), which continues to apply also when multi-queen polygyny has evolved secondarily. Only under these conditions can there be selection for the obligate traits that mediate forms of adaptive suicide during disease defense, a kind of altruism that is not or only rarely expected in open (i.e. not informationally closed) societies of cooperatively breeding animals where all or most individuals maintain their own germlines. It is therefore of key importance to distinguish between the highly integrated purely somatic social immune defenses of superorganismal colonies (right) and the cooperative hygienic behaviors among totipotent parents, offspring, or siblings living in societies (left). The latter unfold as public-good games of mutualistic give-and-take between partners and generations (Duarte et al., 2016; Nuotclà et al., 2019) and lack ultimate self-sacrifice elements because cooperation is almost always condition-dependent. While the perennial superorganismal ants and termites with their mutually exclusive two-dimensional territories have evolved truly advanced social immune systems, promiscuous foraging

(Continued)

of immune defenses requires fine-tuned and mutually inhibitory checks and balances (Eberl and Pradeu, 2018), higher-level analogies of which have been abundantly found in ants that engage in proactive hygiene (Morelos-Juárez et al., 2010), choose between curing and destroying sick workers or brood (Tragust et al., 2013; Frank et al., 2018; Pull et al., 2018), practice self-removal of diseased workers from the colony (Chapuisat, 2010; Heinze and Walter, 2010; Giehr and Heinze, 2018; Conroy and Holman, 2022), and actively manage and disinfect corpses of nestmates (Hart and Ratnieks, 2002; López-Riquelme and Fanjul-Moles, 2013; Kesäniemi et al., 2019). These examples illustrate that a new level of organizational complexity, realized by a higher level of MTE closure, can be kin selected to evolve a matching layer of higher-level immune defenses (Pull and McMahon, 2020). While it is generally true that denser populations can, all else equal, maintain more epidemic and endemic diseases (Anderson and May, 1985; Hamilton, 1987a), this correlation might be reversed when high densities are structured in non-overlapping superorganismal units whose homeostatic control extends into territories that can be monopolized because there is no "routine air traffic". This implies that the ecological success of ants and termites relative to the perennial corbiculate bees is perhaps not surprising.

A third topic of interest is whether superorganisms *have a higher-level developmental biology*, as Wheeler (1911) predicted. If closure and germline–soma differentiation are conceptually comparable across metazoan bodies and colonial superorganisms, it seems reasonable to expect that caste phenotypes and cell phenotypes may differentiate according to similar mechanistic rules (Patalano et al., 2012), despite mismatches related to germline versus somatic aging and the presence or absence of higher-level stem cell analogues as summarized above. As Haig (2016) noted, an animal zygote produces a "germ-stem" of cells that both have daughter cells that become soma (i.e., segregate) and cells that remain in the germ track to ultimately produce gamete daughter cells (i.e. differentiate). Considered that way and loosely defining ploidy as haplotype participation so that mitochondria are included, an informationally closed ant or termite royal pair is a 5–6n merger of lower-level zygotic information (see Figure 5.3), producing 3n or 2n offspring destined for lifetime next-generation mergers. What distinguishes the organismal and superorganismal complexity levels is that differentiation into a gyne or worker caste phenotype

Figure 6.5 (*Continued*)
in three-dimensional space appears to have constrained the superorganismal bees and wasps to either remain annual (bumblebees and yellowjacket wasps) or to replace single-queen colony founding by swarming and thus to adopt regular germline replacement, as we find in stingless bees, honeybees, *Provespa* wasps, and even non-superorganismal epiponine wasps (Boomsma et al., 2014). The extant annual superorganisms appear to merely out-reproduce their pathogenic burden (Boomsma et al., 2005b; Ruiz-Gonzalez et al., 2012), consistent with a recent study showing that invasive yellowjacket wasp colonies that do survive beyond a single year quickly perish from viral loads even when temperatures and food availability are favorable (Loope and Wilson Rankin, 2021). As it appears, these systems have the plasticity to become perennial (Dyson et al., 2022), but such abilities are not elaborated by natural selection because disease pressure cannot be met by adequate social immunity. Reprinted from Social immunity in insects, S. Cremer, *Current Biology*, Volume 29, R458-R463 (2019), with permission from Elsevier.

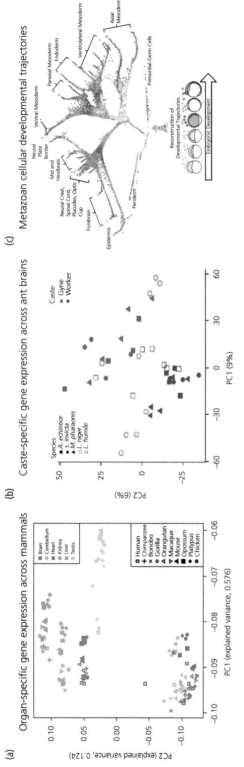

(a) Organ-specific gene expression across mammals

(b) Caste-specific gene expression across ant brains

(c) Metazoan cellular developmental trajectories

Figure 6.6 Organ-specific and caste-specific development. (a) Comparative messenger RNA expression level analysis has shown predictable variation among homologous organs nested within vertebrate species. Reprinted by permission from Springer Nature: *Nature*, The evolution of gene expression levels in mammalian organs, D. Brawand et al., (2011). (b) Similar segregation patterns were obtained between caste-specific brain transcriptomes of adult gynes and workers of five ant species. This study reconstructed the ancestral GRN dating back to the common ancestor of the ants that realized the MTE towards permanent caste differentiation. It further showed that only some of the differentially expressed genes overlapped with the GRN that segregates gyne and worker brains in the honeybee, and that directions of caste-biased transcription between ants and honeybees were uncorrelated as expected when the ant and corbiculate bee MTEs have independent origins. It further emerged that the ant GRN for caste differentiation did not predict any segregation of breeders and helpers in *Polistes canadensis* (Patalano et al., 2015), consistent with paper wasps merely having adult reproductive role differentiation and no pre-imaginal caste differentiation. The brain transcriptomes of ant species that secondarily evolved gamergate (inseminated worker) replacement reproductives no longer express the original ant GRN for caste differentiation; they use an alternative and likely older GRN for reproductive role differentiation triggered by insemination (Nagel et al., 2020). Reproduced from *Towards reconstructing the ancestral brain gene-network regulating caste differentiation in ants*, B. Qiu et al., *Nature Ecology and Evolution*, Volume 2 (2018), with permission of Springer Nature. (c) Developmental trajectories of cell lines during zebrafish embryogenesis, showing that primordial germ cells are the first to enter differentiated and canalized development from pluripotent stem cells, after which somatic differentiation proceeds along other trajectories that are, at least initially, less developmentally canalized. Republished with permission of AAAS, from *Single-cell reconstruction of developmental trajectories during zebrafish embryogenesis*, J.A. Farrell et al., *Science*, Volume 360 (2018); permission conveyed through Copyright Clearance Center, Inc. The original superorganism hypothesis by Wheeler (1911) should imply that similar canalization processes characterize the developmental trajectories of gynes and workers in colonies of ants, bees, wasps, and termites that have evolved from independent MTEs to superorganismality, which appears to be the case in at least some ants (Qiu et al., 2022).

has to happen separately in each meiotically produced egg or larva, while differentiating cell lines in multicellular organisms can suffice with mere division and epigenetic differentiation. Despite this difference, Williams offered a Darwinian definition of ontogenetic development that was accurate enough to also apply to the superorganismal level. As he wrote, "[o]ntogeny is often intuitively regarded as having one terminal goal, the adult-stage phenotype, but the real goal of development is the same as that of all other adaptations, the continuance of the dependent germ plasm" Williams (1966a, p. 44). All that differs is that we are now talking about development towards an integrated and differentiated colony phenotype, as noted by Wheeler in the opening quote of this chapter (see also Lehtonen and Helanterä, 2020).

Trusting their naked eye common sense, naturalists never doubted that superorganismal colony development—from founding, via ergonomic growth, and then (unless annual) iteroparous alternation between reproductive and somatic investments (e.g., Oster and Wilson, 1978, Figure 2.2)—has been shaped by natural selection and continuously finetuned during adaptive radiations. More than half a century after Williams' (1966) book, we now investigate aspects of ontogeny in metazoan bodies with genome-wide transcriptomes, which has generated unprecedented resolution in mapping early embryo development in zebrafish (Farrell et al., 2018; Wagner et al., 2018) and frogs (Briggs et al., 2018). Such studies have shown that organ-specific gene expression spectra have remained remarkably conserved (Figure 6.6a). When exploring possible analogies in superorganisms, the obvious organs to sequence are the brains that coordinate development (McKenzie et al., 2016; Godfrey and Gronenberg, 2019). A recent study comparing gyne-worker gene expression differences in brains showed that genome-wide

GRNs (Table 3.1) for caste differentiation are likewise conserved across five ant species (Qiu et al., 2018) (Figure 6.6b), but that there was no such correspondence with the honeybee (an evolutionary convergent MTE) and a paper wasp (which experienced no MTE). Single-cell sequencing has also increasingly revealed the developmental canalization of vertebrate cell lines following the first cell divisions (Figure 6.6c), consistent with early epigenetic landscape visualizations by Waddington (1956, 1958). Because GRNs are the key informational matrices for understanding developmental differentiation (Wagner, 2005, 2011), and are instrumental for different developmental programs across body parts (Wagner, 2014), we should expect analogous patterns to hold up for caste differentiation trajectories if Wheeler's (1911) argument is correct. A recent study has shown that this appears to be the case, at least in ants (Qiu et al., 2022).

The pursuit of gene regulatory networks that mediate major transitions

The empirical evidence compiled in this chapter is broadly consistent with the hypothetical explanatory framework for fraternal MTE origins that I offered in Chapter 4 and with the neo-Darwinian organismal biology tradition reviewed in Chapters 1–3. However, much further work is needed to challenge these contentions. That research will have to combine comparative natural history and molecular biology, both to obtain more complete phylogenies with higher resolution of nodes, and to exploit genomic technologies to reconstruct and compare the ancestral GRNs that forged the independent fraternal MTEs. I predict that these ancestral GRNs for permanent germline–soma differentiation will all be unique innovations that lack network homologies backward in time and that will have conserved their key novel network homologies as synapomorphies forward in time. Reconstructing these ancestral GRNs will be less challenging for superorganismality MTEs than for multicellularity MTEs, because the latter's coalescence times are much longer. However, once these reconstructions have succeeded, I expect these MTE-defining GRNs to become the foundation for understanding the stepwise functional innovation of life's organization, an assignment that has already been pioneered for metazoan GRN homologies (Wagner, 2014). Advances of this kind may allow metazoan biology to proceed beyond the phenotypic gambit (Grafen, 1984) in a way that has already been partly laid out in a Tinbergen complementary questions framework (Springer et al., 2011). Obtaining the germline-soma division of labor GRNs behind MTEs may thus turn out to carve nature at its most fundamental proximate causation joints for understanding hierarchical and convergent evolution of organizational complexity.

Large-scale comparative transcriptome studies exploring the multicellular domain (e.g., De Mendoza et al., 2013; Sebé-Pedrós et al., 2017) have shown that animals and embryophyte plants have richer transcriptional regulation toolkits than fungal and algal lineages (Figure 6.2). However, once studies start focusing on GRNs, toolkit terminology may become outdated, because gene participation per se is unlikely to predict the direction of differential gene expression within a germline-soma GRN. Although still an abstraction of a very compound reality, the GRN concept (see Table 3.1) maintains that complex phenotypic traits are not encoded by homologous genes per se but by plastic assemblies of metabolic pathways and their interactions, mediated both by conserved and novel genes (e.g., Wagner, 2005). Applying this approach to the superorganismal level has only just began, but the proof-of-concept

study by Qiu et al. (2018) (Figure 6.6b) indicated that network-based causation may subjugate toolkit gene assessments just like statistical path analysis may reverse single correlations when new predictors are added. Deeper comparative understanding will also need to address that MTEs have domesticated and functionally modified GRN homologies of previous domains. Complications of this kind have remained unrecognized in social insect studies using toolkit terminology (e.g., Rittschoff and Robinson, 2016), because "eusociality gradient" thinking failed to acknowledge the discontinuous superorganismal MTEs (blue dots in Figure 6.1c). Lineages such as *Apis* bees, *Polistes* wasps, and *Solenopsis* ants have been assumed to possess parallel patterns of toolkit gene expression, but these proximate comparisons have produced little confirmation (e.g., Hunt and Amdam, 2005; Page and Amdam, 2007; Toth and Robinson, 2007) even when considering pathways rather than single genes (Berens et al., 2014).

Once more, we are drawn back into eusociality semantics having blurred the distinction between Darwinian *understanding* and the phenological *interpretation* of pattern. For example, *Drosophila*, *Dictyostelium*, and *Arabidopsis* are all multicellular in the unspecified sense, but *Drosophila* and *Arabidopsis* represent two independent MTEs towards permanent germline–soma organismality, functionally similar to the convergent origins of *Apis* and *Solenopsis* superorganismality. Likewise, *Dictyostelium* represents a cell society lineage that did not realize an MTE, and is thus functionally analogous to *Polistes* societies that were not ancestral to any superorganismal clade. We need to ask, therefore, why social insect researchers have continued to pursue a reproductive ground-plan hypothesis (West-Eberhard, 1987) as if higher-level germline–soma differentiation must involve the same genes and pathways across all social insects, while geneticists studying transitions towards multicellularity would be reluctant to make such claims when lineages are separated by independently evolved MTEs. If facultative and obligate reproductive altruism always evolved independently, as my distinction between societies and (super)organisms implies, a generalized reproductive ground-plan appears to be an invalid concept for large scale phylogenetic comparison.

No matter what level of organizational complexity one studies, questions of convergent proximate causation should remain relevant for Darwinian understanding of adaptive design because, as Darwin phrased it, "natural selection is continually trying to economise in every part of the organisation" (Darwin, 1859, p. 186). Such comparisons need to have a precise embedding of phylogenetic history rather than being based on phenological terminology that has, in its original Greek meaning of "bringing to light," no ambition beyond interpreting patterns of variation. Phenology-based approaches cannot explain social adaptations unless they are combined with the Popperian hypothesis-testing discipline of statistically controlled comparative analysis to evaluate predictions from inclusive fitness theory. As it appears, toolkit and reproductive ground plan thinking represents zoomed-out least common multiplier thinking, while a comparative understanding of evolutionary innovation and the preservation of homologies requires the pursuit of greatest common divider explanations (Box 1.3).

Throughout this book, I have used a Darwinian framework to pursue testable conjectures for phenotypic social traits that can respond to selection—that is why Table 6.1 took the form it did. There are alternative approaches, for example, a key figure in Queller and Strassmann (2009) that plots organisms on axes of low versus high conflict and low versus high cooperation without distinguishing between the four fundamental levels of organismality (see Figure 3.2). Such an approach produces intriguing patterns but does not bring us closer to understanding Darwinian causation because MTEs and other lineage-specific idiosyncrasies are glossed over. This does not give conceptual incoherence as long as it is made clear that the very purpose of the analysis is a zooming-out survey to capture pattern and has no bearing on any process. However, an example of a similar figure is found in a book by Godfrey-Smith (2009) where it becomes misleading because the title is *Darwinian Populations and Natural Selection*. The book does not report a single pattern that was not already understood in a more parsimonious way by Darwinian argumentation and neither does it offer novel testable predictions. Nevertheless, it aims to upgrade a version of implicit and fully environmentalized Darwinism to become a new scientific philosophy without any basis in first-principle theory. These general problems ultimately connect to equivocal definitions of genes and environments, as I will return to in Chapter 8. But before considering adaptation as an abstract information accumulation process, it is important to evaluate our empirical knowledge of social evolution in the unicellular domain. In Chapter 7, I will summarize the organismal biology of single cells, the Hamiltonian principles by which cell societies evolve, and our present insights in internal division of labor within free-living cells, a topic once more already recognized as pivotal by Huxley (1912).

The free-living prokaryotic and eukaryotic cells

Summary

The last universal common ancestor of cellular life (LUCA) and the last eukaryote common ancestor (LECA) were unique events that committed previously independent replicators to joint vertical transmission although occasional horizontal gene acquisition remained possible. Here I elaborate the explicitly organismal narrative that the organizational complexity of simple prokaryote cells and complex eukaryote cells was based on subsequent major transition events of cell closure and division of labor among cellular elements. My arguments reconcile Huxley's (1912) principle that nothing alive can be functionally homogeneous with Williams' dictum that genetic homogeneity is a necessary condition for maintaining organizational complexity. Echoing arguments brought forward intermittently since the late 1800s, I question the validity of the cell as machine metaphor because that reductionist approach addresses only proximate aspects of causation and cannot explain the self-organized and self-referential aspects of cellular maintenance and reproduction as they are shaped by natural selection. I also review the substantial insights obtained from studies of societies of bacterial and protist cells. They vindicated inclusive fitness theory and are now being extended to address viral social behavior, but they have no direct relevance for understanding the major cellular and multicellular transitions in evolution (MTEs) that irreversibly enhanced organizational complexity. Finally, I evaluate the scattered evidence for germline–soma differentiation within unicellular organisms, which increasingly confirms that these domains also have forms of reproductive division of labor and differential rates of aging.

> It is customary to think of the egg cytoplasm as the soma of a zygote, but there may be elements in this cytoplasm that are more or less a part of the genetic message. [. . .] The smallest protist is an endlessly intricate machine, with all parts contributing harmoniously to the ultimate goal of genetic survival. Any such distinct individual is expected to be genetically homogeneous. We do not expect to find genetically different individuals cooperating in a single somatic system.
>
> **Williams, *Adaptation and Natural Selection* (1966a, pp. 61, 221)**

These quotes illustrate how incisively parsimonious the concept of organismality used to be. Writing just before Lynn Margulis correctly conjectured that the eukaryote cell had a symbiotic origin in which different prokary-

otes had joined forces to produce what we now refer to as the last eukaryote common ancestor (LECA) (Sagan, 1967), Williams identified what truly mattered in a functional sense. He saw that there is no reason to deny that

Domains and Major Transitions of Social Evolution. Jacobus J. Boomsma, Oxford University Press. © Jacobus J. Boomsma (2022).
DOI: 10.1093/oso/9780198746171.003.0007

single cells can have germline–soma segregation and he anticipated that every independent cell should have non-redundant genetic homogeneity within each of its co-replicons (see Table 3.1; Box 2.2), a concept that needed explicit development after the inheritance of organelle DNA was discovered (Cosmides and Tooby, 1981). Williams' emphasis on a cell's harmoniously cooperating parts echoes Huxley's (1912, p. 61) poetic narrative of life finding "in the cell the ground-plan for her first mansion—a one-roomed hut," a metaphor he used to establish the principle that "living matter always tends to group itself into [...] closed, independent systems with harmonious parts." Williams supplemented that narrative with the insight that internal division of labor must be associated with genetic homogeneity, a point elaborated in this book via the non-redundant commitment principle of lifetime enclosed genetic information (LTC) as a necessary condition for the origins of major transitions in evolution (MTEs). While capturing a huge amount of fundamental biology, Williams did not overplay his hand of understanding. He left open that germline and soma may not be fully segregated and that they may differentiate via dissimilar mechanisms across lineages of varying ancestry and organizational complexity. For all these reasons, this epigraph seemed a proper bridge between the somatization of cells and multicellular individuals in Chapter 6 and exploration of the possible validity of developmental somatization of cell parts in the unicellular domains of social evolution.

There have been two uniquely singular MTEs, one to cellularity itself and another to complex cellularity. We now identify the results of these events as the last universal common ancestor of all cellular life (LUCA), and LECA, the ancestor of all extant eukaryote lineages. LUCA's descendants split into Archaea and Bacteria, a divergence that did not involve an MTE but produced the lineages of which single representatives would later join their complementary forces in LECA (see Figure 3.2a). The singular origins of the LUCA and LECA cells preclude the horizontal comparisons that helped understanding the convergent fraternal MTEs and their sister lineage societies that failed to realize MTEs (see Chapter 6). Here I will argue that inclusive fitness theory has been very successful in explaining the social dynamics of cell societies, but that it has remained underappreciated that chimeric redundancy must have prevented that such societies could realize MTEs towards irreversibly higher organizational complexity. I will also briefly highlight that Hamiltonian logic for explaining condition-dependent cooperation in cell societies can also be applied when viral particles form societies. In the final section I will tentatively explore the proximate aspects of ontogenetic development in prokaryote and protist organisms, including the remarkable fact that unicellular life does not appear to have precluded aging. These explorations will suggest that perfect symmetry in cell division is as illusionary as what came to be known as Weismann's dogma of the immutable germline, and that the two phenomena may well be connected. These reflections will finally make it natural to consider how and why cytoplasmatic elements could somaticize as adaptive parts of LECA's organismality, and how LUCA's origin might have been analogous.

As in some of the previous chapters, it is useful to briefly dwell on deeper history, this time of organismal cell biology. The comparative study of unicellular eukaryotes represented cutting-edge biology for Ernst Haeckel and August Weismann in the 19th century (Reynolds, 2008; Churchill, 2015), and the prominent position of protist biology continued into the first decades of the 20th century. In those days, the global scientific elite of zoology was small and generally well aware of each other's work, be it on cell biology, comparative embryology, vertebrate systematics, or social insects. The *Stazione Zoologica* in Naples was an international hub where Weismann spent time, as did Wheeler and Huxley (see Chapter 3), and where Charles Otis Whitman, Wheeler's mentor, met Weismann (Churchill, 2015). In the 1880s, Weismann also paid visits to the UK, where his stringent neo-Darwinism received active support from Wallace and academic biologists, including some of the Oxford scholars (Churchill, 2015). The bewildering variation in protozoan nuclear cell division inspired Weismann in this period, for it allowed him to support his germ plasm theory and to break with the static way in which Haeckel had interpreted protozoan biology (Reynolds, 2007, 2008). It is therefore no surprise that young Julian Huxley's undergraduate training in Oxford made him well versed in the comparative neo-Darwinian study of protozoan cells and metazoan embryos, and that he was able to integrate much of that in his *Individual in the Animal Kingdom* book (Huxley, 1912, 2022) (see Chapter 3). As it appears, many of the professional organismal biologists active between 1880 and 1914 were broadly read natural historians able to recognize functional organismal agency all the way from ciliates to ant colonies. While Wheeler (see Box 3.1) and Huxley (see Box 3.2) stood out, they were also representative of their times.

While protist biology thrived, it was not until the 1930s that prokaryote microbiology grew to maturity, facilitated by improved microscopy and culturing techniques that also allowed viruses to be identified as macromolecular particles (Harold, 2014). An explosive descriptive understanding of the staggering diversity of microbial symbiont diversity followed (Buchner, 1953), but it took another decade before a foundational synthesis of organismal prokaryote biology was provided by Stanier and van Niel (1962) in an article entitled *The concept of a bacterium*. This paper offered stringent mutually exclusive definitions of a bacterium, a virus, and a eukaryotic protist, based on their distinct structural and functional organizational complexity. The Stanier and van Niel (1962) definition of Bacteria also seamlessly fits the later discovered Archaea (Woese and Fox, 1977), which justifies that we continue to use the term Prokaryotes for both in spite of the fundamental differences between them (Harold, 2014). Stanier and van Niel made the important points that (1) the circular bacterial chromosome is haploid and spatially segregated from the ribosomes; (2) new genes can be inserted in such chromosomes at any point, but reciprocal recombination is absent, precluding any deeper comparison with eukaryote sex and meiosis; (3) bacteria practice binary fission of cells, but mitosis is a derived eukaryote trait that would not work for a prokaryote chromosome attached to a cell membrane; (4) the bacterial cell as a whole constitutes the irreducible site of respiratory and (in cyanobacteria) photosynthetic metabolism; and (5) there are clear functional homologies between free-living bacterial and intracellular mitochondrial cell membranes on the one hand, and between the intracellular lamellae of cyanobacteria and eukaryote chloroplasts on the other. However, bacterial flagella and axial filaments are not homologous with the contractile locomotor organelles of eukaryote cells.

THE FREE-LIVING PROKARYOTIC AND EUKARYOTIC CELLS

Reading Stanier and van Niel (1962), one almost expects them to propose the symbiotic origin of the eukaryote cell, but they failed to do that. This major insight that forever changed our perspective on the early evolution of life's cellular complexity occurred to Lynn Margulis a few years later (Sagan, 1967), reviving earlier experimental understanding by Ivan Wallin (1927) whose work had been dismissed by Buchner and ignored by Stanier and van Niel. However, their review makes it unsurprising that Margulis' conjecture of the symbiotic origin of the mitosing cell would prove to be correct for mitochondria and plastids, as Wallin had originally proposed as well, but not for flagella (Harold, 2014, p. 120). More broadly, these flashbacks emphasize that the 1960s were a highly productive decade for organismal biology, not only because of Tinbergen's (1963) four complementary angles of biological inquiry and the gene's eye view of adaptive evolution by Hamilton and Williams (see Chapters 1 and 2), but also because scholarship illuminated the principles of organization in the microbial domains. The fundamental nestedness of life's organizational complexity, as first proposed by Huxley (1912), remained so canonical through to the 1960s that it rarely seems to have required diagrammatic illustrations like Figure 3.2. It is puzzling that this profound understanding was partly overwritten by ad hoc definitions of complexity, sociality, and individuality that were never shown to be valid concepts derived from first-principle understanding of natural selection and descent with modification. A case in point is the erroneous idea that bacteria and viruses have an equivalent of meiotic sex just because there is frequent horizontal gene transfer (HGT) or hybridization. The cellular domains of social evolution thus appear to have experienced reduced conceptual clarity of organismal thinking that resembles

the semantic confusion about multicellular organisms and colonial superorganisms (see Chapter 6), and also here the adoption of mere phenological perspectives seems to have been a major reason.

In Darwin's time, the scientific understanding of cells was exclusively structural and largely driven by German scholars, leading to what has been referred to as an atomistic concept of the cell (Reynolds, 2007; Nicholson, 2010). This early thinking about fundamental indivisible cell units only addressed eukaryote protists and used Virchow's principle of *omnis cellula e cellula* (every cell emerges from a cell) as justification for atomism (Reynolds, 2007; Harold, 2014). It was Charles Otis Whitman who first challenged this view because it did not acknowledge that eukaryotic cells in multicellular bodies also represent higher levels of organization already contained in the egg and fertilized zygote before they unfold during development (Whitman, 1893). Inferences like this were often seen as metaphysical and even dangerous because Virchow and Haeckel (but not young Julian Huxley) had mistakenly argued that human states should resemble multicellular bodies (Reynolds, 2008, p. 131). However, all Whitman and his mentee Wheeler (Evans and Evans, 1970) asked for was due attention to development as a phenomenon that could not be reduced to atomistic structural predictors. It is in this light that we need to understand Wheeler's (1910, p. 7) analogy between multicellular animals and colonial superorganisms as "[t]he queen mother of the ant colony displays the generalized potentialities of all the individuals, just as the Metazoan egg contains *in potentia* all the other cells of the body." It is unfortunate that appreciations of this kind were often dismissed as inappropriately Kantian/Platonic (Reynolds, 2008, p. 140) in the years between the World Wars, because they clashed with the perceived all importance of reductionist transmission genetics as established by T.H. Morgan (Amundson, 2014).

As Whitman (1893) pointed out, cell theory should imply that "the organism is fancied to carry at least two distinct organizations, the organization of the separate cells and that of the cells united. The higher organization thus differs, *qualitatively*, from the lower, so that we may have analogies, but no homology of organs between unicellular and multicellular organisms." He maintained that presently observable nestedness of structural levels tells us nothing about how distinct functional levels of organization came into existence, because "organization precedes cell-formation and regulates it, rather than the reverse." Crucial parts of the organizational complexity that unfolds during development can thus not be reduced to lower-level explanatory paradigms, consistent with my definition of the four domains of organizational complexity having mutually exclusive homologies of organization because they are separated by MTEs. Whitman was right in establishing this principle, but wrong in dismissing cell theory altogether as pointed out by Huxley (1912) (see Box 3.2). Much later, Whitman's mentee Wheeler was the first to conceptualize the organizational differences between any domains of social evolution by arguing that there are (1) ultimate characteristics of organismal order that cannot be deduced from what applies at a lower level of order; (2) reducible characteristics of order that can, at least in principle, be reduced to the properties of lower-level order; and (3) neutral properties, which are simply shared between all levels of order (Wheeler, 1926). The first category implies that a fraction of domain-specific organizational complexity should become irreversibly complex at each MTE origin, consistent with the nested white zones in Figure 3.2b and the

fact that MTE-specific complexity cannot be secondarily reversed to lower-level order (see also Chapter 8).

I have related this early history because the organismal perspective of cells remains highly relevant today. In a recent monograph, *In Search of Cell History* by Harold (2014, p. 10), we find the tension between top-down organismal and bottom-up molecular biology spelled out when he writes that "[g]enes and genes alone, are replicators; the information they contain is potentially immortal, and it dictates all aspects of biological structure and function. By contrast, organisms are transient and disposable, vehicles in which genes ride and confront the judgement of natural selection." But just a few pages later we read that "I am but one of a cohort of skeptics who suspect that the gene-centered view is incomplete and fundamentally misleading" (Harold, 2014, p. 12), and this contradiction is then shown to be a mere paradox by the statement that "[e]clipsed in the 20th century by their molecular constituents, organisms have come to be seen as little more than epiphenomena of their genes, and relegated to the backwaters. Some have even questioned whether biology still needs the concept of organisms! To my mind, this is reductionism gone berserk. [...] Organisms are indeed central to what Richard Dawkins called the 'greatest show on earth': their nature and origin are what we strive to explain" (Harold, 2014, p. 36). These passages illustrate the common misconception that the gene's eye view of adaptive evolution is fundamentally reductionist, a notion that careful reading of Dawkins (1976, 1982a, 1982b) quickly dismantles. There is very little in the gene's eye perspective that demands phenotypic agency of single genes. The reductionism criticized here came almost entirely from molecular genetics and was a prime driver of the disappearance of evolutionary thinking from cell biology after the 1950s (Reynolds, 2007; Nicholson, 2019). This change of focus had the unfortunate consequence that the "baby" of natural selection as adaptively organizing force was thrown out with the organismal "bath water".

When Williams (1985) defended reductionism in evolutionary biology, it was from an organismal point of view that seeks first-principle understanding (see Table 3.1) of how natural selection acting on overall genetic information shapes complex phenotypes including their development—not to defend single-gene agency. Meanwhile, however, purely proximate "model system" views of life penetrated organismal biology while some used Darwin's tenet of evolution by infinitesimally small steps to argue that reductionist approaches would ultimately prevail no matter what. Written as an attempt to reboot organismal biology in parallel with molecular biology (Gibson, 2013), Wilson's (1975) *Sociobiology* offered to its readers a vision in which all forms of social complexity fitted into a single smooth gradient from microbes to men, driven by—as the first chapter suggests—"[t]he morality of the gene." That idea misinterpreted the gene's eye view of Hamilton and Williams and is at odds with this book's analyses. As Amundson (2014) explained, T.H. Morgan's approach founded a school of genetics averse to development—a clear contrast to the teleonomic neo-Darwinian perspective that continued to acknowledge the importance of organismal ontogeny. Morgan's focus decisively influenced how genetics developed in the US—as a novel reductionist philosophy embraced by many, including Huxley (1942), and rejected by scholars like Wheeler (see Chapter 3). This may well explain much of what I described as the conceptual divide across the Atlantic (see Chapter 2), with UK Darwinians adhering to Fisher's statistical and quantitative (i.e. codex) understanding

of genetics to explain adaptive evolutionary change, and US geneticists following Sewall Wright in studying total change in genotype frequencies. As developments accelerated during the second half of the 20th century, the reductionism–holism contrast became particularly visible in cell biology.

A reappraisal of organismal cell biology and naturally selected organizational complexity

When evaluating organizational complexity in the unicellular domains, it is important to conceptualize the prokaryote and eukaryote cells as organisms in their own right, descendants of unique subsequent MTEs that established closed, self-sustaining systems with viable non-equilibrium thermodynamics (Atkins, 1984, 2003; Pross, 2012). They share that fundamental characteristic with the permanently multicellular organisms and the colonial superorganisms reviewed in Chapter 6, but I postponed a brief review of the fundamental internal characteristics of such systems because it is easier to extrapolate from cell biology upwards (see Figure 3.2b) than to take the opposite route. Sometimes referred to as autopoietic, closed systems maintain themselves by necessity, not choice, and their dynamic homeostatic stability continues to secure their own closure, social autonomy, and immunity via self-generated information that allows them to cope, as systems, with environmental challenges (Luhmann, 1992). This implies that, outside the human domain, any functional operators within (i.e., as parts of) autopoietic systems must have become obligatorily altruistic upon the completion of the MTE that defined the irreversible beginning of their distinct organizational complexity. But as nothing in nature is altruistic by default, we need inclusive fitness theory to understand why this could happen. In biology, the idea of closure is generally believed to go back to the start of systems biology (Maturana and Varela, 1980), but it was, as we saw, introduced as a neo-Darwinian concept by Huxley (1912, 2022). Making this explicit is important because Huxley's phrasing allowed the formulation of testable predictions that could easily join forces with the later gene's eye view that gave rise to inclusive fitness theory (see Chapters 3–5). In contrast, systems biology appears to have consistently denied the importance of natural selection for the origin and maintenance of adaptive design as teleonomically defined by Pittendrigh (1958). Systems biology thus remains stale unless this crucial omission is acknowledged (Corning and Kline, 1998a, 1998b).

Moreno and Mossio (2015) recently offered a useful introduction to the biological autonomy that emanates from closure. They conclude (p. 101) that "in order to be considered an adaptive agent, a system must satisfy two conditions. Firstly, it must be able to constrain its boundary conditions to ensure its self-maintenance. This in turn requires that it possess an internal organization that is the causal source of the interactive processes ensuring its identity. Secondly, it must be able to discriminate between specific processes or structures in its environment, and to functionally act on them." This is consistent with how Kauffman (2000) and Atkins (1984, 2003) identified energy flow through an operationally closed system, and with such systems maintaining themselves through recursive reproduction (Luhmann, 1992; Kauffman, 1993; Moreno and Mossio, 2015) and immune defense (Pradeu, 2012). It also implies that closure and nested causality cannot be extended beyond the core family (Darwin, 1859; Weismann, 1893; Williams, 1966a), as I illustrated in Figure 3.2b and provided evidence for in Chapter 6. Adaptive agency arguments thus precluded that Moreno and Mossio could make the mistake of claiming that agency can also evolve at higher levels through what Williams (1966a) coined "biotic adaptation" of population-level groups, communities, ecosystems, and Gaia as abstraction of our planet (e.g., Lovelock, 2000; Jagers op Akkerhuis, 2008). Moreno and Mossio (2015, p. 198) conclude that only three interrelated characteristics—organization, closure, and functionality—matter, because they all refer to the same emergent causal regime. However, their analysis does not offer insight into how and why natural selection shaped the emergence of new autonomous agency—the fundamental question that Wheeler, Huxley, and Williams addressed (see Chapter 3). As I have tried to clarify throughout this book, novel levels of closure can be predicted to breed new layers of organization defined by higher-level adaptive functionality, but only when driven by gene's eye natural selection.

Moreno and Mossio (2015) focused primarily on cell membranes as proximate instruments for realizing physical closure (see Figure 5.1), as did other reviews of the autopoietic free-living single cell (e.g., Harold, 2014), but they also extended their analysis into the multicellular domain, emphasizing that nested causation must imply that new elements of higher-level organization should emerge without being reducible to the lower-level properties of the single cell. However, this is Whitman's (1893) dilemma revisited and Huxley's (1912) solution confirmed, as illustrated in Figure 3.2b where the four fundamental closure levels are made explicit (see also Figure 5.3). It is no surprise, therefore, that Moreno and Mossio's "functionality" criterion is complementary to the Hamiltonian $b > c$ condition, that is, to the sufficient MTE condition for producing—when long lasting—a new irreversibly more complex level of organization, adaptive agency, and autonomy via directional kin selection (see Chapters 4 and 6). Functionality and adaptation thus remain two sides of the same coin, as Williams (1966a) stressed time and again throughout *Adaptation and Natural Selection*. This match between contemporary philosophy and traditional neo-Darwinian logic is gratifying, although one cannot help wondering why the early 20th-century insights were ignored. One reason is, I believe, that many came to consider Monod's (1970) and Jacob's (1974) machine concept of the cell as an "end" rather than a useful "means" for deterministic cause and effect understanding, which contributed to the already mentioned abandonment of the organismal evolutionary perspective in cell biology. As I will address in more detail in Chapter 8, physics envy may also have played a role, while forgetting, paradoxically, that the deeper realms of thermodynamics are not deterministic, but statistical, similar to Pricean inclusive fitness theory.

The logic of considering unicellular organisms as molecular machines (Box 7.1) carries the (often implicit) assumption that biologists should ultimately be able to take cells apart and reassemble them from scratch— a conjecture that inspires present-day synthetic biology (e.g., Schwille, 2011). An apt critique of the machine metaphor was recently provided by Nicholson (2019), who acknowledges the tremendous merits of molecular cell biology but maintains that machine thinking fails as formal proximate theory of organization. This conclusion extends earlier arguments (Nicholson, 2013, 2014) that incoherent machine concepts of the organism in general have precluded proper appreciation of the fraction of biological organization that must remain irreducibly complex. Nicholson (2019) makes explicit that there is a fundamental difference between an extrinsically purposeful machine designed by an engineer and an intrinsically purposeful organism that is self-organizing, self-maintaining, and self-regenerating. A machine is a reductionist device where all parts have independent meaning, both in isolation and as part of the machine, but organismal parts (operators) exist only in network relationships of collective interdependence and maintain overall functionality and agency without proximate external design. However, they need ultimate, natural selection design, so it is puzzling that the same Nicholson papers claim their analyses are at variance with the "gene's eye" view of adaptation. As it turns out, no effort is made to understand that the inclusive fitness paradigm is not about inappropriate reductionism, as shallow reading of book titles may erroneously suggest, but about explaining complex phenotypic traits that are cumulatively shaped by many genes of small phenotypic effect (Grafen, 1984; Fromhage and Jennions, 2019; Queller, 2019) (see also Box 2.3).

The inadequacy of neglecting or dismissing natural selection arguments by many organism-focused developmental and cell biologists, emerges clearly from two quotations that explain the emergence of LUCA in almost identical terms, one formulated in the 1960s as an evolutionary conjecture of how natural selection would likely have shaped life's origin (without data being available) and the other written half a century later in an up-to-date review of organismal cell biology. The first, by Williams (1966a, p. 136) says the following: "Selection eventually produced genetic systems that were so effective in promoting the homeostasis of surrounding somata, and so specialized in their manner of producing this effect, that the fusion of somewhat different systems would produce adaptively inferior combinations. The maintenance of individuality was then selected for, and life became fragmented into physiologically isolated individuals and genetically separate lines of descent." The second, by Harold (2014, p. 55)—with decades of empirical data as guidance—concluded that "[s]ome lines became sufficiently successful that further incorporation of foreign genes was more likely to disrupt existing functions than to enhance the systems' capacity to reproduce and garner resources. At this point, natural selection would begin to favor the erection of barriers to the import of genes from the outside; consolidation and refinement come to prevail over foreign novelties." One would be hard pressed to find a better example of the predictive power of first-principle gene's eye thinking about what adaptations can and cannot evolve. It therefore seems worthwhile to explore how far we might progress by adopting a Tinbergen approach (see Figure 1.1; Box 1.3) of asking complementary questions of putative somatic function in the unicellular domains, as I will do in the final section of this chapter.

Another argument in favor of a partial return to the classical organismal perspective of cell biology is that the functioning of cell membranes cannot be reduced to information in prokaryote genomes. Harold

captured this dilemma in a few well-chosen sentences when he wrote that "[g]rowing cells are not organized by genomes alone; they model themselves upon themselves, and therefore the smallest entity that can be truly said to organize itself is a whole cell. It follows that when biochemists grind cells into a pulp, something irretrievable is indeed lost: the organization that makes the cellular system work." The implication is that cell heredity is a complementary result of reducible effects of coding and regulating genes and more elusive higher-level cell architecture, because "all instructions that can be expressed in linear, digital form, end up in genes [and t]hose concerned with three-dimensional order, such as topology and polarity, are carried by cell heredity" (Harold, 2014, pp. 13–14) (see Boxes 7.1 and 7.2 for details). This conclusion echoes that Pittendrigh (1958) considered naturally selected adaptation and organization to be synonyms (Box 1.2) and that Wheeler (1926) formulated three complementarity categories of organismal organizational order that match Harold's inference, as

---

Box 7.1 The limits of the *cell-as-machine* metaphor

Nicholson (2019) recently argued that ever since Descartes' theory about organisms there has been a split between bottom-up mechanistic approaches and top-down concepts that emphasize organismal agency (see also Reynolds, 2018). Earlier, Amundson (1996) evaluated the contrast between those who emphasize *function* and others who focus on *form*, going back to Russell (1916). Both comparisons are valid, but Amundson's contrast is about ultimate versus proximate explanation (Mayr, 1961) and Nicholson's contrast is about Tinbergen's (1963) subdivision within the proximate domain, that is, about ontogenetic development versus reductionist molecular mechanisms (see Figure 1.1). Nicholson (2019) focused on

the unicellular organism to evaluate how far bottom-up molecular biology can reach until the network functions of intracellular operators can no longer be independently explained. A practical consideration supporting his arguments is that molecular constituents within cells work under Brownian motion conditions that would not, for example, affect ant workers that somatically serve a superorganismal colony, a fundamental difference between the micro- and macro-world reviewed earlier by Went (1968). Nicholson (2019) argues that no biochemical approach will be able to fully explain a phenomenon like the self-organized and self-referential mitotic spindle that combines robustness with flexibility while remaining

Box 7.1 *Continued*

in a state of permanent but obviously controlled steady state far from thermodynamic equilibrium. Despite these limitations, the machine metaphor (Monod, 1970; Jacob, 1974), addressing both form and function, increasingly became the trademark of molecular biology.

The problem is that machines have external designers and the cell obviously does not, so the machine metaphor can at best be a heuristic simplification and never a theory. In fact, emphasizing machine analogies makes cell biology an easy target for intelligent design creationists who are a priori convinced about an external designer (Nicholson, 2013). However, while this may all be correct, any critic of reductionism needs to ask herself whether one can offer testable predictions of where the realm of reductionism reaches its limits. It is fascinating to read that self-organized cell functionality is an energetically more demanding process than dynamic self-assembly as practiced by some externally designed machines, nanoparticles, and viral capsids (Nicholson, 2019), but one needs to explain how natural selection has produced and maintained self-referential phenomena when they incur such extra costs—there cannot be organismal agency without natural selection having shaped it (Pittendrigh, 1958). The only appropriate explanation for costly phenotypic traits is that they used to enhance the inclusive fitness of the organisms that carry the genes mediating these complex phenotypes, and that they may well continue to do so right now. It is this contrast between empirical approaches that I tried to capture with the white irreducible complexity zones in Figure 3.2b, where deterministic explanations of biophysics and biochemistry operate below the white zones, possibly assisted by thermodynamic self-organization, but where the theory of organismal natural selection is needed for top-down understanding of costly adaptations. Hence, Nicholson's analyses cannot offer a theory either, even though the cellular facts are intriguing. As Nicholson (2019) writes, "[d]ue to its dynamic nature, what persists in a cell over time is its form, not its matter:

the individual molecules that make up a cell come and go, but its overarching organization remains." The dynamics that define cellular life thus assemble and disassemble molecular structures continuously to maintain a low-entropy state of organizational complexity.

Apart from highlighting a rather recent statistical understanding of the fluidity of cytoplasmatic processes, Nicholson argues that other findings have also eroded the conceptual validity of the machine metaphor in cell biology. The first of these is structural—the now well-established fact that intrinsically disordered proteins are abundant in all complexity domains, which must imply that ordered conformation is not always necessary for protein function. The second category of challenges is functional—many proteins are now known to be promiscuous in a stochastic and context-dependent manner. This strongly suggests that most protein-level interactions within cells are contingent, which is incompatible with predetermined machine-like design derived only from DNA and RNA sequences. Protein subunits are soft and fluid, not hard and rigid, and their interactions transient and opportunistic rather than stable and predictable (Nicholson, 2019). However, while the rewards of self-organized plasticity without loss of stability are tangible, their high operational costs relative to hard-wired alternatives need justification because the prokaryote LUCA and the eukaryote LECA cell could only evolve as MTEs by realizing superior reproductive success in the face of natural selection.

Continuing Nicholson's (2019) argumentation, brownian motion dynamics within cells suggests that the adaptive information-accumulating force of natural selection may only be slightly more powerful than the stochastic processes that would tend to annihilate organizational information. Probabilistic molecular adaptations appear to move enzymatic processes forward in a discontinuous rather than a continuous manner, via innumerable ratchet-like switches, while genetics tinkers with the overall efficiency of the associated metabolic processes as far as their

---

Box 7.1 *Continued*

phenotypic effects are "visible" to natural selection. This often makes somatic metazoan cells respond in a binary, rather than a gradual deterministic manner, similar to the way that workers of superorganismal bee hives have different thresholds for accepting specific tasks even though all can do them (Richardson et al., 2021). Such conclusions strike most organismal biologists as unsurprising, but they may be upsetting to molecular cell biologists—as Nicholson (2019) argues. His analysis follows Kauffman (1993) in replacing the anthropomorphic machine metaphor by self-organized processes that generate

order. However, the insufficiency of the externally engineered machine analogy does not obviate the hard ultimate condition that natural selection works towards maximizing cellular inclusive fitness within the settings of the Williams epigraph at the start of this chapter. No matter how self-organized intracellular processes may be, those governed by the most appropriate selection-shaped set of self-referential gene regulatory networks (GRNs) (see Table 3.1) will prevail. And they will do so irrespective of whether significant fractions of cellular functioning can or cannot be reduced to independent causal factors.

---

outlined in the first section of this chapter. These arguments undermine the sufficiency of the proximate machine concept of the cell but become unproblematic when also adopting an ultimate teleonomic agential perspective (Reynolds, 2018). In Chapter 6, we saw that the original Weismannian principle of a permanent and immutable germline and ephemeral soma is in the process of being replaced by a concept of reproductive division of labor that optimizes germline maintenance depending on trade-offs with other life-history traits. This broadened germline concept appeared to be at least partly extendable upwards to the colonial superorganisms, so it seems legitimate to ask whether it can also be extended downwards, with due proximate modification, to the unicellular domains of social evolution. In the final section of this chapter, I will argue that answers to this question are closely related to scattered but intriguing evidence for senescence in unicellular organisms.

A Tinbergen approach, in the spirit already pioneered by Huxley and Wheeler more than a century ago (see Chapter 3), always considers questions of adaptation in a comparative phylogenetic context. This must imply

that we cannot ultimately understand the organizational complexity of LUCA and LECA as organisms without an appreciation of what their respective non-organismal or lower-level organismal sister lineages represented. As Harold (2014, p. 161) writes, "[w]hat the historian needs to know, first of all, is what happened and when; only after that does it make sense to inquire into the how and why." Box 7.3 summarizes the factual information that we have today. It suggests that both MTEs likely have (had) sister lineages with social lives characterized by forms of redundancy (see Table 3.1) that LUCA and LECA abandoned via novel cell membrane closure. Research of the last decade has made huge progress in identifying the Loki Archaeota as clade-of-origin for the eukaryotes, thereby establishing the notion that there are only two phylogenetic domains of life—the Bacteria that remained the prokaryotes that they always were and the Archaea that, next to many extant prokaryote lineages, also produced LECA which then spawned all macroscopically visible forms of life. It thus seems straightforward to explain LECA's MTE origin in the same LTC terms that I applied to the

Box 7.2 The origin of the first cell membrane

There is little doubt that cell membrane closure around proto-ribosomal self-replicating RNA had been achieved when LUCA emerged in what is now generally appreciated to have been an RNA world (Gilbert, 1986). Compartmentalization of this kind has long been argued to have fostered cooperation (D.S. Wilson, 1975; Maynard Smith, 1979; Szathmáry and Demeter, 1987; Grey et al., 1995), consistent with the logic of Figure 5.1, and with recent experimental evidence supporting closure's relevance for the origin of life (Matsumura et al., 2016). However, the steps preceding cell closure are harder to capture—whether the genetic code arose by chance or by selection-driven necessity remains unclear, and how a largely error-free translation system could evolve without complex proteins being around is also enigmatic (Wolf and Koonin, 2007; Boza et al., 2014). The genetic code is universal (but see Keeling, 2016), and the earliest forms of cellular life must have shared the basic tools of RNA polymerase, messenger RNA, transfer RNAs, and ribosomes, which in turn ensured directional transfer of information from nucleic acids to amino acids to proteins—known as the central dogma (Harold, 2014, p. 183). As Wolf and Koonin (2007) argue, the small ribosomal subunit could have been a polymerase and the large subunit could then have functioned as a ratchet to produce the triplet code (Harold, 2014, p. 185). In effect, ribosomal RNA may be "the vestigial remnant of a primordial genome that may have encoded a self-organizing, self-replicating, auto-catalytic intermediary between macromolecules and cellular life" (Root-Bernstein and Root-Bernstein, 2015). However, a deeper problem emerges from the fact that cell membranes appear to have an independent origin.

According to the Virchow principle (Reynolds, 2007), every cell that has been in existence so far emerged from a previous cell, so the structural origin of the cell membrane cannot be genetically explained. As Harold (2014, pp. 175–176) writes, "[t]he functions of the plasma membrane go far beyond those of boundary and barrier. In prokaryote cells (and to a lesser degree, in eukaryotic ones), the plasma membrane is the locus of energy transduction [. . .]; it is also deeply enmeshed in cell division and morphogenesis. And in today's world, membranes never arise *de novo*. Each membrane, complete with its complement of carriers and catalysts, is generated by growth of a preexisting membrane of the same kind. In principle, such membranes are as much a heritable feature of cells as the genome, passed continuously from one generation of cells to the next from the very dawn of life." That there are no genes to code for the complex construction of cells (see also Johnson and Lam, 2010) nourished the idea that LUCA, and every cell that came from it later on, must have had elements of irreducible organizational complexity where machine analogies would invariably break down (Box 7.1). One could of course argue that irreducibility is a matter of present technology limitations, to be overcome as science proceeds, but the nagging problem is that, as one follows the major journals of science, the non-understood complexity in cell biology appears to increase faster than reductionist insights can track, consistent with the conjecture that nested level-specific layers of irreducible complexity will always remain (see Figure 3.2b).

Nonetheless, exciting new insights in how, where, and when LUCA happened have been accumulating in recent years. If the first cell membrane arose independent of genomic information, this must have happened through a physical process of self-assembly (Box 7.1), so that natural selection was not involved, at least at the very beginning (but see Pross, 2012). This scenario was explored by Chen et al. (2004), who encapsulated RNA in fatty acid vesicles to show that the ensuing osmotic pressure resulted in membrane growth relative to vesicles without RNA. The authors hypothesized that this process, which they implemented with realistic nucleic acid concentrations, could have been instrumental in establishing the enigmatic connection between genome and membrane in an informationally closed one-gene one-vesicle protocell. In a follow-up study, Budin and Szostak (2011) addressed whether a leaky single-chain lipid membrane could gradually transition to the typical phospholipid bilayer of limited permeability by a gradual correlated increase of metabolic proto-ribosomal efficiency on one hand and the production

Box 7.2 *Continued*

of phospholipid precursors that increase cell closure on the other. Yet another study focused on naturally formed iron sulfide compartments as hatcheries of cellular life, for they might have combined closure with multiple catalyst functions that could have produced the first prokaryote cells (Koonin and Martin, 2005). These studies offer plausible hints towards an intuitive understanding of the key characters of LUCA, but we are still far from a beginning of consensus on the interrelated processes that were of primary significance for the origin of organized cellular life (see also Harold, 2014, pp. 76–78; West et al., 2017).

The ancestral RNA world (Gilbert, 1986; Pressman et al., 2015) is now almost universally accepted. It seems likely that RNA arose rather quickly after the Earth became habitable ca. 4.3 billion years ago (BYA), and new fossil discoveries have pushed the likely origin of LUCA back in time to ca. 3.8 BYA (Javeaux, 2019). Connections with marine hydrothermal vents as cradles of cellular life have also become increasingly explicit (Brazelton, 2017; Dodd et al., 2017). A hot and relatively stable ancestral habitat for LUCA matches

reconstructions of the thermal stability of elongation factor proteins (Gaucher et al., 2008) and nucleoside diphosphate kinases (Akanuma et al., 2013) in the earliest forms of cellular life. The natural ion gradients in alkaline hydrothermal vents with life spans of up to 100,000 years appear to be particularly promising candidate sites for the start of organized cellular life (Martin et al., 2014), because they provide a chemiosmotic gradient of the correct polarity and magnitude (Harold, 2014, p. 81). However, as De Duve (2005, 2011) argued, echoing Pittendrigh (1958), no organized complexity can have arisen without the constant "supervision" of natural selection, and the targets must have been social traits that accumulated this complexity, either fraternally by gene duplication (see Chapter 4) or via HGT (see Table 3.1). It is timely, therefore, that studies have begun to address the evolution of cooperative RNA replicators (Vaidya et al., 2012), co-dispersing weakly autocatalytic protoplasm units (Baum, 2015), cooperative networks of self-replicating RNA lineages (Mizuuchi et al., 2022), and cooperative enzymes (Levin et al., 2020).

multicellular and superorganismal lineages (see Chapter 6). The only difference is that we need inclusive fitness theory for mutualistic cooperation (see Chapter 5) rather than inclusive fitness theory for intraspecific altruism (see Chapter 4) to capture the non-redundant egalitarian merger with mitochondria. However, while the necessary non-redundant LTC condition for LECA's MTE was obviously fulfilled (see Figure 5.3), LECA's subsequent differentiation was directed inwards to somaticize the cytoplasm (Roger et al., 2021), rather than outwards to add somatic units externally, as in the later fraternal MTEs.

Relative to LECA, it has been much more challenging to reconstruct what LUCA was like (Box 7.3), but progress in understanding

this enigma has also been significant, not least because it is clear now that the sister lineages of the first ever cell were virus-like particles (Koonin, 2017). Harold (2014) reviewed the evidence for this contention in the symmetrical perspective of three domains of life (Figure 7.1a), but recent phylogenies (e.g., Nasir and Caetano-Anollés, 2015) are increasingly confirming that there have only been two domains (see also Doolittle, 2020). Redrawing the tree of life from this perspective (Figure 7.1b) suggests that the origins of LECA and LUCA were functionally more analogous than so far appreciated. Both MTEs apparently arose as closure events out of a swarm of simpler and more promiscuous viral replicators (LUCA) and binary dividing prokaryote cells

Box 7.3 LECA and LUCA as organismal concepts relative to their putative sister clades

Mitosis and meiosis appear to have interconnected origins (Archibald, 2015; Koonin, 2015a; Baum and Baum, 2020), with meiotic sex initially being a stress response that possibly arose by some combination of mitotic cell division and ancestral prokaryote transformation (Wilkins and Holliday, 2009; Bernstein and Bernstein, 2010). The process—often infrequent and facultative in protists—is extremely accurate in endowing the diploid nucleus with the same complement of maternal and paternal genetic information—very different from prokaryotes who, as Harold (2014, p. 25) expressed it, can "exchange genes by a kind of sexual congress." It is thus straightforward to conceptualize a protist zygote as a double pairwise LTC phenomenon: a non-redundant fraternal commitment between a conspecific male and female gamete and a similarly non-redundant egalitarian commitment (Queller, 2000) with a clone of mitochondria, increasingly considered to have been the key innovation that made the sexual eukaryotes so successful (Lane and Martin, 2010). There is some controversy about whether the acquisition of mitochondria was adaptive right from the start (Booth and Doolittle, 2015; Lynch and Marinov, 2015), but prokaryotes were obviously unable ever to evolve higher organizational complexity without this symbiotic organelle (Lane and Martin, 2015, 2016). The irreversibly enhanced organizational complexity of eukaryote cells thus appears to be clearly related to the excess energy that mitochondria provided, a phenomenon that is unlikely to be captured by a purely reductionist approach, as both Harold (2014, pp. 124–140) and Koonin (2015a) have implied.

The archaeal cell had become the "informational" ancestor and the bacterial symbiont the "operational" ancestor by the time LECA emerged (Harold, 2014, p. 115; Roger et al., 2021). A similar pattern can be observed in secondary protist mergers where the endosymbiotic partner is retained for specific (usually photosynthetic) services while its nucleus and mitochondria normally disappear (Keeling, 2013). This underlines that mutualistic symbiosis, also in a long-term phylogenetic perspective, is always mutual exploitation where the "informational partner" prevails but remains constrained by the need for streamlined performance of operational team members (Keeling and McCutcheon, 2017). An information/operation distinction is a clear characteristic of germline–soma differentiation in the multicellular domains, so it appears logical to explore its applicability for understanding intracellular differentiation, even in prokaryotes (Shi and Huang, 2019). If we apply that view, it seems likely that a fundamental division of labor was unambiguously established at LECA's origin (Williams et al., 2013, 2020), that is, by the time the ancestral Lokiarchaeota FECA cell (Spang et al., 2015; Akil and Robinson, 2018; Imachi et al., 2020) had irreversibly completed its conversion process (Roger et al., 2021). That eukaryote metabolic enzymes are primarily bacterial seems consistent with the mitochondria always having been part of cytoplasmatic soma, but better understanding of the fundamental divergence between archaeal and bacterial bioenergetics, their complementary biochemistry, and later integration in the symbiotic eukaryote cell still has a long way to go (Sojo et al., 2014; Koonin, 2015b; Gould, 2018; Doolittle, 2020).

Immune defenses can be considered as accurate markers of operational closure in permanently multicellular lineages and superorganismal colonies, with the former being clearly nested within the latter and the latter having added a novel layer of self-organized and partly irreducible organizational complexity (see Chapter 6). That nestedness is much less apparent between prokaryotes and protists, quite possibly because the LECA MTE was not fraternal but egalitarian. Prokaryote cells have both constitutive Argonaute (restriction modification) and adaptive (CRISPR-Cas) defenses (e.g., Dupuis et al., 2013; Koonin, 2017). Both systems

Box 7.3 *Continued*

likely originated from RNA-guided defenses against parasitic replicators that had evolved to produce DNA-guided Argonaute (pAgo) defenses by the time LUCA arose (Koonin, 2017). In contrast, eukaryote immune defenses are based on RNA interference, a defense system derived from prokaryote Argonaute but having become highly diversified and elaborated, possibly concurrent with the emergence of the RNA viruses (Koonin et al., 2015) that mostly remain specialized on eukaryotes (Figure 7.1c). The CRISPR-Cas system was irreversibly lost in eukaryotes while pAgo defenses achieved adaptive characteristics. However, these defenses did not reach CRISPR-Cas sophistication until protein-based extracellular immune recognition systems arose in the jawed vertebrates (Koonin, 2017, including reviewer comments). Protist immune defenses also include phagocytosis as a key element, which sets the unicellular eukaryotes apart from the prokaryotes that lack this organic feeding ability (Fenchel and Finlay, 1994).

After LECA arose, internal forms of HGT started to move genes from mitochondrial and chloroplast organelles to the nucleus (Brandvain et al., 2007; Harold, 2014, pp. 136–144) while also recruiting additional prokaryote genes (Ku et al., 2015). This came to imply a more explicit separation between informational nuclear germline functions and operational cytoplasmatic soma functions—the numbers of remaining organelle genes are often still appreciable in protists but down to small percentages of the total genome in multicellular lineages (Bell, 1989; Björkholm et al., 2015; Janouskovec et al., 2017; Roger et al., 2017). This internal process was complemented by substantial HTG from other forms of life, most significantly in prokaryotes, less so in protists, and rarely in multicellular lineages (Ku and Martin, 2016). A final source of abundant HGT across all domains are the viruses. Recent research is increasingly showing that virus-like replicators evolved before the first prokaryote cell and that viruses have been co-evolving with their hosts (Figure 7.1b) as they went through their MTE ratchets towards higher organizational complexity (Koonin et al., 2006; Finnegan, 2012; Koonin et al., 2015; Nasir and Caetano-Anollés, 2015). The eukaryote protists and particularly the

multicellular lineages have provided substantial novel opportunities for RNA viruses and retroviruses which rarely use prokaryote hosts (Malik et al., 2017). Overall, horizontal exchange of genetic material between viruses and hosts and vice versa has been tremendous (Krupovic and Koonin, 2017; Malik et al., 2017) as genomes are speckled with elements that originated elsewhere.

Only phospholipid bilayer membranes, ribosomes, and actin/tubulin cytoskeletons appear to be truly homologous between prokaryotes and eukaryotes. This is consistent with LECA's organizational complexity not being reducible to mere additive gene expression (Harold, 2014, pp. 111–136) and also with the egalitarian MTE origins having required fundamentally novel division of labor GRNs. However, while the bigger picture for LECA's early history has become much clearer, obtaining similar understanding of LUCA's origin remains a huge challenge. As Harold (2014, p. 52) summarized, LUCA combined the universals shared by all of life (proteins, nucleic acids, ribosomes for transcription and translation, metabolic pathways, and possibly an elementary cytoskeleton) within an energy-transducing membrane as closure device, able to often produce an additional cell wall. However, how LUCA's self-referential agency evolved remains mysterious (Harold, 2014, p. 53) despite credible hypothetical scenarios (Sojo et al., 2016) and novel experimental data (Mizuuchi et al., 2022). The viral nature of LUCAs hypothetical sister lineage now seems somewhat better understood (Figure 7.1b), and we can assert that LUCA represents the origin of the first ever phenotype and the establishment of vertical inheritance by default, in which a duplicated membrane-bound chromosome is passed on to two daughter cells at every cell fission. It also seems reasonable to assume that LUCA could only achieve a competitive edge in a world of naked RNA and DNA replicators after its own DNA replicators had become linked as "pearls on a double-stranded chromosome string" (Maynard Smith and Szathmáry, 1993) to preclude competition for differential representation in daughter cells, a kind of replicator version of the "all in the same boat" principle outlined in Chapter 5.

(LECA). The enhanced organizational complexity of LECA (Boxes 7.2 and 7.3) thus seems ultimately associated with its soma having achieved much higher agency from the cytoplasmatic mitochondrial symbiont than had any previously existing cell type (e.g., Lane and Martin, 2015). Taming the agency of its formerly independent bacterial mitochondria required genome streamlining from ca. 1500 to a mere 200 genes (Archibald, 2012) and selection pressure to compartmentalize transcription and translation, possibly triggered by the emergence of introns (Martin and Koonin, 2006; Brandvain et al., 2007). None of that applied when LUCA arose amidst the ancestors of extant double- and single-stranded DNA viruses and positive-sense (+) RNA viruses, the lineages with the simplest replication biology that remain most diverse today (Figure 7.1c; Box 7.3).

The cell societies

While single prokaryote and eukaryote cells are clearly organismal by themselves, each can also form multicell societies. As we saw in Chapter 6, society life is characterized by condition-dependent forms of cooperation between individuals whose own bodies practice obligate forms of lower-level somatic altruism to a germline. Classic inclusive fitness theory has always been about societies in which the three variables of Hamilton's rule (r, b, c) vary independently; they do not when we extend the theory to explain the organismal MTEs (see Chapters 4 and 5). In societies, particularly those of cells, these variables can often be experimentally manipulated, but we only learned around the turn of the millennium that microbes are excellent testbeds for inclusive fitness theory (Pál and Papp, 2000; Crespi, 2001). The first studies focused on *Dictyostelium* cellular slime molds and *Myxococcus* Proteobacteria, both of which had been studied for decades

from an organismal, rather than a society perspective (Bonner, 1967; Dworkin and Bonner, 1972) and primarily by molecular biologists. Until the end of the 20th century, researchers considered the transient, condition-dependent multicellularity of *Dictyostelium* to be a model of incipient ontogenetic development (e.g., Ennis et al., 2000; see also Buss, 1982), which was unfortunate because it helped establish the idea that conflict reduction, rather than single diploid cell foundation, was the essential condition for MTEs to permanent germline–soma segregation in multicellular organisms. While Bonner (see Figure 3.1) never developed an interest in questions of social conflict in cell societies, Joan Strassmann and David Queller did (Strassmann et al., 2000), similar to Greg Velicer and Richard Lenski (Velicer et al., 2000). It soon emerged that the cell societies of *Dictyostelium* and *Myxococcus* (Figure 7.2a,b) were endemically plagued by gene's eye conflicts when their social aggregations form asexual fruiting bodies.

Because of their lower organizational complexity (see Figure 3.2), the social traits of microbes are more likely to be encoded by single genes than those of multicellular organisms. In addition, the habitat patches microorganisms live in may be sufficiently isolated for fitness to become frequency dependent when selection is strong (Ross-Gillespie et al., 2007). This possibility had not emerged in Hamilton's original kinship models where additivity of quantitative polygenic traits was a reasonable assumption (see Box 2.3) so advantages of rare cheaters tended to cancel against cooperators doing better in other patches with many cooperators. These differences notwithstanding, social interactions in the unicellular domains are also essentially always about kinship and communication, supplemented by responses to omnipresent disease pressure. However, the much smaller scale of the cellular domains implies that social

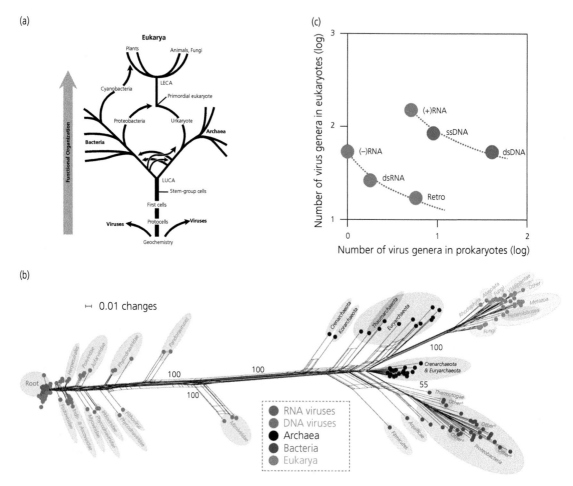

Figure 7.1 Towards the roots of the tree of life. (a) Harold's (2014, p. 219) hypothetical tree of life, proposing (based on his book-long review) that viruses originated in the precellular RNA world and evolved in parallel with the cellular domains of life. In this perspective, it is conceivable that viral attack may actually have selected for the type of cell membrane closure that marks the origin of LUCA. This view is consistent with huge amounts of subsequent viral HGT to and from cellular hosts (Koonin et al., 2015) across the domains of social evolution, and with extant viruses all being highly derived (Box 7.2). Republished with permission of The University of Chicago Press, from In Search of Cell History, F.M. Harold (2014); permission conveyed through Copyright Clearance Center, Inc. (b) A distance-based phylogenomic network reconstructed from more than 400 protein-fold superfamilies across almost 400 proteomes, highlighting that both RNA and DNA viruses predate cellular life, while also confirming that LECA arose from archaeal stock. Working with proteins is attractive because their three-dimensional-folding carries signatures of homology between viral and cellular life that RNA and DNA sequences cannot capture (Malik et al., 2017). Reprinted with permission of AAAS from *Science Advances*, A phylogenetic data-driven exploration of viral origins and evolution, A. Nasir and G. Caetano-Anollés, Vol. 1:e1500527 (2015). © The Authors, some rights reserved; exclusive licensee AAAS. Distributed under a CC BY-NC 4.0 Licence (http://creative commons.org/licenses/by-nc/4.0/). (c) The relative biodiversity abundances of the main categories of extant viruses (number of known virus genera) attacking prokaryotes and eukaryotes, plotted on a double-log scale using combined information from (Koonin et al., 2015, figure 1) and Nasir and Caetano-Anollés (2015, figure 3). There appear to be two overall abundance categories. The three most abundant classes are double-stranded (ds) DNA, single-stranded (ss) DNA and positive-sense (+) RNA viruses, of which the DNA viruses primarily exploit prokaryote cells (blue dots). The three less abundant classes (negative-sense (−) RNA, dsRNA, and retroviruses) are mostly associated with multicellular animals, plants, and fungi, like the +RNA viruses that belong to the abundant category (red dots). The dsRNA viruses, commonly associated with protist and fungal hosts, seem generally little diverse, but these numbers could also reflect unequal research effort across the major host

(Continued)

interactions are not only about cell aggregation or cell adherence but can also concern the release of public good molecules in the environment. Quorum sensing to assess the density of neighboring cells (Lerat and Moran, 2004; Keller and Surette, 2006; Diggle et al., 2007a) is now known to be fairly universal across the Gram-positive and Gram-negative bacteria, and generally assumed to be an evolutionarily conserved form of social communication that regulates the production of public good molecules (Diggle et al., 2007b; Darch et al., 2012; Whiteley et al., 2017; Smith and Schuster, 2019). With relatedness being variable, social communication systems being available, and the contribution of public good molecules being costly, the unambiguous prediction is that kin selection can maintain conditionally expressed altruistic traits, provided average relatedness is sufficiently high to more than compensate for the cost of non-clone mates getting away with faster cell division when they do not contribute public good molecules.

While cell–cell interactions in the fruiting bodies of *Myxococcus* and *Dictyostelium* resemble how social paper wasps and vertebrates interact physically, the production of public goods does not require that cells ever touch each other. In spite of much greater social working distances, *Pseudomonas aeruginosa* Gammaproteobacteria were shown to preferentially produce public-good molecules such as iron-scavenging siderophores in the presence of clone-mate cells (West and Buckling, 2003; Griffin et al., 2004). Something similar, but subtly different, applies to conditionally produced bacteriocins (Riley and Wertz, 2002) that spitefully harm competing strains

(Gardner et al., 2004). As predicted, the level of relatedness in cell cultures turned out to be decisive also for the expression of cooperative virulence in this human pathogen. Conceptual reviews (e.g., West et al., 2006) went on to overturn the often-implicit idea that microbial interactions are benign or even harmonious by default. Rather, they showed that tragedies of the commons arise when strains of microbial symbionts associated with the same host are genetically variable—a fitting confirmation of Williams' intuition in the final sentence of the epigraph for this chapter. As we saw in Chapter 5, such redundant chimerism often triggers the evolution of complementary mechanisms for repression of competition (Frank, 2003; West et al., 2006). In a more general sense, the cooperation and conflict studies of microbial societies have supported the rephrasing of inclusive fitness theory via the Price equation (see Chapter 4) because relatedness in unicellular organisms is mostly a binary variable—a weighted average of 100%-related clone-mate cells and unrelated other cells—without a direct connection to pedigrees that apply in classic kin selection theory.

Failure to express cooperative traits for selfish reasons has often been referred to as cheating (e.g. Smith and Schuster, 2019), even when such responses are prudently "rational" from a gene's eye inclusive fitness perspective. While some continue to find this view of sociality disturbing, for most this just asserts that all naturally evolved strategies are ultimately self-serving for the genes encoding them, irrespective of whether they look nice or nasty. Although the term cheating

---

Figure 7.1 (*Continued*)

clades. The viruses associated with Archaea are remarkably different from all other known viruses—morphologically, genomically, and in terms of proteins (Prangishvili, 2013; Krupovic et al., 2018), consistent with early adaptive diversification with very different lineages of host cells. The bacteriocins of Bacteria and Archaea are also very different (Riley and Wertz, 2002).

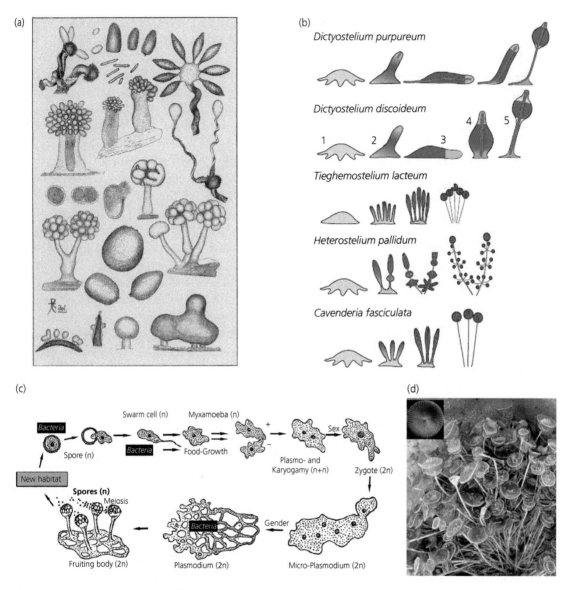

Figure 7.2 Bacteria and protists that form societies of cells or differentiated cytoplasms. (a) The Myxobacteria (delta-Proteobacteria) express a remarkable diversity of dynamic multicellular phenotypes. These produce durable dispersing myxospores when cells aggregate in multicellular fruiting bodies, consisting mostly of rod cells that altruistically support the reproductive myxospores while dying by suicide to release their nutrients to the developing fruiting body (Bretl and Kirby, 2016). The best studied species, *Myxococcus xanthus*, has relatively simple globular fruiting bodies, while *Chondromyces* species have differentiated stalks and *Stigmatella* species have branched stalks. The condition-dependent altruism of these bacteria can be explained by Hamilton's rule in a straightforward manner (Smith et al., 2010), and their social dynamics include both self-sacrifice in response to collective starvation and multiple modes of exploitative cheating mediated by cell signaling and recognition systems that are increasingly well understood (Box 7.4). The image from Bretl and Kirby (2016) was originally drawn by Thaxter (1892).

(Continued)

has been commonly applied in the behavioral ecology literature, its generalized use in the unicellular domains elicited the need for a precise definition, which was provided by Ghoul et al. (2013). Irrespective of cheating being facultative or obligate, and of the behavior happening within or across species, this formalism forced researchers to make sure that a conjectured cheating trait can in fact respond to selection, a necessary requirement when it is hypothesized to have evolved to exploit an otherwise honest signaling system.

Evidence for cheating was indeed found in natural *Pseudomonas* habitats (Andersen et al., 2015; Bruce et al., 2017; Butaite et al., 2017). In a comparative analysis, Kümmerli et al. (2014) further showed that iron-scavenging siderophore molecules tend to be diffusible public goods when bacteria live in structured habitats inside hosts or in soil cavities, while they remain attached to the cells that produce them when bacteria live in unstructured habitats such as water columns where neighboring cells are unlikely to be clone-mates. While

---

Figure 7.2 (*Continued*)

(b) A similar but more elaborate diversity of social reproduction based on spore-bearing fruiting bodies has evolved in cellular *Dictyostelium* slime molds, which are heterotrophic eukaryotes belonging to the same clade as the animals and fungi (Schaap et al., 2006; Glöckner et al., 2016). Also here, one species, *D. discoidum,* has been studied in much greater detail than all others (Box 7.4; see also Benabentos et al., 2009; Romeralo et al., 2013; Strassmann, 2016; Noh et al., 2018). As in *Myxococcus,* cooperation and conflict in *Dictyostelium* matches expectations from inclusive fitness theory, and to such extent that behaviors of cell groups that aggregate in response to starvation can be quantitatively predicted and verified by experimental evolution under controlled conditions (Box 7.4) (Mehdiabadi et al., 2006; Hirose et al., 2011; Kuzdzal-Fick et al., 2011; Ho et al., 2013). In *D. discoideum*, the stalk cells are the unambiguous reproductive altruists, but the proportion of cells sacrificing themselves for their kin is much lower (ca. 20%) than in the Myxobacteria, where almost all cells end up adopting active altruistic support functions or providing passive suicidal service. Fruiting body formation is analogous to the Myxobacteria in the sense that spore formation is asexual. However, being a eukaryote, *Dictyostelium* has multiple cell types and a (resting spore-producing) sexual cycle which is now fairly well known in *D. discoideum* (Douglas et al., 2017; Bloomfield et al., 2019). Another remarkable difference with the Myxobacteria is that *Dictyostelium* cells have evolved ways to husband and disperse bacteria, trading lower digestive yield for a possible head start via inoculation of new patches with preferred prey (Box 7.4; see also Brock et al., 2016). Republished with permission of Springer Nature, from The multicellularity genes of dictyostelid social amoebas, G. Glöckner et al., Nature Communications, Volume 7 (2016) but with new nomenclature (Sheikh et al., 2018); permission conveyed through Copyright Clearance Center, Inc. (c) A third clade of bacterial predators in which both fruiting bodies and bacterial farming evolved are the plasmodial slime molds (Kutschera and Hoppe, 2019). These myxomycetes (or myxogastrids) have an integrated life cycle with alternating haploid gametophytes and diploid (karyogametic) sporophytes (plasmodia) that grow sexual fruiting bodies to produce dispersing haplospores after meiosis—asexual slugs forming fruiting bodies are absent. However, whereas *Dictyostelium* and Myxobacteria have multicellular phases, the plasmodial slime molds extend their clonal cytoplasm acellularly by forming syncytia with many diploid nuclei that may reach diameters of a few cm without any outside or internal cell walls. A well-known clade of myxomycetes is the genus *Physarum*, whose best-known species *P. polycephalum* forms bifurcating syncytia reminiscent of macroscopic fungal hyphae with advanced levels of problem solving agency (Nakagaki et al., 2000; Dussutour et al., 2010; Alim et al., 2013, 2017; Boisseau et al., 2016; Gao et al., 2019). This mode of syncytial differentiation and sexual reproduction is similar to *Caulerpa* green algae, mentioned already by Huxley (1912) and now proved to be sexual (Phillips, 2009). Reprinted by permission from Springer Nature: *Theory in Biosciences*, Plasmodial slime molds and the evolution of microbial husbandry, U. Kuschera, T. Hoppe (2019). (d) For the sake of comparison, I have added an image and close-up of an *Acetabularia* green alga. This genus has evolved germline–soma differentiation by cytoplasmatic extension of a single cell. Its large nucleus, produced as a sexual zygote, remains in the rhizoid base of each "stem" that produces a remarkably differentiated and patterned cytoplasmatic soma on which cysts with new gametes form (Hämmerling, 1953; Nishimura and Mandoli, 1992), a fitting example of the potential of germline–soma differentiation within a single free-living and singly-nucleated cell. Photo credit Patrick Keeling.

kin selection based on proximity of clone-mate cells is now increasingly well established (e.g., Lyons et al., 2016; Özkaya et al., 2018), much remains to be discovered about the ways in which prokaryote cells communicate, for bacterial traits are affected by multiple hierarchical quorum-sensing systems of which we remain mostly ignorant (West et al., 2006).

The evidence for competition, rather than cooperation, being the default setting of the microbial world is now quite overwhelming (e.g., Granato et al., 2019; Smith et al., 2020), and this paradigm seamlessly extends into mutualisms where hosts engineer microbiomes, both conceptually (e.g., Scheuring and Yu, 2012; Coyte et al., 2015) and mechanistically (e.g. Lories et al., 2020; Worsley et al., 2021) (see also Chapter 5). It appears indisputable that we need an inclusive fitness perspective to explain that microbial cooperation can exist despite nature's omnipresent competitiveness, no matter whether bacteria remain planktonic, make biofilms, hunt collectively, or engage in multicellular structures to maximize dispersal of resting spores, as the opening quote of Chapter 5 asserts. I have summarized the advances in the Hamiltonian study of cell societies in more detail in Box 7.4, starting with the bacterial examples and then addressing the more complex eukaryote slime molds, to highlight the substantial detail they have added to our dynamic understanding of cooperation in the unicellular domains. The same studies have also clarified that none of these almost exclusively aggregative social interactions have anything in common with the obligate somatic altruism that characterizes permanently multicellular organisms—nothing in the social life-histories of *Pseudomonas*, *Myxococcus*, *Bacillus*, or *Dictyostelium* is higher-level organismal. If anything, these microbial interactions are reminiscent of Hobbes' dictum that the original human condition was a life that was nasty, brutish, and short, so any apparent harmony should trigger a microbial researcher's suspicion and prompt her to dig deeper to discover how such phenomena are de facto self-interested at the level of the genes encoding their expression. An illustrative case was recently provided via an experimental evolution study of cross-feeding *Escherichia coli* strains towards more cooperative cell clusters (Preussger et al., 2020), which was accompanied by a critically probing commentary (Figueiredo and Kümmerli, 2020) suggesting that gene's eye self-preservation may well have been the driving force. While many explicit tests with microbes could potentially prove inclusive fitness theory wrong, that has not happened yet (Box 7.4).

There are also dynamically synergistic cell phenomena that are not cell societies. The plasmodial slime molds (Figure 7.2c; Box 7.2) have a biology similar to *Dictyostelium* but consist of a single syncytial cell with many clonal nuclei, and mononucleate *Acetabularia* green algae (Figure 7.2d) realized advanced nuclear–cytoplasmic division of labor within a single cell membrane. Another exception are viruses (Figure 7.1), dependent as they are on symbiotic interaction with closed cells for expressing their agency (e.g., Ge et al., 2020). Outside their host environments, viruses lack irreducible, self-organized complexity, so they cannot independently maintain organization above thermodynamic equilibrium, consistent with lone viral particles having specific half-lives of environmental decay. In terms of sheer abundance, viruses dominate the global near-surface atmosphere with ca. 2–40 million particles per cubic meter of air, leaving bacteria in second place with two to four times lower numbers (Whon et al., 2012). The closure-based autopoietic definition of organized life implies that viruses cannot be considered alive, a contention that was articulated in considerable detail by Moreira

Box 7.4 Cooperation and conflict in prokaryote and eukaryote cell societies

Three main model systems have been extensively studied since the turn of the millennium, in all cases with the explicit aim of testing and potentially refuting predictions from Hamiltonian inclusive fitness theory. Essentially without exception, the theoretical predictions have been confirmed. At the same time, standard assumptions of the theory often turned out to matter and were made more precise, enriching the overall conceptual framework and often generating new counterintuitive predictions. These social adaptation studies showed that the Hamiltonian study of microorganisms—an unrecognized application domain of gene's eye evolutionary theory before the turn of the 21st century—amply confirmed what both Monod (1975) and Popper (1962, 1975) had identified as the hallmark of a good theory because more and more would fall into its lap (see Chapter 1). The paragraphs below outline in some detail the achievements of these three research programs, starting with the most mobile *Pseudomonas* bacteria, followed by the substrate-bound *Myxococcus* bacteria, and ending with the eukaryote amoeboid and cellular *Dictyostelium* slime molds. I will also briefly mention some remarkable unicellular systems with similar division of labor, realized either in a single multinucleate cytoplasm or in a single uninucleate cell.

Pseudomonas

Using "signal negative" and "signal blind" quorum-sensing mutants, various early studies showed that public goods controlled by quorum sensing indeed have group-level benefits and that cheating, by not producing the signal or not responding in kind when detecting the signal, pays off in higher cell division rates—these experiments used high and low mixing of cell cultures and confirmed that quorum-sensing wild types prevail at high relatedness whereas cheaters prosper at low relatedness (Griffin et al., 2004; Diggle et al., 2007b). After identifying pyoverdine as a key public good, Kümmerli et al. (2009a) showed that the

production of this iron-scavenging molecule is a plastic response to cell density variation, environmental iron concentration, and the presence of cheats. Another study showed that viscous media, reminiscent of the soft host tissues that this microbe operates in when infecting the lungs of cystic fibrosis patients, increase relatedness among interacting cells and thus impose selection for siderophore production (Kümmerli et al., 2009b; Weigert and Kümmerli, 2017). Cell–cell cooperation tends to have unambiguous positive effects on overall growth rate, so that quorum sensing is usually expressed when relatedness is high. However, conflict over the production of public goods reduces population-level fitness of *P. aeruginosa* and more so in biofilms than in planktonic cultures (Rumbaugh et al., 2009), which implies that cheater-dominated biofilms become more susceptible to some antibiotics (Popat et al., 2012).

Pseudomonas cheats prosper most when environmental iron levels are low, but this threat to cooperation is attenuated when patches vary in the need for siderophore production, that is, in the cost of cooperation (Jiricny et al., 2010; Luján et al., 2015; Dumas and Kummerli, 2012). It also appeared that partial cheats (low public good producers) are often fitter than complete loss-of-function cheats (Dumas and Kummerli, 2012; Andersen et al., 2015), which explains why natural soil populations are complex assemblies of cooperators and exploiters (Vasse et al., 2015) and why cooperators may also have evolved defenses in the form of secreting private rather than public goods (Dandekar et al., 2012; Andersen et al., 2018). Together, these social interactions between bacterial cells have offered compelling and largely consistent confirmation of the predictions of inclusive fitness theory (Crespi, 2001; Keller and Surette, 2006; West et al., 2006; Diggle et al., 2007b; West et al., 2007c). However, it is important to note that most of these interactions are actions at a distance. They may involve impressive examples of cellular altruism (e.g., Ackermann et al., 2008), but there is no

Box 7.4 *Continued*

organized division of labor because multicellular structures for the sole purpose of reproduction and propagule dispersal are lacking. In other words, all social behavior of *Pseudomonas* is meant to maximize resource acquisition of individual cells, but there is no joint purpose of resource allocation beyond public goods and quorum sensing (Raymond et al., 2012). This is different in the Myxobacteria where cell aggregations form spore-dispersing fruiting bodies.

Myxococcus

The Myxobacteria (Figure 7.2a) are special because they produce resting cells (myxospores), a trait shared with only a few other bacteria such as actinomycetes, cyanobacteria, and *Bacillus* that all practice gliding mobility, as Dworkin and Bonner (1972) realized decades ago. This may have predisposed these prokaryotes to develop simple forms of filamentous multicellularity to aid resting spore dispersal or three-dimensional structures as in the Myxobacteria. The best studied model system, *Myxococcus xanthus*, predates on other bacteria by cooperative "amoeboid" swarming (Kaiser, 1979; Velicer and Yu, 2003), a genetically variable and phenotypically plastic collective trait that is likely to keep clone-mate cells together (Vos and Velicer, 2008). The first *M. xanthus* studies explicitly focusing on social behavior revealed that solid growth media are essential to maintain population structures conducive to cooperation in experimental laboratory lines (Velicer et al., 1998), and that multiple cheating clones could be identified in such cultures (Velicer et al., 2000). Most cells altruistically die when up to 100,000 of them form a fruiting body mediated by intercellular signaling (see also Strassmann, 2000). Chimeric fruiting bodies tend to favor cheater lineages (Fiegna and Velicer, 2005), which may drive local populations extinct (but see Fiegna et al., 2006). Remarkably, natural strains from distant locations often refuse to mix in the same fruiting body while sympatric lines tend to form dominance hierarchies of sporulation success, mediated by a substantial arsenal of self/non-self recognition mechanisms (Fiegna and Velicer, 2005; Vos and Velicer, 2009).

In field populations of *Myxococcus*, social environments relevant for local adaptation are mosaics of tiny square-millimeter-sized patches whose cells jointly form fruiting bodies (Kraemer et al., 2016). Mean relatedness among interacting cells thus remains high even though a variety of social phenotypes can often be isolated from a single fruiting body (Kraemer and Velicer, 2014). High local relatedness is reinforced by positive frequency-dependent elimination through interference competition and by molecular kin discrimination mechanisms acting upon cell contact (Rendueles et al., 2015a, 2015b; Bretl and Kirby, 2016) rather than by action at a distance. A similar result was obtained for swarming *Bacillus subtilis* colonies (Lyons et al., 2016), suggesting that killing non-kin bacteria is often more advantageous than socially interacting with them (Velicer and Plucain, 2016). Overall, these studies emphasized that high relatedness is crucial for benign and mutually synergistic interactions. This seems true more generally for swarming bacteria such as *Myxococcus* and *Bacillus* and less so in *Pseudomonas* whose cells do not have direct membrane exchanges to mediate antagonism or positive kin discrimination (Wielgoss et al., 2018). Recent studies confirmed that synergistic cooperation in single fruiting bodies is in fact restricted to clones with recent common ancestry that subsequently diversified into sets of genetically complementary social phenotypes (Wielgoss et al., 2019). This implies that relatedness of strains collaborating in making a fruiting body is even higher than previously appreciated (Pande and Velicer, 2018), and that incompatibility mechanisms may evolve readily and rapidly.

Dictyostelium

The evolutionary study of eukaryote cell societies has focused primarily on *Dictyostelium* cellular slime molds, a rather diverse genus (Schaap et al., 2006) in which a variety of fruiting body shapes have evolved (Figure 7.2b). Like *Myxococcus*, *Dictyostelium* is also a bacterial predator, but rather than having swarm foraging by default, *Dictyostelium* operates as solitary amoebae and only aggregates upon starvation.

Box 7.4 *Continued*

Altruistic reproductive suicide by becoming part of the stalk of a fruiting body was first documented by Strassmann et al. (2000), together with the frequent occurrence of chimeric fruiting bodies in which cheating for higher shares of spore production can occur. Follow-up studies concentrated on showing that sympatric clones often form linear dominance hierarchies (Fortunato et al., 2003a), and that chimerism occurs in the wild (Fortunato et al., 2003b) and normally carries a fitness cost (Foster et al., 2002; Castillo et al., 2005). This established *D. discoideum* as a prominent model for testing inclusive fitness theory in eukaryotes, because these amoebae could also be shown to practice homophylic cell adhesion in a green-beard manner independent of overall relatedness (Queller et al., 2003) (see also Crespi and Springer, 2003). Comparative phylogenetic research established that *D. discoideum* belongs to an evolutionary derived branch of dictyostelid social amoebae characterized by larger solitary fruiting bodies rather than a clustered set of smaller sorocarps (Schaap et al., 2006) (Figure 7.2b). However, ca. three-quarters of *D. discoideum*'s genes still have orthologues or homologues in *Physarum*, the closest unicellular sister taxon (Glöckner et al., 2016).

It turned out that relatedness of interacting *D. discoideum* cells in natural habitats is almost invariably high, although never maximal in the sense that no *Dictyostelium* population produces obligatorily clonal fruiting bodies. This outcome is very similar to what was found for *Myxococcus*, underlining that inclusive fitness predictions for societies of cells are the same no matter whether they are prokaryotic or eukaryotic, and irrespective of whether they have collective reproduction or remain free-living cells interacting at a distance as in *Pseudomonas* and *Bacillus*. Under high relatedness, kin selection theory would predict that cheating *Dictyostelium* phenotypes that produce more than their fair share of spores would remain rare (Gilbert et al., 2007). That expectation was borne out because a genome-wide screen of laboratory-induced mutants showed that many cheaters were facultative in the sense of only having disproportional success when they were part of chimeras while they cooperated normally when clonal (Santorelli et al., 2008). As it appeared, cheaters can either work through self-promotion or by coercing other clones into a reduced share among the spores (Buttery et al., 2009), and their dishonesty will select for resistant phenotypes (Khare et al., 2009), once more analogous to *Myxococcus* (Fiegna et al., 2006). These insights were gained in cumulative steps—more than a decade of increased understanding can be appreciated in more detail by comparing reviews by Shaulsky and Kessin (2007), Medina et al. (2019), and Ostrowski (2019).

The total of 150 induced cheating loci identified by Santorelli et al. (2008) were later shown to have elevated polymorphism, unusual patterns of linkage disequilibrium, and unexpectedly low differentiation across populations, consistent with recurrent invasion of new alleles and negative frequency-dependent selection preventing fixation so that cheating genes remain endemic and diverse (Ostrowski et al., 2015). However, around the same time, Wolf et al. (2015) cautioned that what seems to be social winner and loser strategies may in fact be rather neutral in terms of inclusive fitness when the expression of social traits has hidden costs. The trait they emphasized was spore size because cheaters produced smaller and less viable spores than cooperating strains in the same chimeras. This is an interesting finding, but it could not be replicated in another study (Votaw and Ostrowski, 2017), which found a positive correlation between spore number and spore size. However, these ambiguous results could be due to differences in resource levels across populations, a general problem in field studies of life-history trade-offs. As shown by Van Noordwijk and de Jong (1986), a series of negative trade-off correlations across a population-level gradient of resource richness tends to produce an overall positive correlation. Taken together, the *Dictyostelium* studies have been illuminating in showing that predictions of inclusive fitness theory are equally powerful when studying societies of eukaryote microbes as they are when the model systems are cooperatively breeding insects or vertebrates (see Chapter 6).

Box 7.4 *Continued*

Bacterial husbandry

Given the abundance of interactions with bacteria, it was to be expected that some of them would have entered into mutualistic associations with *D. discoideum*, but that this would involve a form of husbandry-like inoculation of new sterile habitat was a surprise. As Brock et al. (2011) showed, farming strains of these amoebae do not exhaust all food bacteria in their patch at the onset of starvation but stop feeding earlier to keep some bacteria alive and vector them to the spores in fruiting bodies. Food bacteria (e.g., *Klebsiella* and *Escherichia*) then co-disperse with host spores to inoculate a new substrate upon germination. Bacterial husbandry carries various fitness costs which are only recovered when spores land in habitat without food, so farmer and non-farmer strains coexist because their respective reproductive success is context dependent (Brock et al., 2011). Farmer cells vector a variety of bacteria, of which about half are just food, but they also transfer *Burkholderia*, a non-food bacterium, one of which appeared to be a defensive mutualist that suppresses non-farmers and thus helps to privatize crops (Brock et al., 2013). Other *Burkholderia* strains facilitate their own transmission by infecting non-farmer strains and turning them into vectors for both food and non-food bacteria (DiSalvo et al., 2015). It thus appears that there are coadaptations to farm and be farmed, but the farmer-crop associations remain facultative throughout, as one would expect in an aggregative society of eukaryote cells and mutualistic bacteria. This system is therefore fundamentally different from the closed non-redundant farming practices of fungus-growing ants, termites, and ambrosia beetles (see Chapters 5 and 6).

Similar looking but not analogous microbial models

The less well-studied *Dictyostelium* sister clade of acellular (myxogastrid) plasmodial slime molds evolved independently from prostelid ancestry but retained unicellularity, albeit with massive multinucleate syncytia whose tubular mycelium-like networks can reach macroscopically visible dimensions (Alim et al., 2013; Hillmann et al., 2018) (Figure 7.2c). Plasmodial slime molds move around by peristaltic contractions, which transfer signals (Alim et al., 2017) to mediate habituation and optimization of resource acquisition (Dussutour et al., 2010; Boisseau et al., 2016). Fruiting bodies are never chimeric and always sexual, and spores carry specialized adhesive symbiont bacteria that may be essential for fruiting body formation (Kutschera and Hoppe, 2019).

Other examples of eukaryote cell societies are the unicellular *Chlorella* algae that aggregate in response to predation pressure (Kapsetaki and West, 2019), and unicellular volvocine algae (Ratcliff et al., 2013) and budding yeast (Ratcliff et al., 2012). However, in unicellular organisms where both clonal and nonclonal group formation is possible, it is important to clarify the specificity of adhesion genes. Such genes may either mediate homophilic (ligand-to-ligand self-recognition) adhesion that guarantees clonality or they encourage catholic forms of recognition via less specific cell surface molecules—only the former might therefore express altruistic traits, while the latter are best interpreted as green beard phenomena (Queller, 2008). Molecular mechanisms mediating these interactions appear to be better known in *Saccharomyces cerevisiae* yeast than in other model systems (Smukalla et al., 2008; Ratcliff et al., 2012; Regenberg et al., 2016; Fisher and Regenberg, 2019).

---

and López-Garcia (2009). The issue remained controversial, however, because an alternative definition of life is based on the capacity to respond to natural selection when propagating

heritable variation. The latter reductionist definition admits viruses to the realm of life (Harold, 2014, p. 47), but that merely implies that viruses are part of the universal phylogenetic tree (Figure 7.1b). That is not the same as being part of the tree of organized life, which requires physical (cellular) or functional (higher-level) closure to maintain independent organizational complexity away from thermodynamic equilibrium (see Chapter 5). Critics of Moreira and López-Garcia (2009) often confused secondary reductions (e.g., in genome size) with the impossibility of reversal—no LUCA descendant ever gave up cellularity. An apparent exception of an archaeal virus changing morphology outside the host (Häring et al., 2005) appears to be a form of adaptive phenotypic plasticity and does not require independent organismal agency.

Extant viruses have very little in common with their ancestors from the RNA world, for they have been subjected to antagonistic coevolution with their unicellular and multicellular hosts for up to billions of years (Koonin and Makarova, 2018; Marino et al., 2018). These tremendous time spans and the very high mutation rates of viruses (Bonhoeffer and Sniegowski, 2002; Gago et al., 2009) make it interesting to ask whether viruses have evolved social traits when they interact with other viruses in the same host. An early study documented that an RNA virus produces intracellular public goods upon clonal infections but selfishly limits such investments when mixed with unrelated strains (Turner and Chao, 1999), a result that was confirmed for an *Escherichia coli* bacteriophage (Kerr et al., 2006). The emerging field of sociovirology (Diaz-Muñoz et al., 2017) recently gained momentum after Hamiltonian predictions of cooperation and conflict could be shown to apply, mediated by quorum sensing in a way similar to cellular *P. aeruginosa*. Social cheating traits in viruses were recently documented to be abundant as well (Leeks et al., 2021). Of particular note is the discovery of a bacteriophage-produced peptide fragment that enables viral communication across infected host cells. This allows for collective decisions to either lyse a host cell (at low concentrations) or to lysogenically integrate in the host's genome so the infected cell survives and will henceforth propagate the viral genes indirectly (Erez et al., 2017). Because these peptide fragments are strain-specific, this form of communication with ancestral clone mates tells new viral infections whether uninfected host cells are available to make the lytic strategy rewarding. The fact that viral social life depends on organized cellular life implies that these phenomena must be secondary developments that could not exist in the RNA world.

Somatization inside free-living cells

Huxley (1912) assumed that closure of the first cell preceded its internal division of labor, a conjecture that remains consistent with some strain-specific origin-of-life scenarios currently considered realistic (e.g., Chen et al., 2004; Sojo et al., 2016). However, division of labor is not necessarily only a post-MTE phenomenon—societies of cells have achieved some cell-type differentiation in spite of lacking the potential to realize an MTE, as, for example, in budding yeast (Ratcliff et al., 2012; Regenberg et al., 2016). More specifically, *D. discoideum* has five (Schilde et al., 2019) and *B. subtilis* has six cell types (Aguilar et al., 2015). That division of labor can evolve in nonclonal cell societies is consistent with a model by Cooper and West (2018), even though the permanently multicellular lineages that arose clonally almost invariably have more cell types (Fisher et al., 2020). However, what matters is not that some division of labor may be in place before an MTE becomes irreversible, but that partnerships remain non-redundant in the sense of not competing for the same organismal functions as, for example, multiple strains of mitochondria within the same cell would do. If Huxley's principle that nothing homogeneous can be alive and Williams' rephrasing of that insight in terms of genetic information (see opening quote of this chapter) hold, the origin of organized cellular life (LUCA) must have been the result of a series of non-redundant internal differentiation steps, not unlike an analogous sequence that is now becoming better understood for LECA (Roger et al., 2021). Given that the deepest division of labor at fraternal MTE origins is germline–soma differentiation (see Chapter 6), it then seems legitimate to ask whether an analogous principle might have applied to the early egalitarian MTE origins, so that intracellular differentiation resulted in functional germlines with informational genes and somatic elements with organizational functions (Box 7.3).

Contemplating possible organizational analogies between LUCA and LECA inevitably leads to asking how proto-ribosomes shaped LUCA's origin (Szabó et al., 2002). We know that ribosomal RNAs are among the 100–200 prokaryote genes (3% of the genome) that are essentially always vertically transmitted, strongly suggesting that this set must represent the core informational genes of early cellular life (Harold, 2014, pp. 26–27; Root-Bernstein and Root-Bernstein, 2015). This then begs the question how the circular chromosome became a co-carrier of informational genes and which other agents might have affected early cytoplasmatic somatization. Candidates for this function might be horizontally acquired and subsequently vertically transmitted proto-plasmids or plasmid-like virions (Krupovic, 2012), particularly if they could have mediated coupling between transcription and translation (Irastortza-Olaziregi and Amster-Choder, 2021). LUCA subsequently produced Bacteria and Archaea, adaptive radiations of comparable but different complexity, each with their own plasmids (Shintani et al., 2015). After this diversification, a single representative of these two lineages merged in the hierarchical LECA arrangement whose informational and organizational complementarity is now

known in more (Box 7.3) but still insufficient detail (Harold, 2014, p. 85). It is interesting that Archaea have rarely become intracellular pathogens or mutualists of eukaryotes, while bacterial pathogens and endosymbionts abound (Abedon, 2017). This seems to suggest that bacteria were somehow preadapted to contribute somatic operational genes to LECA whereas the Loki Archaeota may have had preadaptations to function as informational hosts. However, the largely unresolved question is whether hierarchical differentiation between informational and operational functions may also have characterized LUCA's origin, that is, whether the concept of germline–soma differentiation can be extended backwards to the very start of organized life.

These inferences are perhaps too speculative to be real arguments, but they serve the purpose of loosening our perspective on germline–soma division of labor. Once we accept that Weismann's original germline concept was an oversimplification of what may in reality be a dynamic form of reproductive versus somatic division of labor, that is, a variable life-history syndrome, it becomes natural to ask whether unicellular individuals and their clonal copies suffer from aging by default, similar to multicellular organisms and superorganismal colonies (see Chapter 6), and whether such patterns would then be connected to asymmetric cell division (Ackermann et al., 2007a). This asymmetry condition was proposed by Partridge and Barton (1993) who conjectured that aging should be expected in any organism where cell division is asymmetric in the sense of there being an older mother cell and a younger daughter cell, rather than two identical cell copies. A further argument for including aging in the comparative MTE framework of this book is that the metazoan logic of germline–soma differentiation could at least in part be extrapolated upwards to apply to colonial superorganisms, provided we took into account that there are fundamental differences in how germline aging and soma aging proceed (see Chapter 6). When going the other way, downwards to the unicellular levels, we should therefore be prepared not to find precise analogies, but patterns that may allow parallel Darwinian interpretation across the complexity domains. In another review paper, Partridge and Gems (2002) used the terms "public" and "private" for mechanisms of aging across distantly related metazoan lineages to discriminate between general and lineage-specific explanations. The question then becomes whether aging as a phenomenon is public in the sense of being general, while each of the four domains of social evolution (see Figure 3.2b) have distinct private aging principles related to their organismal complexity levels.

As mentioned at the beginning of this chapter, the organismal study of protists was a key research area in the early 20th century (see Buchner, 1953 for an historical overview). In his monograph *Sex and Death in Protozoa*, Bell (1989) provided an exhaustive review of this early research and reinterpreted its findings (30% of his references are from before 1925) in light of modern neo-Darwinian concepts. He focused in particular on the intriguing question whether laboratory cell cultures of protists always or only sometimes senesce, a question that had been pursued by many, among them Jennings and Lynch (1928) whose studies of asexual rotifers would be continued by Lansing (1947) which led to the term 'Lansing effect' that we encountered in Chapter 6. Bell's (1989, p. 43) finding that lineage-level senescence (measured as decline in cell division rate in laboratory cultures over time) was almost universal appears to offer a dilemma because later cell copies cannot be considered as somatic altruists to previous and no longer existing ancestral cells. Yet, referring in particular to the ciliates, there is no doubt that cell lines often have three life stages. In

Bell's words, "an initial period of immaturity, during which conjugation does not occur; a period of maturity, during which mating occurs regularly, and leads to the production of viable exconjugants [offspring cells after sex]; and a period of old age, signaled by the decreasing vitality of exconjugants and eventually, vegetative senescence" (Bell, 1989, p. 131). In addition, he argued that the evolution of large cell size in ciliates required division of labor between a somatic macronucleus and a germline micronucleus. Both accumulate deleterious mutations over time, but only the germline micronucleus is involved in meiotic sex and can thus purge genetic load (Bell, 1989, p. 123).

As I write 30 years later, we know more, but still remarkably little, about germline–soma segregation in ciliates. The micronucleus has been confirmed to be a truly sequestered, transcriptionally silenced germline with a very low mutation rate, while the macronucleus genome is indeed responsible for the cell phenotype (Sung et al., 2012; Long et al., 2018). A substantial, albeit variable, fraction of the non-transcribed micronucleus genome is deleted at each reproductive event (Chalker et al., 2013), suggesting a genuine Weismannian segregation from the somatic macronucleus functions. However, little new research with an organismal focus seems to have been added. For example, Bell's mutation accumulation hypothesis for the sexual propagation of the two nucleus types remains to be tested, even though its core idea—that observed periods of immaturity in a life cycle are "a compromise between a longer period that would reduce the viability of exconjugants too far, and a shorter period that would entail too great a loss of opportunities for sexual reproduction" (Bell, 1989, p. 133)—seems highly relevant now that multicellular germlines are increasingly conjectured to be such optimal life-history compromises (see Chapter 6). Later studies of asymmetry in cell division have remained limited in number as well, even though asymmetrically budding yeast and symmetrically dividing fission yeast are obvious comparative model systems. For example, fission yeast cells in laboratory cultures with optimal resources divide symmetrically and no aging is apparent, but stress-induced asymmetric division concentrates protein complexes in one of the daughter cells that dies almost instantaneously (Barker and Walmsley, 1999; Coelho et al., 2013). In budding yeast, such protein aggregations characterize older cells and induce inability to engage in meiotic sex that would allow recombination (Zhou et al., 2014; Schlissel et al., 2017). If aging in protists is as common as Bell (1989) suggests, it becomes an interesting question whether ancestral lineages to permanently multicellular MTE origins might have had symmetric or asymmetric cell division.

Prokaryotes also offer interesting models for testing evolutionary hypotheses of differential aging based on asymmetries in cell division. Early studies indicated that cell aging does occur, both in *Caulobacter* where asymmetric mother–daughter division can be directly observed (Ackermann et al., 2003, 2007b), and in *E. coli* where cells appear to divide symmetrically but decline in vigor depending on the age of cell poles (Stewart et al., 2005). Also here, protein waste accumulaties in older cells (Lindner et al., 2008; Boehm et al., 2016), similar to what has been found in budding yeast. However, the Stewart et al. (2005) result was later argued to be reminiscent of stress-induced aging as it could not be reproduced when cultures were kept under unambiguously benign conditions (Wang et al., 2010), a result remarkable parallel to the fission yeast results by Coelho et al. (2013). Many proximate aspects of asymmetric cell division show striking parallels between unicellular and multicellular organisms as well, particularly for

stem cells (Inaba and Yamashita, 2012; Katajisto et al., 2015), so one wonders whether asymmetric but growth-independent inheritance of harmful protein aggregates may be a general principle by which cells age across the complexity domains (see also Rang et al., 2011). It is puzzling that this field has primarily developed in proximate molecular biology directions, while opportunities to test general evolutionary principles seem so obviously worthwhile (Florea, 2017). If aging would turn out to be common across bacterial cell lines, it might shed novel light on the challenges that domesticated organelles met when they no longer had options for HGT and conjugation, mechanisms that otherwise mediate considerable innovative potential (Norman et al., 2009). Recent studies, discussed by Shi and Huang (2019) have indicated that even prokaryote cells have specific and mutually exclusive positions for the nucleoid (chromosome) and the cytoplasmatic ribosomal proteins so that informational replication and operational protein synthesis are fundamentally compartmentalized.

Further suggestive evidence for a consistent distinction between germline and soma in unicellular organisms was provided by Aanen and Debets (2019). They interpreted a comparative data set across bacteria and yeasts obtained by Krašovec et al. (2017), showing that mutation rate appears to be a phenotypically plastic reaction norm where high cell density induces low mutation rates, and low density induces high mutation rates. Using modeling scenarios with exponential and linear growth, Aanen and Debets found that patterns were best explained when asymmetric cell division at high densities resulted in aging mother cells that continue to divide and in rejuvenated daughter cells that stop dividing until conditions improve. Dividing mother cells would thus assume somatic functions while securing uncompromised germline templates via dormant daughter cells. In *E. coli*, this plasticity is mediated by simple quorum sensing (Krašovec et al., 2014), suggesting that the negative correlation between mutation rate and population density could be a generally adaptive life-history trait in the unicellular domains. Asymmetric cell division for positional retention of template DNA also applies in some multicellular fungi, but here the mother stem cells stay dormant (Aanen, 2019). These results suggest that there are two ways to protect germlines from mutation accumulation: reducing the number of mutations per se by minimizing cell division rate as in animals and possibly plants, or maximizing copying fidelity via asymmetric division as in unicellular microorganisms and fungi (Aanen, 2019). The specific contributions of mitochondrial mutations remained implicit, but were explicitly modeled for multicellular organisms to explain that early germline sequestration is adaptive in animals while plants with lower mitochondrial mutation rates are better off without sequestered germlines (Radzvilavicius et al., 2016).

This chapter has highlighted the power of inclusive fitness theory for understanding adaptation in the unicellular domains and argued that the origins of LUCA and LECA as organisms depended on informational closure around lifetime committed genetic information carriers, similar to the fraternal MTEs (see Chapter 6). I explored aspects of germline–soma differentiation and aging that can be qualitatively compared across the domains of social evolution, with due modifications as we should expect when comparing different levels of organizational complexity. My analysis confirmed that nature has produced strictly symmetrical forms of commitment in gametic mergers such as zygotes (see Figure 5.3), but that splitting up during binary cell fission or mitotic division may never be completely symmetrical. This would predict that aging is generally unavoidable albeit hard to detect when cell division approaches symmetry. Although modern molecular work has given fascinating insights, it is remarkable how relevant neo-Darwinian organismal biology remains to ensure that biology continues to get its fundamental concepts right. Classic definitions of the autonomous organism (Reynolds, 2007, pp. 71, 76) and of organization-specific immune defense (Pradeu, 2016) stand as tall today as they did previously, just like T.H. Huxley's notion of the principal difference between cell aggregation and cell adherence, with only the latter enabling epigenetic development (Reynolds, 2007), and thus MTE potential towards permanent multicellularity as conjectured in this book. Also Julian Huxley's insistence that the survival value of a phenotypic trait is intimately connected to an organism's development, that is, to the continuance of individuality over time, remains generally relevant. It underlines that individual agency and independence of the outer world apply to lone cells in analogous but simpler terms than we need to use in the multicellular domains, as Huxley, (1912, pp. 16, 25, 28) clearly appreciated. Viewing the cell as a naturally selected social agent is therefore at least as powerful a metaphor as viewing the cell as a machine (Reynolds, 2018)—all this is asking for a better balance between top-down organismal and bottom-up reductionist approaches.

CHAPTER 8

Adaptation, control information, and the human condition

Summary

Following the inductive logic of Hume's fork, this book has partitioned Hamilton's rule so that it offers a general necessary and concrete sufficiency condition for the origins of major transitions in evolution (MTEs). My conjecture could in principle have been formulated decades ago, but several forms of confusion have constrained its development. First, Darwinism was "environmentalized" throughout much of the 20th century, often becoming a mere extension of ecology or genetics, which allowed researchers to avoid questions about adaptive design. This problem disappears only when adhering to a strict definition of what an environment really is, as George Williams did in 1966 and Helena Cronin updated in 2005. Second, many have failed to appreciate that the gene's eye view of adaptive evolution is a concept of genetic information and phenotypic agency rather than of genetic reductionism. Making that perspective explicit helps to connect inclusive fitness theory to Shannon's information theory, because both are about the use of information for prediction and control. It then emerges that inclusive fitness theory, captured by Price equation logic of selection and transmission, is the only biological theory that might ultimately be reconciled with theoretical physics via quantum Darwinism. Third, anthropomorphic bias has been rampant in biology, particularly in conceptualizing the human MTE. Our own transition is exceptional in not having originated via a new level of lifetime commitment and obligate reproductive division of labor, but from an already complex society that became domesticated by cumulative culture produced through endosymbiosis between individual brains and sets of self-organizing algorithmic symbionts. In that perspective, a dual Price equation framework may be the best unbiased tool for unraveling the extent of cooperation and conflict between our genetic endowment and the reproductive interests of our Lamarckian brain symbionts. Dual host–symbiont agency parsimoniously removes the "self" from anthropomorphic concepts such as "self-domestication" and "self-deception". The epilogue summarizes the main predictions of the kin selection theory of organismality outlined in this book.

> *When I was writing the manuscript in the early 1960s, I was convinced [. . .] that adaptation was pervasive in biology, essentially defining the subject; that natural selection could explain all examples of adaptation; that adaptations, with few exceptions, were the properties of individual organisms and not groups thereof. I fully expected that the perspective I urged would ultimately be accepted as orthodox; however, I did not at that time expect it to prevail so soon.*
>
> **Williams, *Adaptation and Natural Selection* (1996, second edition, p. x)**

Much of this book has elaborated this retrospective quote. I have updated the evidence and made the implications explicit by keeping groups (societies) and (super)organisms apart, because only the latter can evolve novel higher-level adaptations. I have confirmed that

Domains and Major Transitions of Social Evolution. Jacobus J. Boomsma, Oxford University Press. © Jacobus J. Boomsma (2022).
DOI: 10.1093/oso/9780198746171.003.0008

genetic relatedness is an exclusive predictor of condition-dependent reproductive altruism in societies, and I have argued that MTE origins towards new levels of organismality are something entirely different because they have universally required non-redundant informational closure via strict lifetime commitment (LTC)—and thus maximal possible relatedness among an individual's offspring propagules (see Table 3.1). My explanation of the nonhuman MTE origins used Hamilton's rule, a fully sufficient heuristic for understanding the continuous dynamics of social evolution and adaptive design (Levin and Grafen, 2019), and I showed that LTC allowed partitioning the rule into a necessary and sufficient condition for how life's discontinuous, static levels of nested organizational complexity must have evolved. This conjectural framework has the structure of Hume's fork. It first identified an analytical a priori condition (LTC) that has to be true in an abstract, invariant sense and whose origin is independent of previous social evolution, that is, LTC can be a functional exaptation analogous to the structural concept of Gould and Vrba (1982). It then, by the necessity of Hamilton's rule, identified the contingent sufficiency condition (long-term $b/c > 1$) that captures the remaining a posteriori explanations for all concrete hierarchical MTE origins outside the human domain. The logical implications of this conceptual framework were summarized in Figure 3.2b while Figure 8.1 encapsulates the conclusions reached in Chapters 4–7 and makes explicit the key differences between societies and (super)organisms and between relatively common evolutionary reductions and reversals, which are impossible after MTEs.

Figure 8.1 conveys, firstly, that all expressions of nonhuman social life have been shaped by adaptive, hierarchical MTEs, and that society life always evolved within these MTE domains but not across them. Secondly, it makes clear that there is a natural upper limit for Darwinian explanation of organizational complexity, set by the lifetime committed monogamous founding members of a closed family. I have added another diagram (Figure 8.2) to illustrate that there are four levels of neo-Darwinian explanation where Tinbergen's four complementary angles apply, and that there are higher levels where complexity needs to be explained ecologically because these levels cannot generate adaptations themselves—they are just populated by organisms that do. Ignoring this distinction (Williams, 1966a, pp. 247–248) has caused significant confusion by nourishing the idea that loose ecological pattern hierarchies are the universal basis of organismal biology, rather than natural selection having established the four fundamental MTE levels of hierarchically nested organizational complexity (see Figure 3.2). This confusion has made some infer that non-molecular biology is just about ecological "population-thinking," phylogenetic "tree-thinking," and developmental "homology-thinking" (e.g., Wagner, 2016) with the implicit suggestion that evolutionary change just happens when it happens, similar in fact to the mid-20th-century convenience Darwinism that Williams (1966a) showed to be inadequate and misleading (see Chapter 1). More than 50 years later we need to continue insisting that Darwin's (1859) seminal book was explicitly about the origin of adaptations and that there is ample modern evidence that the only process that can maintain, over evolutionary time, organizational complexity beyond thermodynamic equilibrium is natural selection (Pittendrigh, 1958; Dennett, 1995; Atkins, 2003), not environmental or mutational variation itself. Believing in forms of environmental stochasticity or genetic drift as organizing forces is like maintaining one's household by investing in lottery tickets rather

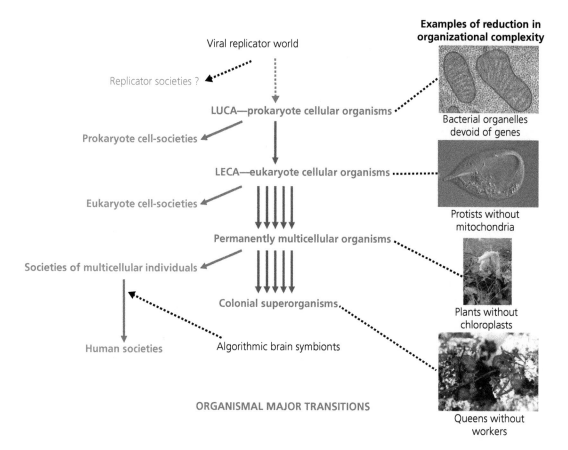

ORGANISMAL MAJOR TRANSITIONS

Figure 8.1 The major transitions in a nutshell. The hierarchical levels at which life has experienced ratchet clicks towards irreversibly enhanced organizational complexity over evolutionary time (red), relative to sister lineages that remained in the previous level of complexity and lower-level agency, where societies can form but new levels of germline–soma-differentiated organization cannot (each single blue arrow represents several to many such developments). As illustrated in Figure 3.2, all four nonhuman domains of social evolution, and the transitions between them, can be conceptualized to have arisen by non-redundant hierarchical closure in terms of genetic information, forms of LTC that always ensured full alignment of interests in resource acquisition, collective defense, and reproducing jointly (see Table 3.1 for definitions of terms). Organized life (bold-faced print and solid arrows) started with LUCA, but extant viruses are increasingly known to have evolved social traits shaped by kin selection just like those in societies formed by cells (see Chapter 7), hence the dotted black arrow at the top. However, viruses are never independently organismal no matter how big viral genomes may become—outside host cells virions decay with fixed half-lives similar to radioactive atoms. In this respect, there is a remarkable parallel between the beginning and the end of the arrow diagram, because the algorithmic cultural brain symbionts that forged the human MTE (black dotted arrow at the bottom) are not independently alive or organismal either—archaeological human artifacts also decay with specific half-lives unless we inhibit that process by active preservation. Images towards the right illustrate that organizational complexity at all four fundamental levels can be secondarily reduced when lineages specialize on parasitic or mutualistic lifestyles, as first noted by Huxley (1912), but that these reductions never implied reversals to a previous level of organizational complexity. Thus, prokaryote symbionts of reduced complexity (mitochondria; top image; photo credit Mariya Zhukova) never came to resemble acellular viral particles, parasitic protists without mitochondria (*Saccinobaculus ambloaxostylus*; second image; photo credit Patrick Keeling) never returned to prokaryote cell complexity, parasitic plants lacking chloroplasts (*Monotropa hypopitys*; third image; photo credit Hans Henrik Bruun) never lost their multicellularity, and socially parasitic inquiline ants that lost the somatic worker castes that characterize the free-living sisters lineages that they exploit (*Acromyrmex insinuator* queen next to its *A. echinatior* host queen; fourth image; photo credit David Nash) never came to resemble solitary digger wasps.

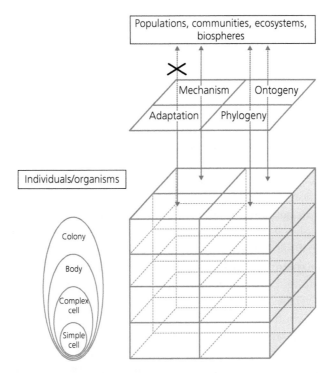

Figure 8.2 Delineating the domains and research agendas of the neo-Darwinian synthesis. The four complementary questions of Tinbergen (see Figure 1.1; Box 1.3) apply (red downward arrows) in the same way in each nonhuman domain of social evolution (red text in Figure 8.1). This deep analogy is rooted in the fundamental nestedness of the four hierarchical levels of organizational complexity (concentric ellipses to the left; see also Figure 3.2b), that have been uniquely or recurrently initiated by MTEs. The adaptation quadrat is always about the maximization of inclusive fitness, and the ontogeny quadrat applies not only in the multicellular (somatic body) domain, but also in the superorganismal (somatic colony) domain and, quite likely, in the two unicellular domains (see Chapter 7). Within the 4 × 4 cube, phylogenies are trees of organized life, and mechanisms can be any proximate factor. The 4 × 4 cube thus defines and delineates the entirety of non-human organized life to which aspects of individuality and organismality apply as first clarified by Huxley (1912). More encompassing levels (populations, communities, ecosystems, biospheres) are defined by their ecological properties, but these are irrelevant for Darwinian investigation because they are not defined by closed, non-redundant (sets of) genetic information, a lack of organizational stringency that precludes that adaptations can evolve for the exclusive benefit of these higher levels. As the upward red arrows indicate, these higher levels offer crude reminiscences of ontogeny (ecological succession), phylogeny (historical contingency of ecosystems), and mechanism (any proximate biotic or abiotic factor), but not of adaptation (crossed out). This does not mean that communities are not full of well-adapted organisms, but only that these organisms always evolve their adaptations at the four lower levels of individuality/organismality towards the left. Neither does it mean that individuals residing in the 4 × 4 cube have no ecologies—ecology provides the context for natural selection in all four quadrats, but nothing else of first-principle significance. And because higher-level adaptation is impossible beyond the core family, we should not pretend either that Tinbergean physiological ontogeny applies to ecosystems, or that aspects of historical contingency (e.g., continental drift) or mechanism (e.g., soil fertility or salinity) are comparable to evolutionary explanations where organization is synonymous with organismal adaptation. Throughout the 4 x 4 cube that specifies the domains of social evolution, all organization is relative and end-directed, so that adaptations are never perfect but always the converse of randomness (Pittendrigh, 1958). That statement has no analogy at the higher ecological levels, which are defined by the abundances of populations and species, energy flow, spatial structure, and degree of stability over time. This organismality scheme does not cover societies, which precluded the inclusion of the human condition.

than pursuing incremental income directly or indirectly based on labor.

Thirdly, Figure 8.1 sets the human MTE apart as the only one that could happen without somaticizing and "germlinizing" its participants. Instead, it uniquely evolved from ancestral societies that were already complex without cumulative culture. This suggests a form of human exceptionalism that has nothing to do with our well-documented gradual descent with modification from a great ape, but with an exceptional MTE forged by self-organizing algorithms gaining a foothold in, and starting to coevolve with, individual human brains as host substrate. Figures 8.1 and 8.2 therefore show that explaining the human MTE requires asking different questions than we needed for all other MTEs.

Given the frequent confusion between Darwinian arguments of adaptation and sociobiological, ecological, or biogeographical arguments of pattern (see end of Chapter 6), this chapter will first ask what environments really are, which will lead to the conclusion, drawn by Helena Cronin (2005), that outside the human domain they are nothing else than gene-discovered opportunities. A second natural question will then be how organisms have stored adaptive genetic information about environments since the dawn of life, and translated it into phenotypic agency. To obtain tentative answers, I use insights from cybernetic systems biology that are exclusively applicable to the four neo-Darwinian domains of organizational complexity (Figure 8.2). Exploring informational aspects of adaptive design allows comparison with how other natural sciences understand order and disorder, analogies that suggest that inclusive fitness theory will ultimately be reconciled with at least some of the other grand theories of science. After these two sections, which are complementary efforts to eliminate as much anthropomorphic bias as possible, I attempt a brief evaluation of the human condition

within the MTE context that I have used throughout this book.

Genes discover environments and identify them as opportunities to replicate

Williams' (1966) book *Adaptation and Natural Selection* was subtitled "a critique of some current evolutionary thought." This is rarely made explicit although the opaque argumentation that he challenged is a phenomenon of all times. Cronin (2005) recently refreshed our memories of Williams' logical principles while adding that his "concept of adaptation has again become abused, even traduced," necessitating a new early 21st century critique. Her starting point was to posit that environments do not exist until they are discovered by genes, which implies that environments are fundamentally disinterested entities without purpose or meaning unless genes start exploiting them, similar to what Pittendrigh (1958) emphasized (see Box 1.2). As he wrote, the statement "Organism X is fit for environment Y" is not equivalent to "Environment Y is fit for organism X", because such synonymizing "obscures the fundamental asymmetry of the relationship, the fact that the essential nonrandomness of adaptation is due entirely to the organism's (not the environment's) capacity to accumulate and retain information both phylogenetically and ontogenetically." In this context, Cronin's "discovery" must therefore be stripped of its anthropomorphic meaning of finding or learning something for the first time—it just means the initiation of an informational connection. When adaptive design makes organisms act as if they maximize inclusive fitness (Levin and Grafen, 2019), Cronin's gene-discovery definition of environments implies that human consciousness is explicitly defined by our exceptional ability, not shared with any other organism, to reflect on environments that are irrelevant for our own inclusive fitness. As Cronin (2005) argued, the fact that genes exploit environments to replicate implies that genes and environments do not compete for deep causal explanations in biology. Instead, the two connect "in such a way that the richer and more complex the adaptations, the richer and more complex the environments they rely on." Allelic gene-discoveries can be gained and lost over time, but the ones that become part of robust gene regulatory networks (GRNs) (see Table 3.1) prevail (Wagner, 2005, 2011). Each MTE origin can thus be conjectured to be unique, because it must have fixed massive and robust developmental gene expression networks for irreversibly enhanced, hierarchically nested, division of labor throughout development (see Figure 3.2b; Figure 6.6).

Cronin (2005) confronted notions of environmentally induced phenotypic plasticity by West-Eberhard (2003) with the stringency of Williams' teleonomic theory of adaptation. She found them wanting because Cronin's environment definition leaves no room for the idea that environments can ever be in the driving seat of evolutionary causation. As so often, the key point relates to Williams' (1966a) fundamental distinction between the function of an adaptive trait and any mere byproduct effects that remain without adaptive consequence unless discovered by new genes. She argued that if a rarely expressed phenotype is within the naturally selected design of an adaptation, it is inconceivable that it is really new and there is no reason to assume that its expression will allow evolutionary responses that were not already understood with standard theory of phenotypic plasticity and reaction norms (e.g., Crozier, 1992; Stearns, 1992; Delgado et al., 2020). Alternatively, a new phenotype may be due to a recombination event that merely reflects a disruption, that is, a byproduct effect. But then it cannot be part of any adaptive design, so its haphazard genotype is vanishingly improbable to become

important for adaptive evolutionary change. In the end, it does not matter which of these scenarios applies, because new gene mutations will always be required to make an organism become "responsive to new aspects of the world-at-large and in such a way as to offer new solutions to adaptive problems" (Cronin, 2005). This sobering critique in the style of Pittendrigh and Williams (see Box 1.2) returns the phenomenon of phenotypic plasticity to the canon of neo-Darwinism, which offers a clear conceptual understanding of reaction norms and of the ways in which trial-and-error learning may connect to novel gene discovery via an adaptive assimilation process known as the Baldwin effect (Baldwin, 1896; Dawkins, 1982a; Dennett, 1995) (see Chapter 3).

Cronin's (2005) critique exposes that implicit "environmentalized" Darwinism—a problem of which I have given other examples earlier—is incorrectly argued in a principal manner because environments cannot drive any adaptation by themselves. In addition, concepts about how genes involved in phenotypic plasticity may mediate adaptations need to follow the parsimony of Occam's razor. Here again, Williams (1966a, p. 81) led the way when he wrote that "[t]he origin of a fixed adaptation is simple. The population merely needs to have or to acquire some genetic variation in the right general direction. The origin of a facultative response is a problem of much greater magnitude. Such an adaptation implies the possession of instructions for two or more alternative somatic states or at least for adaptively controlled variation of expression. It also implies sensing and control mechanisms whereby the nature of the response can be adaptively adjusted to the ecological environment." In this view, plasticity is as onerous a concept as adaptation itself (see Box 1.4), so it should not be lightly invoked or interpreted by adopting increasingly inclusive interpretational concepts of the least common multiple type (see Box 1.3). Faithful to Hamilton's (1996) narrow roads of gene land and its greatest common divider arguments, Williams concluded that "[o]nly when adaptive adjustment to uncertain conditions would be important would one expect facultative control. Since the obligate is more economical of information, it can always be expected in situations in which a facultative response would not be significantly more effective" (Williams, 1966a, p. 82–83). Of course he knew, as Tinbergen (1965) had made explicit, that benefits of plasticity and learning often exceed their costs. His point was that such cases need to be proven by comparative research—one cannot assume that plasticity is a kind of natural null hypothesis of life.

While plasticity arguments merely need a clear interpretational context to be properly Darwinian (e.g., Levis and Pfennig, 2016, 2019), recent epigenetic challenges of Williams' understanding violate the crucial distinction between proximate and ultimate explanations more seriously by suggesting that Lamarckian mechanisms can drive adaptive change (e.g., Jablonka and Lamb, 1995, 2002, 2005). It is striking that such arguments are always presented with phrases like "is possible," "may be relevant," "could be important," or "can reflect," and that testable predictions are never offered. This is as expected when epigenetic factors are only proximate effects. Committed enthusiasts often ignore that epigenetic mechanisms must also represent naturally selected adjustments in the expression of genes. These genes originated by random mutation but came to offer higher inclusive fitness to phenotypes when their effects became extended across two or three generations, a parsimonious ultimate explanation consistent with all evidence (Tooby et al., 2003; Haig, 2007, 2012; Scott-Phillips et al., 2011; Uller et al., 2015; Bonduriansky and Day, 2018; Henikoff, 2018; Adrian-Kalchhauser et al., 2020). Why then is it that some like to believe that magic epigenetic bullets exist? Despite the first-principle

confidence that Williams expressed in the epigraph of this chapter, it appears that recent decades have instilled a false belief in the all-sufficiency of proximate data (Stern, 2019; Nurse, 2021), a kind of postmodern Baconian myth (see Chapter 1). We have seen the emergence of bottom-up developmental genetics to study epigenetic mechanisms without any relationship to the top-down adaptive epigenetic concepts developed by Waddington (Nijhout, 1990; Jablonka and Lamb, 2002; Haig, 2007, 2012), and a similar divide was diagnosed for microbiology in a *think before you sequence* essay (Prosser, 2013). Molecular biology that pretends not to need Darwinian understanding remains a field whose ad hoc hypotheses limit its significance.

As Cronin (2005, p. 20) noted, "[t]he means by which genes gain access to the world's properties is through adaptations. In shaping adaptations over evolutionary time, natural selection has seized on regularities, features of the world that are stable, recurrent, and dependable, features that can be relied on, generation after generation, so that adaptations can do their work." Williams (1966a, p. 259) categorized those feature-providing environments as being *genetic* (other genes in the same genome), *somatic* (e.g., organelle genomes in the same cell), *social* (directly interacting other agents), or *ecological* (see Box 2.3; Chapter 3). The implication, across the four domain levels in Figures 8.1 and 8.2, is that other cells are "social environment" for a cell society of *Dictyostelium*, but somatic environment in a fruit fly body. Similarly, other wasps are social environment in a *Polistes* society, but somatic environment in a yellowjacket colony, the terminal wasp branch that became superorganismal (see Chapter 6). Novel genes thus seem comparable to one-person start-up companies introducing a single new discovery in a market of much larger and well-established competitors who work with entire networks of patents that generate consistent profits. Start-ups might go all the way under their own steam but are often bought up when their new discovery can be integrated into a larger context with profits for all parties (note that forms of spiteful elimination also occur). It is in this way that novel genes and the new opportunities they identified can turn invention into phenotypic innovation (see Schumpeter, 1926). As Hamilton (1987b) put it, "complex traits of altruism can be built up by accumulation of genes that satisfy the criterion [of Hamilton's rule]. Each adds to a set already there and this set – to which, in effect, the new gene is applying for admission – is part of the [genetic] 'environment' over which gene effects must be averaged in determining whether selective forces will allow it to rise in frequency."

Natural selection has long been known to be algorithmic because it operates on genes carrying digital information (Kimura, 1961; Monod, 1975; Williams, 1992; Dennett, 1995; Frank, 2012c; Sherwin et al., 2017). As Cronin (2005) wrote in summarizing what adaptations are for, "natural selection accumulates in the species' gene pool an appropriate store of information about environmental regularities [and f]rom this archive, the organism can pick out the current instantiations that it encounters in its own lifetime of that long, unbroken succession of past environments." This implies that gene pool information is directly related to individual agency and adaptive organismal design (Frank, 2018), and that Hutchinsonian (1957) ecological niches represent the total of interrelated opportunities that sets of organismal GRNs utilize to maximize inclusive fitness. Because MTEs are discontinuous enhancements of adaptively designed organizational complexity (see Figure 3.2), it should imply, as Huxley (1912) was the first to note, that unicellular bacteria and protists have less adaptive potential than multicellular organisms, which may in turn be surpassed by colonial superorganisms—in spite of the rate of adaptive evolution being inversely proportional to generation time. In Cronin's (2005) words: "In the beginning, natural selection created genes that could barely exploit the heavens or the earth; their adaptations were meager and their environments commensurately without form and void. But, down evolutionary time, genes have bootstrapped themselves from inchoate nakedness into magnificent dwellings, organisms of great sophistication." This is, and without a trace of undue reductionism, a modern particulate genetics version of Darwin's (1859) statement that "from so simple a beginning endless forms most beautiful and most wonderful have been, and are being, evolved." It is the genome-wide statistical genetic information perspective (see Box 2.3) that allows such general conclusions when focusing on adaptive change rather than total change. Models based on theoretical material gene loci cannot do that.

The major transitions in terms of information, agency, and order

Digital allelic information increases by naturally selected trial and error, a form of slow "learning" that becomes accurately archived in a population's gene pool across the generations (Dawkins, 1982a). The question remains, however, how much adaptive order this Fisherian improvement process can actually create? The short answer is a significant but never maximal amount, as Darwin's first-principle theory of natural selection established and all later developments in biology confirmed (Frank and Fox, 2020). Yet, if so much fell into Darwin's germane theory's lap—to reuse Monod's (1975) metaphor from Chapter 1, why is this deep insight so often kept implicit and disguised in correlative gradient analyses that cannot augment first-principle understanding (see Chapter 6, final section)? Are many biologists conceptually confused or just reluctant to stare their own overarching theory in the face? Physicists and chemists have always been ahead of biologists in achieving understanding based on first principles that quickly became almost completely shared by their communities, even when abstractions began to surpass concrete imagination. These insights have greatly influenced, and remain essential for, the molecular and computational branches of biology, but few biologists would think that all they need is physics and chemistry. This issue becomes very explicit when we give center stage to natural selection as the only possible cause of adaptive design, no matter how moving the target may be, because it is that Darwinian theory that gives biology its distinguished status among the natural sciences, not physics and chemistry. The current gap between how physical engineers and organismal biologists understand phenomena of information and order remains wide, so it is legitimate to ask whether that discrepancy is likely to persist.

The road towards an encompassing understanding of the principles of life and matter has been one of increasing abstraction away from concrete day-to-day meaning. In the sciences, this has always implied that theory became statistical rather than deterministic. For example, Bohr's correspondence principle showed that the "certainties" of classical Newton mechanics are special cases of quantum uncertainties in the microworld (Wicken, 1986), and Claussen's concrete steam engine entropy was generalized by Bolzmann and Gibbs into statistical entropy based on energy-dependent probabilities of microstates (Corning and Kline, 1998a; Atkins, 2003). As Bohr (1933) phrased it, quantum theory gave atoms individuality and forced physicists to accept statistical probability answers while renouncing complete causal description. Likewise, we have seen in Chapter 4 that Hamilton's original inclusive fitness theory for family pedigrees was generalized by the statistical Price equation that works with an abstract measure of relatedness. As Frank (1995b) ably

portrayed, it was George Price who recognized for the first time that a general mathematical theory of selection—on par with Shannon's theory of information—was necessary to connect biology's logic of natural selection to the abstract achievements of the other natural sciences. Price understood that Shannon had only been able to develop his groundbreaking theory by using a purely mathematical definition of information deprived of anthropomorphic meaning, and he followed Shannon's example when deriving his equally abstract covariance equation. The Price equation therefore purged inclusive fitness theory of any anthropomorphic elements it might have had, for it only considers the statistics of selection and transmission. A crucial merit of such abstractions is that they carefully separate subject and object and impose clear definitions that are likely to carve nature at its fundamental joints.

Hamilton's rule ($br_x > cr_o$) (see Box 2.1) could be rephrased in Price equation terms because it met the criterion of purity of definition, but other remarkable connections have recently emerged as well. By establishing that natural selection maximizes a statistic known as Fisher information, Frank (2009, 2012c) showed that the environmental information accumulated in gene pools can be quantified. Fisher's fundamental theorem of natural selection then emerges as a natural consequence of maximizing Fisher information, which gives a direct relation with the acceleration of Shannon (1948) information. In a series of further papers, Frank (2017, 2018, 2020) continued to unravel other aspects of mathematical convergence. He argued that all fundamental equations of change have a similar form, including a first term of force that induces change in frequency of population members while holding the properties of members constant. A second term, usually of opposite sign, does the reverse, that is, it captures the change in frame of reference while keeping the frequencies of population members constant. As it turned out, natural selection and Bayesian updating (e.g., learning algorithms) increase information in analogous ways (see also Crewe et al., 2018), so the Price equation can be used to show that natural selection and information have an identical basis (Frank, 2017, 2018). These explorations also revealed that an invariant geometry perspective (probably first explored by Grafen, 1985, see also Charnov, 1993) may have the potential to ultimately show that all fundamental equations of science have a similar root (Frank, 2020). It thus appears that a connection between the Darwinian theory of adaptation through natural selection and Shannon's information theory is likely to become more firmly established fairly soon. However, convergences with theoretical physics appear to be more remote as yet.

While anticipating that analogies with information science look promising, Corning and Kline (1998a) concluded that biology's complexities cannot be understood with purely physical methods, which prompted them to develop a cybernetic "thermoeconomic" approach that explicitly requires teleonomic natural selection (Corning and Kline, 1998b). Their conceptual framework maintains that the structures and mechanisms that mediate the capture and use of energy for goal-directed work introduce bioeconomic criteria into thermodynamic processes that are reminiscent of biologically and economically familiar concepts such as costs, benefits, and efficiency. Their logic applies to organisms and superorganisms alike, but not to ecosystems or inert inanimate objects (Corning and Kline, 1998a), consistent with the early arguments by Huxley (1912) and the extensions developed in this book (Figure 8.2). The cybernetic implementation of thermodynamic reasoning by Corning and Kline (1998b) also identified a set of key characteristics of living systems that cannot be reduced to the laws of physics. These include closure, teleonomic drive, hierarchical organization, and the synergistic interactions between functionally interdependent organismal parts—a set that is almost identical to Huxley's (1912) criteria for defining an individual. Their logic remained focused on energy as joint currency and on the efficiency of its use by organisms. This appears compatible with gene copies in future generations as joint downstream fitness currency after natural selection and transmission have proceeded. As their arguments unfold, Corning and Kline (1998a) also showed that Schrödinger's (1944, p. 73) heuristic (i.e., not formally derived) argument that life feeds on negative entropy is either incorrect or without practical meaning because it is incompatible with adaptation-focused definitions of life in the sense of Pittendrigh (1958).

Corning and Kline (1998a, 1998b) defined the term "control information" as "the capacity (know-how) to control the acquisition, disposition and utilization of matter/energy in purposive (teleonomic) processes," referring to Pittendrigh (1958), and proposed this measure as a proxy for the cybernetic work accomplished by an organism. Control information has no independent material existence, consistent with Cronin's (2005) conjecture that genes accumulate virtual information about the environments that they define. This implies that control information shares with Shannon information that it requires a recipient user, but it differs in not needing a sender or being conceived of as a message. The accumulation of control information is determined by cost/benefit ratios associated with the fitness differentials that make adaptive information accumulate in population gene pools. Natural selection remains decisive throughout because a potential increase in control information will fail to substantiate when its energetic benefit/cost ratio remains unfavorable. Corning and Kline (1998b) concluded that "latent structural information becomes control information if and when it is utilized, and [that] its 'power' is a function of its organizing ability – the organizing 'work' that it can do with the available energy at hand in relation to a given system." To understand societies, it thus seems logical to connect selection for control information with Hamilton's rule, that is, with the b/c ratio scaled by a relevant relatedness ratio (see Box 2.1). Closure and non-redundant informational LTC as necessary condition for MTE origins (see Figure 4.6; top row Figure 5.1; Figure 5.3) then makes that scaling variable cancel so that symmetrical symbiosis (as in the last eukaryote common ancestor (LECA)) or directional kin selection (as in the fraternal MTEs) can breed irreversibly advanced organismal control information, consistent with a credible link between inclusive fitness theory and Shannon's (1948) information theory.

As a concept, control information extends broadly to cover both adaptations that are reducible to gene-level causation and irreducible self-organized complexity. While self-organization can create some physical order for free (e.g., Turing, 1952; Prigogine and Nicolis, 1971; Prigogine, 1978; Kauffman, 1993), this notion is insufficient for explaining organized life (Dennett, 1995). The origin of individuality and agency always implies what Dennett (2017) called *competence without comprehension*—abilities that need to be maintained by natural selection. As Johnson and Lam (2010) argued, self-organization is a spontaneous intrinsic-property process in physics, but the start of natural selection proceeded to domesticate self-organization by regulating and directing its spontaneity. In this view, self-organization is both a powerful instrument to mediate the expression of adaptive trait syndromes and a constraining force that can only secure stable organizational complexity within what is ontogenetically and mechanistically possible. Johnson and Lam highlight that appreciating the role of self-organization is key for understanding the connection between genotype and phenotype

during organismal development, consistent with the diagrammatic representation of the (white) hierarchical layers of self-organized complexity in Figure 3.2b. However, this should also imply the sobering notion that it remains a tall order for reductionist approaches to deeply penetrate these white zones when self-organization is an irreducibly complex phenomenon (see also Nicholson, 2014). They finally concluded that, "although selection does not need to construct an elaborate plan to generate complexity when self-organization is involved, it does have to drive the evolution of elaborate mechanisms for invoking self-organizing processes and controlling their dynamics," even when observed complexity seems spontaneous (Johnson and Lam, 2010).

The cybernetic approach of Corning and Kline (1998a, 1998b) reconciles natural selection and information theory in a system-orientated context not widely known among evolutionary biologists. I have explored its potential in a heuristic diagram that visualizes the MTE trajectories in control information terms (Figure 8.3). My confidence in this visual narrative was nourished by the similarity of Corning and Kline's (1998b) three-pronged thermoeconomic approach to Tinbergen's four complementary questions scheme (see Figure 1.1; Box 1.3) and to its match with the top-down versus bottom-up complementarity of organismal versus molecular biology at both sides of the white irreducible complexity zones in Figure 3.2b. As they write, "[i]n order to understand any complex multi-level system, at least three complementary analyses are necessary, namely: (1) a reductionist study of the detailed structure and functioning of the 'parts' at various levels, (2) a systemic view of the 'whole' and its emergent properties (including its behavior), and (3) an understanding of how the parts (and levels) fit together and interact with one another." This correspondence between logic based on first principles of behavioral and cybernetic

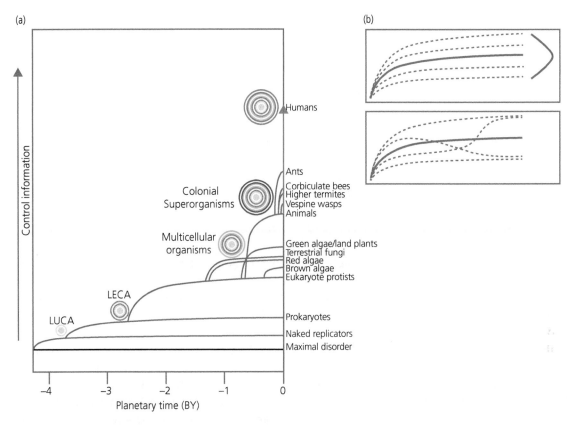

Figure 8.3 The major transitions as accumulations of hierarchical control information sensu Corning and Kline (1998a, 1998b). (a) The Hamiltonian perspective on MTE origins advocated in this book implies that Dawkins' (2006) evolutionary metaphor of *Climbing Mount Improbable* can be conceptualized as a discontinuous hierarchical set of trajectories with, at least initially, very limited overlap on the control information scale. Each MTE started a fundamentally new level of organizational complexity, either uniquely as in LUCA, LECA, and the human domain, or multiple times convergently as in the multicellular organisms and colonial superorganisms. Each trajectory can thus be conceptualized as a convex curve that, in cross section, retained the signatures of nested closure (colored concentric rings) that represent present and previous MTE histories, alternated by the matching hierarchical levels of irreducible self-organized complexity (white rings) (see Figure 3.2b,c for further details). The human MTE is exceptional because its outer (red) ring symbolizes a personalized egalitarian symbiosis with brain algorithms that enabled populations, via natural selection on individuals, to accumulate cumulative culture—a striking difference with adding a new layer of obligatorily somaticized individuals as in the colonial superorganisms. The diagram plots these developments on a horizontal time axis spanning the age of planet Earth (billion years (BY)) and assumes that new organizational complexity, or control information *sensu* Corning and Kline (1998b) (some would consider using the term negative entropy), is acquired steeply during MTE origins and more gradually later on. Approximate dates of MTE origins were derived from references in Chapters 6 and 7, and their putative accumulation of control information was approximated from the number of cell types (Fisher et al., 2020) and the diversity and elaboration of physical caste phenotypes across the four superorganismal lineages (Wheeler, 1928b; Wilson, 1971; Boomsma and Gawne, 2018). No curve has been drawn for the Siphonophora (see Figure 6.1) because the date of their origin is unknown. Although all MTEs achieved approximately global radiations, their species numbers (see Box 6.2) and total DNA content per cell vary enormously (Landenmark et al., 2015; e.g., Larsen et al., 2017). (b) Zooming in, each convex curve in panel (a) consists of a bundle of adaptive radiation curves which, in a null-model arrangement, can be assumed to form a unimodally distributed set of slowly diverging control information curves (dashed) that represent branching phylogenetic trees (upper panel). However, in reality such curves will form less regular control information trees because some lineages gain organizational complexity at a disproportionally enhanced rate while other clades secondarily lose independent control information after adopting specialized symbiotic lifestyles (lower panel; see also images in Figure 8.1). Overall, in terms of control information, *Climbing Mount Improbable* has accelerated over evolutionary time but it has remained a fundamentally discontinuous process.

reasoning appears to confirm that physics and organismal biology cannot presently share a joint philosophical framework. That inference seems much clearer now than after the involved argumentation that Mayr (1982) needed to make the same point (see Chapter 1). This is because the argument has reduced to physics being unable to ask and answer Tinbergian why-questions, and biology having no deeper Darwinian meaning unless it addresses these why-questions. The bigger question that remains is whether this current philosophical incompatibility is a final and inevitable reality or just reflects an underdevelopment of fundamental scientific theory that will likely be overcome in the decades ahead.

The contours of a possible future reconciliation between the theory of natural selection and statistical thermodynamics are conceptually hazy and fairly technical, so I have referred a tentative overview to Box 8.1 to not disrupt my narrative here. A key problem appears to be that Corning and Kline's (1998a, 1998b) analyses indicate that teleonomy cannot be derived from established laws of physics and chemistry, be it the second law of thermodynamics, the concept of dissipative structures (e.g., Prigogine and Nicolis, 1971), or any form of entropy, as famously suggested (but not derived) by Schrödinger (1944). Prigogine and Nicolis (1971) explicitly made "[n]o reference to concepts other than the laws of chemical kinetics and fluctuation theory," which precludes analogies with naturally selected biological adaptation. It is no surprise, therefore, that such approaches by what Corning and Kline call "structuralists" are always explicitly or implicitly anti-Darwinian because they effectively degrade biology to a derived branch of physics. In his organic chemistry account of the origin of organized life, Pross (2012) joined Corning and Kline (1998a, 1998b) in disagreeing with this view, writing that "[s]imply sweeping the issue of teleonomy under the complexity carpet with a shallow

explanation of 'emergent properties of complex systems' will not suffice. Such a response is little more than dressing up the discredited *élan vital* concept in scientifically more acceptable attire." While current physics thus appears unable to be relevant for understanding adaptive design in organismal organization *sensu* Figure 3.2b, a sufficient number of angles have been probed to at least tentatively suggest that attempts to reconcile physical and biological aspects of order ought to continue because the conceptual rewards of such further synthesis are potentially enormous (Box 8.1).

Intriguingly, Grafen (2018) recently noted that Prigogine (1978) interpreted the second law of thermodynamics as a concept of partial change over time, just like the fundamental theorem of natural selection, which Fisher (1930) claimed was analogous to that second law. As Grafen (2019, 2020) recently summarized, the ultimate aim of his formal Darwinism project (see Box 2.3) has been to vindicate Darwin's intuition that natural selection is an improving process, which he achieved by showing that Fisher's fundamental theorem mathematically captures the *dynamics* of Darwin's reasoning in a fully adequate manner. It is therefore not surprising that Frank (2012c, 2018, 2020) argued, and Grafen (2018) implied, that the coming decades may well produce further consilience (the "jumping together of knowledge" as coined by William Whewell, one of Darwin's tutors at Trinity College, Cambridge) among the natural sciences. Whether this will ultimately imply that the philosophies of organismal biology and physics can be reconciled remains to be seen. Three-quarters of a century ago, Schrödinger (1944, p. 81) contemplated that new understanding may have to come from quantum theory, if not a "new physics." We now have a field of quantum Darwinism (Zurek, 2009) claiming that macroscopic material reality requires environmentally induced selection in the microscopic quantum world. It is

Box 8.1 Are adaptive design and statistical thermodynamics incompatible?

Pross (2012) has argued that there is a continuum between kinetic chemical complexity and incipient biological complexity, and that teleonomic principles apply to both kinetic selection and natural selection. Analyses of this kind suggest that there may be more common philosophical ground between biology and chemistry than between biology and physics. His chemical approach to life acknowledged the concept of organization (Ganti, 1975), as did Maynard Smith and Szathmáry (1995) at the beginning of their book when they pondered life's origin. However, de Duve (2005a, 2005b) took an opposite perspective to emphasize the array of organizational singularities that the LUCA MTE has generated while arguing that none of them is explainable without natural selection as organizing force. All these analyses have in common that they remained primarily proximate and did not relate to the general concept of adaptation as synonym of organization (Pittendrigh, 1958). Remembering that Fisher (1930) compared his fundamental theorem of natural selection with the second law of thermodynamics (see Box 1.4), and that Pittendrigh (1958) defined adaptive organization by its deviation from randomness in the sense of maximal disorder (see Box 1.2), it seems puzzling that physics and organismal biology have developed essentially no overarching, agreed-upon interfaces as yet.

Reconciliation of biological (dis)order with informational and physical entropy has been attempted. For example, the informational entropy of allelic polymorphisms has been used to draw parallels between selection in theoretical population genetics and multiplicative weight updating, an established optimization algorithm in computer science (Barton et al., 2014; Chastain et al., 2014). Other studies focused on the analogies between statistical thermodynamics and genetic redundancy (Barton and Coe, 2009; Iwasa, 1988), exploring parallels between free energy (expressed as the energy versus entropy ratio) in thermodynamics and "free fitness" (expressed as the natural selection versus drift ratio) in population genetics (Sella and Hirsh, 2005). However, caveats by the authors indicate that challenges of analytical consistency are far from resolved (see also Van Baalen,

2013). A more empirical approach focusing on the adaptive information content of complex phenotypes as functions of gene expression networks benefitted from information theory (Wagner, 2017), as did cellular signaling entropy research for cancer diagnostics (Teschendorff et al., 2014; Teschendorff and Enver, 2017). These studies appear to confirm that connections between genetic information and Shannon information are realistic, but that attempts to follow in Schrödinger's (1944) footprints to uncover analogies with statistical thermodynamics have had little success. Perhaps this is because the few thermodynamics studies available used natural selection as a scale-free ratio rather than relating to a Fisherian/Hamiltonian force of fitness improvement (Grafen, 2018, 2019).

The continuing lack of analogies between Price equation logic and thermodynamics, in spite of Fisher's (1930) and Schödinger's (1944) early efforts, is intriguing, because there are biological phenomena of extreme order that seem to be begging for zero entropy interpretations. The weak hydrogen bonds between AT and GC base pairs in a double helix molecule is one example. As Watson and Crick (1953) wrote, bonding complementarity "immediately suggest[s] a possible copying mechanism for the genetic material." When an adenine purine always links up with a thymine pyrimidine and a guanine purine with a cytosine pyrimidine there is, barring random mutation, zero uncertainty for these interactions. Copying can therefore, in principle, continue indefinitely across the generations. The double helix thus appears to underline what Schrödinger (1944, p. 77) had in mind when he concluded that life's "existing order displays [both] the power of maintaining itself and of producing orderly events." But do events of maximal possible order deserve special consideration similar to what I proposed for invariably maximal relatedness and strictly non-redundant LTC in Chapters 4 and 5? Meiotic sex, the irreversible commitment between two gametes, is an even stronger example of maximal order and zero uncertainty, similar to the founding of ancestral superorganisms by strictly monogamous parents. These stringently non-redundant LTC events were fundamental for my partitioning of Hamilton's

---

Box 8.1 *Continued*

rule to obtain necessary and sufficient conditions for MTE origins, so it would be gratifying if they would turn out to be analogous to extreme cases in thermodynamics.

If any analogy of this kind can be made, maximal order that tolerates no exception in biology would translate into zero entropy in physics and necessitate zero degrees kelvin (Atkins, 2003). Macroscopically, that would require that no energy fails to be converted to work (Atkins, 2003, p. 115), while the statistical entropy equation of Bolzmann and Gibbs ($S_G = -k \sum P_i \ln P_i$) would give zero only when just one state in an energy-dependent distribution of alternative microstates is possible (i.e., $\neq 0$, so $P_i = 1$) (Corning and Kline, 1998a). Intriguingly, the Bolzmann and Gibbs equation is isomorphic with the Shannon equation ($S_S = -\sum P_i \log_2 P_i$) that specifies uncertainty in predicting the identity of a random element (i) in a message. This would produce a zero outcome when (non-redundant) pairs are always

viable while (redundant) crowds of three or more elements never are, that is, when there is only a single answer to *where can I find a male gamete with reproductive potential?*, which is, *alone in the exclusive company of a single female gamete*. However, the concepts that produced these identical equations address very different aspects of reality, and what the living-system equivalent for thermodynamic entropy in joules per degree kelvin should be remains elusive (Wicken, 1986; Corning and Kline, 1998a, 1998b). One is left with the alluring feeling that there are extreme state analogies across physics and biology that existing theory cannot capture. Could it be that the extreme order of informationally closed zygotic or monogamous founding-pair arrangements were necessary to evolve obligatorily altruistic soma also from a physics perspective because, mechanistically, new levels of non-reducible self-organized complexity under the auspices of natural selection demand zero uncertainty about partner redundancy?

---

intriguing that *monogamy* in the sense of LTC both appears to be a necessary condition for MTE origins and has emerged as a key characteristic of quantum entanglement, in contrast to the promiscuous correlation phenomena that characterize accessible and objective physical reality (Koashi and Winter, 2004; Brandão et al., 2015). Might it be, therefore, that natural selection is just as fundamental in physics as in biology, but only in the most abstract quantum domain?

Conceptualizing the human condition with minimal anthropomorphic bias

Just as theoretical physics aims for first-principle understanding of the universe, so do biologists have the obligation to pursue deep understanding of the axiomatic principles of life. Since Darwin's *Origin*, we know that this requires explicit attention to adaptations

shaped by natural selection, although historical contingency always remains an essential covariable. Physicists resort to mathematical abstraction to avoid anthropomorphic bias, but maintaining such distance has been a challenge for biology. It is hard not to become subjectively involved when theories or data are about "charismatic" birds, mammals, and particularly primates. Yet, a deep analysis of the human condition needs to be Darwinian, not just molecular, genetical, ecological, or epidemiological, let alone political (e.g., Cronin, 1991; Elwick, 2003). As I have mentioned several times, environmentalized versions of Darwinism have been an underappreciated stumbling block towards unbiased understanding of social evolution across the domains of organizational complexity. Such convenience Darwinism often turned adaptation and natural selection into hazy, low-priority concepts necessitating the fundamental criticism of "current evolutionary thought" by Williams (1966a) and Cronin (2005) (see also Stearns', 2011 account of Williams' position). A deeper cause of this confusion may be that it seems natural to think of evolution as a dialectic process of give and take between genetic endowments and environmental factors, while overlooking that this only applies when we project anthropomorphic views of the human condition on the rest of nature. These attitudes are encountered in studies of animal behavior (Box 8.2), and also in unspecified "plasticity" arguments (e.g. West Eberhard, 2003) and dehumanized "niche construction" thinking (e.g. Odling Smee et al., 2003) where environments become agents that could potentially interfere with neo-Darwinian inclusive fitness principles. Scientifically, such approaches appear to mostly be 'emperors without cloths', that is, extrapolations of anthropomorphic bias without first-principle conceptual foundation (see also Gupta et al., 2017).

In a review of 16 common misconceptions about the evolution of human cooperation, West et al. (2011) focused on precise terminology to replace the multitude of definitions that saw the light in recent decades, all emanating from the anthropomorphic perspective that humans are somehow collectively special in an ultimate (evolutionary) sense. They argued that belief in various forms of group selection is mainly responsible for this confusing situation and posited that humans are only special in a proximate sense, that is, because of their extreme—relative to any other animal—cognitive abilities that allow individuals to exploit many subtle maximizing strategies not available to other organisms. These analyses illustrated that distinguishing between proximate and ultimate explanations has been more of a problem in the human behavioral sciences than in nonhuman behavioral ecology (Scott-Phillips et al., 2011). West et al. (2011) went on to challenge the notion that humans are

particularly interesting because of their levels of altruism. Indirect fitness benefits clearly occur, most notably in our universal kinship-based inheritance laws and in individual support networks (Lieberman et al., 2007; Burton-Chellew and Dunbar, 2015; Waynforth, 2020; Tanskanen et al., 2021), but humans are only truly exceptional in their enormously diverse ways of engaging in cooperation with nonrelatives (Trivers, 1971; West et al., 2011; Kurzban et al., 2014). Moreover, implicit assumptions of instinctive human pro-sociality have been put to the test and found wanting. The results of economic games were shown to be unconvincing because their untested assumptions seem incompatible with real-life situations experienced by pre-industrial and pre-agricultural humans who were "players" in more transparent "games" (Trivers, 2004; Burton-Chellew and West, 2013; Burton-Chellew et al., 2016, 2017). Null hypotheses also generally remained vague if they were specified at all, precluding formal hypothesis testing.

The human MTE is obviously related to the accumulation of cultural syndromes, but evaluations emphasizing the role of language as instrumental mechanism for its origin (Maynard Smith and Szathmáry, 1995) did not allow much progress beyond asking for scientific rigor (Stearns, 2007). Early models of gene–culture coevolution (Boyd and Richerson, 1985; Richerson et al., 2003) have suggested that cooperative traits can increase by group selection but were skeptically received (Lehmann et al., 2007, 2008), and a recent update of this approach (Richerson et al., 2016) also met with considerable controversy (e.g., Barclay and Krupp, 2016; Krasnow and Delton, 2016; Mace and Silva, 2016; Singh et al., 2016; Tooby and Cosmides, 2016). A common cause of disbelief may have been Williams' (1966a, p. 93) argument that "group-related adaptations do not, in fact, exist," to which he added that it is incorrect to assume that "one must look for group functions" in order "to explain the functional aspects of groups" (Williams, 1966a, p. 210)—the danger of interpreting byproducts as adaptive functions looms large. It also emerged that cultural inheritance cannot really be handled with reductionist population genetics models because memes lack stable transmission modes (Claidière and André, 2012) so the interpretation of gene–meme coevolutionary processes becomes ambiguous (Claidière et al.,

2014). As its abstraction captures any process of change without bias, the Price equation has been used to clarify various forms of extended inheritance (Helanterä and Uller, 2010). However, what has been lacking is an explicit exploration of the origin of the human MTE, that is, an individual Darwinian explanation of what made our own MTE unique as a bipartite mutualism with co-adapting Lamarckian algorithms (Box 8.2).

Box 8.2 The uniqueness of the human major transition

Acknowledging that reductionism is a tool rather than an end is particularly relevant in discussions about the human symbiosis with culture because biologists used to agree with humanities scholars that culture is an "organismal" phenomenon (e.g., Alexander, 1979, p. 66). It is only recently that studies of isolated cultural traits transmitted by other animals have abducted the term, so that many animals could become "cultural" (e.g. Whiten et al., 2005; Heyes, 2020), even though these studies shed no light on the unique human domain of *cumulative culture* as we now call it (Mesoudi and Thornton, 2018). It is important to make explicit that human cumulative culture is as far removed from single cultural traits in other animals as Waddingtonian organismal development is from a single differentially expressed genetic pathway. That cultural traits can be studied in isolation is interesting, but such single traits are irrelevant for understanding human culture because the very fact that ancestral humans realized an MTE should preclude retrospective homology comparisons with organizational complexity in a previous domain. Yet, to justify a sense of cultural continuity across the human-great-ape MTE, the 21st century has seen a surge of cognition studies based on non-parsimonious animal intelligence hypotheses driven by "humanizing" confirmation bias, usually after first proposing watered-down definitions of consciousness, episodic memory, and syntax (Wynne, 2004; Suddendorf, 2013, pp. 111, 152–153; Adamo, 2016; Bolhuis et al., 2018). Other studies

made more of pro-social empathy than is supported by hard evidence (e.g., Vasconcelos et al., 2012, Cronin, 2016). Bonner characterized this phenomenon as a "subconscious desire among us to be democratic even about our position in the great scale of being." Williams (1966a, pp. 254–255) had earlier diagnosed that the tendency to humanize groups of organisms may be due to "a desire, unconscious in many and expressed by a few, to find not only order in Nature but a moral order." As he wrote, "[i]n human behavior a sacrifice of self-interest and devotion to a suprapersonal cause is considered praiseworthy. If some other organisms also showed concern for group welfare and were not entirely self-seeking, these organisms, and Nature in general, would be more ethically acceptable." However, such anthropomorphic wishful thinking is inconsistent with the scientific method.

Similar skepticism is warranted to avoid humanizing the great apes (Suddendorf, 2013). It is crucial to keep in mind that an MTE lies between us and that our human social lives are really exceptional. Human societies "consist of families within groups that are part of larger communities," a nested arrangement that no other animal has, and people never live in baboon-like troops or gorilla-like harems (Wrangham, 2019, p. 8), or in isolated and mutually hostile families as colonial superorganisms do (Alexander, 1979). The social insects needed strict ancestral monogamy to evolve superorganismality, but our human ancestors

Box 8.2 *Continued*

were never somaticized and only became fairly monogamous as a secondary cultural development (Alexander, 1979; Fortunato and Archetti, 2010; Kramer and Russell, 2015; Schacht and Kramer, 2019). These inclinations were established during the human MTE transition period, when it became impossible to pass on genes to future generations without being surrounded by a network of cumulative cultural achievements and expressing the mental, physiological, and morphological traits that had coevolved with early cumulative culture (e.g., Pagel, 2012; Henrich, 2016). In *The Goodness Paradox*, Wrangham (2019) argued that the human MTE encompassed the emergence of both cumulative culture and *Homo sapiens* as a species ca. 300,000 years ago and that humans lost most of their reactive aggression within groups (relative to both chimpanzees and bonobos) while evolving more proactive aggression in the form of intergroup warfare. This correlation has long been known (e.g., Muller, 1960; Alexander, 1979, p. 220), but has not previously been so clearly reconstructed as a Darwinian narrative of individual natural selection.

Recent work has also challenged the idea that gene–culture co-evolution (e.g., Boyd and Richerson, 1985; Sober and Wilson, 1998; Richerson et al., 2003; Bowles, 2006) was driven by group selection for parochial altruism (Choi and Bowles, 2007; Bowles, 2009; Bowles and Gintis, 2011). Alternative concepts based on individual benefits of personal reputation have gained increasing support (e.g., Nesse, 2009, 2016; Claidière and Sperber, 2010; Kumru and Vesterlund, 2010; Sperber and Baumard, 2012; Santos et al., 2018). It is this view that Wrangham (2019) corroborated, starting with the observation (known for two centuries) that humans have key characteristics of a domesticated species, a trait syndrome including tall and round skulls, small faces, less protruding brow ridges, and a chin—all consistent with lower exposure to intragroup aggression. Wrangham developed a parallel example of Bonobo domestication, explaining that a vegetarian diet and relatively competitor-free habitat allowed groups to be large enough for females to gain social control of isolated male bullies. In contrast, suppression of dominant, antisocial males in humans was achieved by

coalitions of subordinate males. That change could only take place after language and shared intentionality had evolved to make ingroup conspirational killing frequent enough to impose directional selection against reactive male aggression. Concealed ovulation (Alexander, 1979, pp. 135, 209–216) and higher ingroup cooperation (e.g., alloparenting; Hrdy, 2009) evolved in its wake. This ultimately led to a "moral" norm of egalitarian coalitions of adult males jointly controlling tribes (see also Cofnas, 2018; Wrangham, 2021).

In this view, parochial altruism and intergroup warfare, starting in the late Pleistocene, are byproducts, that is, mere effects *sensu* Williams (1966a), of adaptations to increase reproductive success of cooperating males. This implies that natural "morality" did not emerge from unspecified altruism for group wellbeing but from individual senses of right, wrong, and reproduction, driven by social selection for reputation (Nesse, 2008, 2009, 2016). The ensuing tribal manifestations of internal-control aggression, incompatible with modern society, are just a small part of the so-called human universals (Wrangham, 2019, p. 162), of which the appendix of *The Blank Slate* (Pinker, 2002) lists ca. 375 examples, to which mindreading (Suddendorf, 2013, p. 118), shame (Sznycer et al., 2018), and ingroup conspirational killing (Wrangham, 2019, p. 158) were later added—a long list confirming the tremendous cognitive gap between *H. sapiens* and the great apes. Our MTE was particularly characterized by the emergence of a theory of mind (Penn and Povinelli, 2007; Suddendorf, 2013), without which ingroup conspirational killing would not have been possible (Wrangham, 2021). Overall, Wrangham's (2019) monograph has done the field the great service by analyzing the human condition with minimal anthropomorphic bias. As an aside, it is instructive to note that several components of Wrangham's reconstructions were hypothetically worked out in an essay by Wheeler (1934), in which he compared the colonial social insect solution of the male problem by lifetime imprisonment (dead or alive) with the human habits that leave much to be wished for but allowed our ancestors to evolve societies without obligate reproductive division of labor (see Box 3.1).

Box 8.2 *Continued*

All this implies that human cultural exceptionalism is a fact, not a mere value statement, because the human condition has many irreducible MTE-specific complexities while great apes merely have isolated cultural traits (Abler, 2005). The human cultural "white zone" of irreducible complexity is not comparable, let alone analogous, to the superorganismal white zone (Figure 8.3), because the human MTE could happen in a society setting while the superorganisms needed a new level of organismality (Figure 8.1). This difference sets humans apart from all "Tinbergen organisms" that I specified with the 4 × 4 cube in Figure 8.2. The human condition can only be conditionally cooperative because our societies consist of reproductively totipotent individuals, very different from the fraternal MTE lineages with obligate somatic

altruism of cells or multicellular individuals. The fact that a single force, the acquisition of and co-evolution with cumulative culture, could become *both* a necessary *and* sufficient MTE condition, while bypassing the fraternal MTE rule of clonal or maximal sibling relatedness, raises interesting questions about the true nature of cultural commitment. In the main text, I argue that informational closure also appears to apply to the way in which the individual human mind is moulded during ontogenetic interaction with virtual symbiotic algorithms that pursue their own control information. Finally, it is worth noting that the human MTE was as irreversible as the other MTEs specified in Figure 3.2a. If humans go extinct, it will be as reduced (see Figure 8.1) cumulative cultural beings, not after reversal to great ape cultural complexity.

The Price equation models of El Mouden et al. (2014) and Aguilar and Akçay (2018) appear to have come closest to avoiding anthropomorphic bias by capturing the human condition as a form of dual agency emanating from genetic and cultural inheritance affecting the same phenotypic traits while allowing cultural and genetic fitness to be aligned to varying extents. These approaches match the assumption that the human condition arose as an egalitarian MTE when isolated cultural traits made network connections and acquired independent agency, perhaps not unlike what ribosomal variants may have done before cell membrane closure (Mizuuchi et al., 2022). Further development of this dual Price-equation methodology, based on codex versions of genetic and cultural information, would seem preferable to reductionist group-level concepts of dual material-inheritance (Richerson et al., 2016) or *gene-meme* concepts (Dawkins, 1976; Dennett, 1995), although the latter still offer proof of concept

arguments similar to green beards exemplifying rare single-gene agency. A host brain versus algorithmic symbiont network perspective may also allow culture's unstable epidemiological transmission dynamics (Sperber, 1996; Claidière and André, 2012; Claidière et al., 2014) to be handled (Day et al., 2020), and to replace the contradiction-in-terms idea of "self-domestication" by a dynamic symbiont-host coadaptation narrative. It is striking that the mutualism between a human brain and its algorithmic symbiont complex has an informationally closed foundation that persists, albeit in a peculiar way. As Williams (1992, p. 4) noted, the mental realm is entirely private—hence Descartes used the singular *cogito ergo sum* (I think, therefore I am) rather than the plural *cogitamus ergo sumus* (we think, therefore we are). Conceptualizing the human MTE origin as a LTC event between a host brain and an algorithmically learning, self-organizing virtual symbiont complex would allow drawing inspiration from analogous understanding of parasitic and mutualistic symbiont "virulence" in interactions that evolve only by natural selection of genetically encoded traits (Frank, 1996a; Queller, 2014). In such applications, virulence is conceptualized in abstract codex phrasing, that is, as a quantification of the extent to which a symbiont wastes host resources in proportion to its interests not being aligned with those of the host (Box 5.1).

As we have seen for other egalitarian LTC arrangements in Chapter 5, when host–symbiont commitment is closed at foundation and symmetrically restricted to a minimal non-redundant set of complementary "seeding" haplotypes, resource acquisition interests are completely aligned provided partners have mechanisms to prevent secondary forms of chimerism (see Figure 5.2). However, exploitative and manipulative traits tend to evolve when mutualistic interactions have redundant commitments so that hosts need to evolve countermeasures to retain sufficient control (see also Chapter 7). As a human individual develops, the interactions between her physically isolated host brain and its algorithmic symbiont complex is likely to acquire increasingly asymmetrical characteristics. The symbiont complex is mindfully committed in its "home brain" but has many opportunities to improve its "software" performance by learning from the social and material world outside while competing for horizontally transmitting its innovative acquired characters to other brain symbionts—a form of promiscuous agency that a host brain, closed for genetic information in its founding zygote stage, lacks. As development proceeds, virtual brain symbionts are therefore increasingly ahead in terms of generating control information *sensu* Corning and Kline (1998b). This is because their share of the population's brain–symbiont information pool grows during a host's lifetime while the diploid genetic sample

defining the host brain substrate remains constant.

The following analogy may help to clarify how publicly open human minds are yet uniquely personalized by closure at foundation. Imagine a perennial legume that germinates from a seed with a diploid informationally closed zygote. At some point soon after, the plant will acquire *Rhizobia* mutualists as infections that induce its root system to grow the first nodules to house these nitrogen fixing bacteria. Even though the first symbiont strain obtained may be identical by chance between two neighboring germinating plants, subsequent acquisitions will independently sample from a large population of available *Rhizobia* strains. If we assume that there are ten possible strains, the symbiont community of a single plant with just four nodules will already be as unique as a four-digit pin code. As plants grow, they remain uniquely defined by the diversity, order of acquisition, and mass of *Rhizobia* symbionts while asymmetric redundancy of symbiont strains at the overall plant level keeps increasing. None of this would be stable if all strains coexisted and competed in the same arena (remember that many symbioses highlighted in Chapter 5 tolerated little or no redundancy). However, root symbiont compartments are modular, so host plants have just a few competing strains in each nodule allowing free-riders to be sanctioned with reasonable but not perfect efficiency (Kiers et al., 2003). The same should apply to "newly inoculated" human minds, which appear to be modular as well (Cosmides and Tooby, 1992; Kurzban and Aktipis, 2006). Descartes' *I think, therefore I am*, thus seems analogous to *I receive fixed nitrogen, therefore my symbiotic individuality functions*, even though no two plants will ever acquire their partnerships in the same way. Functional compartmentalization of brain algorithms in complementary brain units can thus be expected to enhance parallel processing and network communication power while reducing potential conflicts, but not to eliminate such conflicts.

These arguments imply that it is reasonable to assume that human brains are not blank slates, that they have inborn operative control systems to manage algorithmic symbionts, and that these start working before children are born (Dennett, 1991; Pinker, 2002; Pagel, 2012). Our brain can then also be expected to be proactively defensive against manipulation by its innovating brain–symbiont complex, when horizontal transmission of its cultural information is constrained by reproductive biological interests, contentions that have been evaluated in detail by Dawkins (1982) and Dennett (1991, 1995). This observation relates to self-deception, another ambiguous term that anthropomorphically assumes single human agency and that would converge on mere deception when we assume dual host–symbiont agency. Self-deception was first conjectured to be real by Robert Trivers in his foreword to the first edition of *The Selfish Gene* (Dawkins, 1976), which he followed up with a full monograph 35 years later (Trivers, 2011). In my present terminology, Trivers' argument is that subtle forms of social deception are such an integrated part of the human condition that our virtual brain symbionts serve our reproductive interests (and their own) by ensuring that we remain unconscious of these forms of alter-ego agency (see also Pagel, 2012). In his words, "[t]he conventional view that natural selection favors nervous systems which produce ever more accurate images of the world must be a very naïve view of mental evolution" (Trivers, 2015, p. 183). Deception will be an innate part of the human condition if it has often paid off, reproductively, in our evolutionary past, and if these gains were higher when it operated unconsciously. It was Richard Alexander who first elaborated some of the general implications of self-deception,

writing that "[i]t is a remarkable fact that humans have not only failed, throughout history, to acquire any understanding that they have evolved to maximize reproduction, but that even today they deny the possibility vehemently. [. . .] By conveying the impression that we do not intend to deceive, and that we are in fact altruistic and have the interests of others at heart, we actually advance our own (evolutionary) self-interest" (Alexander, 1979, p. 134). The fact that humans are culturally exceptional therefore does not imply that mindful harmony or superior morality was a logical consequence, as also George Williams maintained (see Stearns, 2011). Rather, the establishment of incorruptible political and social institutions appears to be the prime route to secure that most citizens maintain a personal interest in agreed upon moral norms.

Deception by brain symbionts may explain why it takes effort to mobilize our rationality. This dilemma was succinctly captured by Daniel Kahneman (2012) when he argued—while providing a broad spectrum of evidence—that we have a cognitive system 1 for instinctive rules of thumb that we effortlessly apply for routine social interactions, and a cognitive system 2 that "can follow rules, compare objects on several attributes, and make deliberate choices between options" (Kahneman, 2012, p. 36). System 1 represents our primate legacy and system 2 appears to be evolutionarily derived. System 2 can, but does not need to, overrule inclinations obtained instinctively via system 1, but that requires time, effort, and training. In daily life, we are continuously confronted with subjective information generated by system 1, although "only few of the activated ideas will register in consciousness" (Kahneman, 2012, p. 52). It must therefore have taken considerable system 2 control for Alexander (1979, pp. 274–275) to expose the real meaning of parental moral education when he concluded that, rather than educating children never to deceive, "parents actually teach their children how to 'cheat' without getting caught." That is, "parents teach their children what are 'right' and 'wrong' behaviors in the eyes of others, and what truth-telling and forthright behavior actually are, so that from this base of understanding children will know how to function successfully in a world in which some deceptions are profitable, some unforgivable, hence expensive, and some are difficult to detect, others easy." He concluded that it then follows logically from inclusive fitness theory that "parents are more likely to punish children for (a) cheating close relatives, (b) cheating friends with much to offer the family in a continuing reciprocal interaction, or (c) cheating in an obvious, bungling fashion, sure to be detected, than they are to punish them for simply cheating."

These dynamics are consistent with Kahneman's system 2 having emerged during the human MTE and having started to coevolve with the ancestral system 1 in what became a perpetual instinct versus rationality tug-of-war. Similar to other evolutionary arms races, the balance between the opposing forces in such confrontations is intrinsically unstable. As Dawkins (1976, p. 212) noted, "nothing is more lethal for certain kinds of meme than a tendency to look for evidence," so also from that perspective we should expect that we cannot count on our consciousness being overly rational. All such inferences are heuristic, defined by Kahneman (2012, p. 98) as providing "adequate, though often imperfect, answers to difficult questions," but combining his arguments with Trivers' reasoning suggests that our system 1 has been under selection to make us deny inconvenient social truths unless it stands corrected by system 2. If that is even partly correct, we are our own worst enemies in scholarly attempts to get to the bottom of the human condition because anthropomorphic bias is likely to be actively maintained. It may even serve our brain symbiont's interest to make us "democratically"

believe that animals are just like us (Box 8.2). However, despite all this justifiable distrust in its rational performance and an evolutionary history of just 300,000 years, the human brain–culture symbiosis was as revolutionary an egalitarian MTE as the last common ancestors LUCA and LECA were. Only after an already sophisticated ape brain started to host self-organizing and co-adapting algorithmic software complexes could our ancestors begin to create the multitude of global niches that human populations became adapted to, combining cultural innovations with the older legacy by which hominin genes used to have the monopoly of carrying and storing adaptive information. Niche construction is therefore a crucial element of our human legacy, but only because we have cumulative culture and are able to consciously reflect about real or imaginary environments without the requirement that this activity is directly relevant for maximizing our inclusive fitness. No other organisms, including great apes, should be assumed to be able to do something even remotely similar.

Epilogue: Towards a kin selection theory for organismality

By starting my survey of neo-Darwinism further back in time than is normally done, I have tried to show that adaptation-focused organismal biology used to be a coherent science, not a field just waiting to be displaced by reductionist molecular biology—a view shared by early molecular biologists (see also Nurse, 2021). For example, Monod's (1975, p. 23) understanding that "all linguists and ethologists now agree that there is a huge gap between animal communication of any sort and actual human language" could have been the opening quote of Suddendorf's (2013) monograph—no GPS tracking device, omics dataset, or mathematical model made that conclusion wrong or irrelevant in the almost five decades that passed, and neither can it be challenged by the occurrence of single cultural traits elsewhere in the animal kingdom. Likewise, the Huxley quote at the end of Box 3.2 could have introduced Wrangham's (2019) book. But apart from emphasizing the often-striking logical coherence of neo-Darwinism as it re-emerged from its decades-long slumber (see Chapter 1), this book has defended the view that there have been discontinuous enhancements in life's organizational complexity that continuous gradient thinking fails to explain and that the origins of these singular MTE events must have a single adaptive explanation. If my conjectures survive further empirical testing and logical analysis, it would suggest that organized life has to be fundamentally Hamiltonian/Pricean no matter where it

evolves and how far it proceeds in realizing its subsequent MTEs (see also Levin et al., 2017). I have summarized the main testable predictions that follow from my arguments in Box 8.3. If they hold up, the implication is not that all evolutionary change was adaptive—just that by and large all change that historically mattered was. But I hasten to add that this claim is far from new, for it was defended by Darwin, Wallace, Weismann, and Tinbergen, and given gene's eye power by Pittendrigh, Hamilton, and Williams.

The ancient Greek philosophers had two concepts of time, *Chronos* and *Kairos*. Chronos is conventional sequential time, which is our familiar basis for reconstructing and interpreting all dynamic change. Kairos is a qualitative

---

Box 8.3 Predictions to be tested and potential further insights to be gained

I started this chapter by highlighting that the central hypothesis of this book has the structure of Hume's fork, a standard form of inductive reasoning. However, reasoning of that kind cannot be used to test conjectures—it can only be corroborated by the kind of consistency evidence that I have reviewed in previous chapters. This challenge can be overcome by developing a hypothetico-deductive framework, offering nontrivial predictions that have the potential to refute the conjectures of my extension of inclusive fitness theory to cover the static and discontinuous transitions in degree of organismality. In *The Rationality of Scientific Revolutions*, Popper (1975, pp. 82–83) established two logical criteria for deciding whether a novel theory will be better than an existing one: (1) its predictions should contradict previous understanding, and (2) their implications should provide a deeper general understanding. For the present purpose, this means that my conjectures should contradict understanding of the MTEs in organizational complexity as they were framed in recent decades, and that deeper understanding should arise from explicitly connecting new first-principle understanding of MTE

origins with already well-established inclusive fitness theory, while highlighting potential for further theoretical generalization. The previous chapters of this book have already formulated predictions in several contexts, so I only make the most important ones explicit here to encourage new efforts at refutation with comparative data, modelling, or genomic reconstruction of the unique GRNs (Table 3.1) for division of labor that should have mediated the irreversible and discontinuous advances in social organization. My most fundamental prediction that needs no further comment is that there exists no adaptation that was not shaped by natural selection at the lowest relevant level of selection. I challenge incredulous critics to refute this foundational principle, which follows directly from Pittendrigh (1958) and Williams (1966a), not by showing that gene frequencies or even the species diversity of entire lineages can change by haphazard historical events, but by proving that even a single complex adaptation could emerge from genetic drift and/or geographic isolation without being shaped by natural selection.

Box 8.3 *Continued*

Predictions that contradict, or do not follow from, current conceptualizations of MTEs

Prediction 1 - MTEs do not just emerge

MTEs have never evolved via emergence from complex society-like ancestry where lower-level selection (MLS1) gradually gave way to higher-level selection (MLS2; acronyms for multilevel selection are from Damuth and Heisler (1988) and Okasha (2006)). Instead, nonhuman MTEs emerged from inconspicuous ancestors with generalist life histories that established, most likely as an exaptation, a novel form of closure between non-redundant carriers of genetic information, which precluded that ancestral open-society habits, often still extant in sister clades that underwent their own adaptive radiation, could continue (Figure 8.1). I thus predict that it will be impossible to document that any nonhuman MTE originated from a group-level process that underwent reduction of social conflict. The human MTE is exceptional because our irreversibly enhanced organizational complexity did not arise from a new level of obligate altruistic soma, but from a closed symbiosis between individual brains and a personal set of virtual algorithms that coevolved to shape our individual minds by natural selection while leaving our populations with increasingly advanced collective cumulative culture. Reduction of social conflict has been a key requirement throughout human societies, as it was in all the nonhuman societies that never were ancestral to MTEs.

Comments: The key point is that non-redundant genetic informational closure by LTC appears to be a universal necessary condition for the nonhuman MTEs. Societies lack stringent closure, irrespective of being simple or complex, and their redundant chimeric nature (Table 3.1) always maintains a lower level of selection, a characteristic that is prohibitive for MTE potential. The human MTE is the only one that arose from a complex society ancestor. It did not involve new closure between carriers of genetic information, but Williams' (1992, p. 4) conclusion that the human mind is entirely private, likely from before birth onwards, seems interestingly analogous with genetic information closure in the nonhuman MTE lineages.

Prediction 2 - MTEs are strictly hierarchical

After terminal differentiation, there are, under natural conditions, no cells left in multicellular organisms that have retained their totipotency and, analogously, totipotent individuals are absent in colonial superorganisms (this does not preclude that tumors and laboratory experiments can secondarily recreate totipotency). In contrast, societies always have at least some and often many replication-potent cells or behaviorally totipotent adult members. Societies thus lack collective germlines and ontogenetic development that somaticizes all non-germline participants into obligate altruists, consistent with cooperation for altruistic or mutual benefit remaining condition dependent for most society members. The presence of some somatic cells or caste individuals (e.g., cellular slime mold stalk cells or lower termite soldiers) therefore does not make societies (super)organismal because condition-dependent and obligate cooperation evolve independently.

Comments: The key point is that MTEs cannot just appear anywhere because their emergence always requires complete hierarchical nestedness of previous levels of organizational complexity (see Figure 3.2b). Nestedness implies that all formerly independent agency is permanently domesticated. Organizational nestedness is discontinuous by definition—it is a defining a priori trait for an MTE, while subsequently realizing a global adaptive radiation appears to be an a posteriori defining trait (Figure 8.3). Societies may form a continuum of organizational complexity, but MTE-induced (super)organismal lineages are not part of such a continuum. Despite their recurrent, strictly non-redundant LTC origins, superorganismal colonies have secondarily evolved chimeric, society-like forms of redundancy, which triggered conflict regulating responses by the foundress queen, such as termination of sperm competition in *Atta* leaf-cutting ants (see Chapter 5) or by the fully developed superorganism as when honeybee workers police worker-laid eggs (Ratnieks, 1988; Ratnieks et al., 2006). The proximate non-viability of polyspermy has precluded such forms

Box 8.3 *Continued*

of evolutionarily derived chimerism in the multicellular domain, although fungi offer interesting partial exceptions that deserve more study (see Box 6.3).

Prediction 3 - Groups cannot evolve MTEs

Darwin's (1859) statement that "[i]f it could be proved that any part of the structure of any one *species* had been formed for the exclusive good of *another species*, it would annihilate my theory, for such could not have been produced through natural selection," can be modified, into: if it could be proved that any part of the structure, *function, or behavior* of any one *group* had been formed for the exclusive good of *descendent groups*, it would annihilate my theory, for such could not have been produced through natural selection, *unless groups were derived from ancestors with non-redundant lifetime closure around strictly "monogamous" gametic or parental carriers of genetic information* (italics highlight modified text).

Comments: Only the four nested domains of organizational complexity specified in Figures 3.2b, 5.3, and 8.2 qualify as having originated from strictly closed LTC around non-redundant carriers of genetic information. As mentioned in Chapter 4, Williams (1966a, p. 93) understood this principle when he wrote that "group-related adaptations do not, in fact, exist," while adding that "[a] group in this discussion should be understood to mean something other than a family and to be composed of individuals that need not be closely related." This confirmed Darwin's family selection concept that has later been, and continues to be, misunderstood as implying that families are merely groups (see Boomsma and Gawne, 2018, for references). Note that the same Darwin quote was used by Wyatt et al. (2013) to argue that mutualistic symbioses cannot exist unless their interspecific cooperation benefits the inclusive fitness of each partner separately.

Prediction 4 - MTEs have unique GRNs

Every nonhuman MTE origin established its own unique network of co-expressed genes (also referred to as a GRN; see Table 3.1) for mediating obligate reproductive division of labor. These GRNs characterize the new levels of organizational complexity, including higher-level germline–soma differentiation, a concomitant novel potential for higher-level ontogenetic development, and a new "layer" of irreducible self-organized agency. The latter needed to evolve seamless homeostatic interactions with the previous, now somaticized levels of agency (Figures 3.2b and 8.1). These GRNs remained homologous throughout subsequent adaptive radiations but have, as novel networks governing reproductive division of labor, no homologies across MTE origins, neither horizontally (within social evolution domains) nor vertically (backwards across such domains).

Comments: MTE-specific GRNs for germline–soma reproductive division of labor may have so-called toolkit genes in common with sister lineages that retained society-level complexity or with evolutionarily convergent MTEs within the same domain. However, the identification of such toolkit genes is likely to be largely irrelevant for understanding a specific GRN that made a particular MTE irreversible and robust. As outlined in Chapter 6, toolkit status or homology within a clade of society-forming organisms should not predict participation in an MTE-inducing GRN. Neither does it necessarily predict the direction of gene expression bias in a new higher level of MTE-induced germline-soma differentiation. The arguments in this book are based on the conceptual restriction of MTEs to irreversible events that created novel levels of hierarchically nested complexity in organismal organization. Other transitions like genome duplications or fundamental shifts in development (e.g., hemimetabolous to holometabolous insects or the emergence of the mammalian placenta) started new adaptive radiations, but they should not be confused with the MTEs as defined here (see Table 3.1; Figure 3.2). For example, the analogous definition of colonial superorganismality and metazoan organismality (sensu Wheeler, 1911) implies the prediction that caste development should be canalized (Flatt, 2005; Qiu et al., 2022), a functional expectation that would not apply in any non-hierarchical use of the MTE concept.

Box 8.3 *Continued*

Prediction 5 - Ectosymbionts and MTEs

While the MTEs that produced LUCA, LECA, and the human condition were all endosymbiotic, there appear to have been no obligate ectosymbioses with the potential of producing egalitarian MTEs. This offers a sharp contrast to the recurrent fraternal MTEs towards obligate multicellularity and colonial superorganismality, which all relied on external adherence of somaticized cells or multicellular offspring. As it appears, a host cannot fully somaticize a symbiont belonging to a different gene pool unless it loses almost all its genes, which could happen to LECA's endosymbiotic mitochondria and various later domestications of plastids and other bacterial endosymbionts, but seems impossible for ectosymbionts as long as they maintain any traits that mediate independent resource acquisition.

Comments: Some ectosymbiotic mutualisms based on LTC are borderline cases of higher-level organismality and could therefore be test systems for analogous understanding of the origins of ancient egalitarian MTEs such as LECA. A crucial Williams (1966a) criterion for considering such systems as higher-level organismal is whether they evolved mutualistic adaptations that were only beneficial for the higher organismality level and do not make sense for any of the partner species separately. The best examples I can think of come from the fungus-farming social insects, because they fulfil the criterion of ectosymbionts having lost all independent resource acquisition potential as they can only grow on host-provided substrate. The fungus-growing ants are (outside the evolutionarily derived leaf-cutting ant crown group; see Figure 5.2) a non-redundant lifetime symbiosis between a single queen inseminated by a single male that maintains a single haplotype of vertically transmitted fungal cultivar. As an exclusively higher-level symbiotic adaptation, the ants evolved cuticular crypts to maintain mutualistic actinomycete bacteria that suppress fungal disease in their gardens (Currie et al., 2006). No other ant is known to have such cuticular crypts for maintaining similarly specialized bacteria, and no other fungus has this specialized pathogen (Worsley et al., 2019).

The fungus-farming termites are another example of a non-redundant lifetime symbiosis between a single queen, a single king, and a single clonal cultivar. Here, the crucial process of positive frequency-dependent selection that precludes cultivar chimerism is jointly maintained by both partners (Aanen et al., 2009) (see Chapter 5). These trait syndromes would make no sense in free-living cockroach or lower termite relatives, and none of these higher-level adaptations can be interpreted as byproducts of selection for unilateral benefit. Neither the fungus-farming ant nor the fungus-growing termite symbioses quite realized global adaptive radiations, but they did spread over much of two continents: South and North America for the fungus-growing ants and Africa and South Asia for the fungus-growing termites. Interestingly, it cannot be objectively decided in either of these cases whether the insect superorganisms domesticated the fungal cultivars or the other way around (Boomsma, 2022), which attests to the perfect non-redundant symmetry of these ectosymbioses, not unlike the nuclear-mitochondrial endosymbiosis that shaped LECA.

Deeper understanding emanating from all MTE origins being products of strict informational closure

The analyses of this book reinforce the conclusion that inclusive fitness theory offers deep general understanding of social adaptations as they are shaped by natural selection. This understanding is independent of the timescale involved and covers both the less committed dynamical social interactions in open societies and the static origins of irreversible MTEs to enhanced organizational complexity. This understanding is abstract, heuristic, and statistical, reminiscent of the ways in which other important scientific theories have become generalized after first having been formulated in deterministic terms. As I have argued throughout this book, inclusive fitness is a theory of ultimate causation. Hence, we need the complementary approaches formalized for the first time by Niko Tinbergen (see Figure 1.1; Box 1.3)

Box 8.3 *Continued*

to understand all major factors that facilitate, modify, or constrain social adaptation. At the same time, we ought to acknowledge Tinbergian inspiration by comparative natural history. For example, Hamilton was (ultimately) interested in social life in general and not, as some maintain, primarily and proximately in haplodiploid Hymenoptera. These insects were just convenient models, similar to other biologists using fruit flies, guinea pigs, or rockcress plants. Hamilton's prime general interests were in open societies, both those of wasps and humans (Caniglia, 2017; Costa, 2013). Only occasionally had he things to say about the organismal colonies of honeybees or ants (e.g., Hamilton, 1964a, 1964b, 1972), and when he focused on these organismal colonies, as he did when reviewing Michael V. Brian's (1965) *Social Insect Populations* book (Hamilton, 1967b), it becomes transparent that he was aware of the fundamental differences between societies and organismal colonies.

A second point worth making is that ultimate causation arguments transcend the idiosyncrasies of proximate causation. The clearest example of ultimate abstraction is the first (1859) edition of Darwin's *Origin* that established the theory of natural selection. What Hamilton's gene's eye rephrasing of Darwinism achieved was making the special position of naturally selected adaptation explicit in terms of particulate genetics. After Darwin, it is no longer possible to imagine life on any planet without having been shaped by natural selection, and after Hamilton it is no longer credible to imagine life anywhere without senescence as an unavoidable byproduct of natural selection (see Chapter 2). The same applies to Hamiltonian inclusive fitness and kin selection theory, because it represents neo-Darwinian natural selection theory writ genetically (see epigraph of Chapter 4). This implies that we cannot be sure that life on other planets is carbon based, because that is a proximate issue, but we can logically defend the ultimate conjecture that phenotypic agency will always value copies of its replicators that are realized by kin. The extension of inclusive fitness theory offered in this book implies the prediction that organized life on any planet will always be based on some form of cell closure and on subsequent MTEs mediated by higher-level closure around heritable units of information. In that view, any planet with intelligent life will have experienced at least three levels of MTE-closure—one to autopoietic cellular life subsisting on inorganic resources, a second to heterotrophic unicellular life, and a third to mobile multicellular life. Only with these minimal MTE-levels in place can brain-like substrates become substantial enough to be colonized by algorithmic symbionts that are not just processing and learning facilitators, but achieve their own self-referential autopoiesis so that cumulative culture can emerge.

In an edited volume on progress in science, Bondi (1975, pp. 3–4) offered the following succinct assessment of how scientific theories also go through major transitions and domestication events. As he wrote, when Newton's theory of gravitation fell "victim to the increasing precision of observation and calculation, one certainly feels that one can never rest assured. This is the stuff of progress. You cannot therefore speak of progress as progress in a particular direction, as a progress in which knowledge becomes more and more certain and more and more all-embracing. At times we make discoveries that sharply reduce the knowledge that we have, and it is discoveries of this kind that are indeed the seminal point in science. It is they that are the real roots of progress and lead to jumps in understanding, but in the first instance they reduce what we regard as assured knowledge. [. . .] It is, of course, important to remember that when a theory has passed a very large number of tests, like Newton's theory [of gravitation], and is then disproved [. . .] you would not say that everything that was tested before – all those forecasts – were wrong. They were right, and you know therefore that although the theory qua general theory is no longer tenable, yet it is something that described a significant volume of experience quite well."

It is this level of progress that biology as a science ought to aim for, and to do that biology should take seriously the Darwinian theory of natural selection in its modern teleonomic gene's eye phrasing that we

---

Box 8.3 *Continued*

owe to Pittendrigh, Hamilton, Williams, and Price. Rather than being envious of reductionist physics, biologists should appreciate how much better we now understand the social world around us compared to half a century ago. And we should cherish the scattered glimpses suggesting that the quantum realm may soon force physicists to start asking *why* questions that biologists began to explore after Darwin's *Origin*, a possibility that Schrödinger (1944) intuited almost 80 years ago. At the same time, we should admit that understanding is always provisional, even when based on biology's presently most coherent Hamiltonian theory of adaptation through natural selection. Ahead of us is always what Popper (1962) captured when quoting Einstein at the beginning of his *Conjectures and Refutations*: "There could be no fairer destiny for any [...] theory than that it should point the way to a more comprehensive theory in which it lived on, as a limiting case." Temporarily truthful scientific theories therefore also survive by nested domestication and likely also with some, albeit elusive, upper limit set by a putative universal theory of life and matter. It is in this sense that neo-Darwinian inclusive fitness theory domesticated original Darwinism and reduced it to a special case that remains valid in a non-social world where relatedness between interacting agents is zero.

While some consider the heuristic nature of inclusive fitness theory a weakness because it implies dynamic insufficiency in the sense of abstractly specified material gene loci, the theory's genetic information focus is in fact a strength because very general predictions of how social adaptations evolve can be derived and formulated in concrete language so they are testable in real-life populations. In fact, the necessity for biologists to pursue complementary

angles *sensu* Tinbergen must imply that no dynamically sufficient model can ever be as general as the inclusive fitness approach even though such models can be practically useful. In physics there are no proximate and ultimate ways to address questions from different angles and it is therefore that mathematical description can be general and both deterministically adequate and sufficient. However, physics has moved on to become a statistical science in its quantum domain (Bohr, 1933) and gene's eye inclusive fitness theory realized a similar statistical revolution in biology.

Discontents may argue that biologists should nonetheless prioritize understanding the fragments of total evolutionary change that are concretely reducible to material gene loci or ecological factors, rather than focus on the overall genetic information necessary for adaptive evolutionary change. Appreciating total variation and trying to capture significant parts of it in deterministic prediction is indeed important. Not all biological change that we see around us can be adaptive here and now, and some of what evolved in the deep past may well not be at present. Understanding specific genetic conditions is crucial for medicine, tree-of-life reconstruction, agriculture, and biodiversity conservation. However, those agendas of immediate urgency should not lead to the erroneous conclusion that proximate biology is the only relevant response to address challenges to our present human condition. First-principle understanding remains essential, also because biology without neo-Darwinian theory is vulnerable to anthropomorphic optimism bias, which may inhibit rather than facilitate trajectories towards personal and collective rationality, particularly when such bias deceptively suggest that nature has moral lessons to offer.

---

concept of time used to mark special events that cannot be directly reduced to Chronos. It is important that biologists realize that MTEs (see Figures 3.2 and 8.1) are Kairos events because we can precisely define them on a qualitative ratchet scale but we can only estimate their chronological age and make

educated guesses about the proximate events that induced each new level of informational closure. Human descent with modification is about a short stretch of Chronos while the human cognitive and cultural MTE is about a unique Kairos event. The neo-Darwinian gene's eye theory that offered these insights has been hugely successful, for it survived numerous attempts at refutation and has only gained conceptual simplicity and depth. However, social evolution theory is psychologically the most unsettling branch of neo-Darwinism because, as Richard Alexander (1979, p. 4) phrased it, "other major scientific theories do not threaten to invade or change our everyday existences, at least not in fashions that we are likely to regard as pernicious. They do not threaten to make our behavior predictable, to expose what we are actually doing in our social interactions, to infringe our concept of free will, or to influence the ways in which we think about right and wrong." However, discomfort can never be a valid reason to avoid a first-principle scientific theory of adaptation through natural selection that is one of humanity's pinnacle achievements and which continues to be the only *raison d'être* for biology's claim to an independent position among the natural sciences. This seems particularly opportune when indications are increasing that the statistical appreciation of naturally selected adaptation by Fisher, Hamilton, Williams, and Price is biology's best interface for ultimate scientific reconciliation with theoretical physics.

References

AANEN, D. K. 2014. How a long-lived fungus keeps mutations in check. *Science*, 346, 922–923.

AANEN, D. K. 2018. Social immunity: the disposable individual. *Current Biology*, 28, R322–R324.

AANEN, D. K. 2019. Germline evolution: sequestered cells or immortal strands? *Current Biology*, 29, R799–R801.

AANEN, D. K. & DEBETS, A. J. M. 2019. Mutation-rate plasticity and the germline of unicellular organisms. *Proceedings of the Royal Society B*, 286, 20190128.

AANEN, D. K., EGGLETON, P., ROULAND-LEFEVRE, C., GULDBERG-FROSLEV, T., ROSENDAHL, S., & BOOMSMA, J. J. 2002. The evolution of fungus-growing termites and their mutualistic fungal symbionts. *Proceedings of the National Academy of Sciences of the United States of America*, 99, 14887–14892.

AANEN, D.K., DE FINE LICHT, H.H., DEBETS, A.J.M., KERSTES, N. A. G., HOEKSTRA, R. F., & BOOMSMA, J. J. 2009. High symbiont relatedness stabilizes mutualistic cooperation in fungus-growing termites. *Science*, 326, 1103–1106.

AANEN, D. K., SPELBRINK, J. N. & BEEKMAN, M. 2014. What cost mitochondria? The maintenance of functional mitochondrial DNA within and across generations. *Philosophical Transactions of the Royal Society B*, 369: 20130438.

ABBOT, P., ABE, J., ALCOCK, J., ALIZON, S., ALPEDRINHA, J. A. C., ANDERSSON, M., ANDRE, J. B., VAN BAALEN, M., BALLOUX, F., BALSHINE, S., BARTON, N., BEUKEBOOM, L. W., BIERNASKIE, J. M., BILDE, T., BORGIA, G., BREED, M., BROWN, S., BSHARY, R., BUCKLING, A., ..., ZINK, A. 2011. Inclusive fitness theory and eusociality. *Nature*, 471, E1–E4.

ABEDON, S. T. 2017. Why Archaea are limited in their exploitation of other, living organisms. *In*: WITZANI, G. (ed.) *Biocommunication of Archaea*. Cham: Springer.

ABLER, W.L. 2005. Evidence of group learning does not add up to culture. *Nature*, 438, 422.

ACKERMANN, M., CHAO, L., BERGSTROM, C. T., & DOEBELI, M. 2007a. On the evolutionary origin of aging. *Aging Cell*, 6, 235–244.

ACKERMANN, M., SCHAUERTE, A., STEARNS, S. C., & JENAL, U. 2007b. Experimental evolution of aging in a bacterium. *BMC Evolutionary Biology*, 7: 126.

ACKERMANN, M., STEARNS, S. C., & JENAL, U. 2003. Senescence in a bacterium with asymmetric division. *Science*, 300, 1920.

ACKERMANN, M., STECHER, B., FREED, N. E., SONGHET, P., HARDT, W.-D., & DOEBELI, M. 2008. Self-destructive cooperation mediated by phenotypic noise. *Nature*, 454, 987–990.

ADAMO, S.A. 2016. Consciousness explained or consciousness redefined? *Proceedings of the National Academy of Sciences of the United States of America*, 113, E3812.

ADASHI, E. Y., RUBENSTEIN, D. S., MOSSMAN, J. A., SCHON, E. A., & COHEN, I. G. 2021. Mitochondrial disease: replace or edit? *Science*, 373, 1200–1201.

ADRIAN-KALCHHAUSER, I., SULTAN, S. E., SHAMA, L. N. S., SPENCE-JONES, H., TISO, S., KELLER VALSECCHI, C. I., & WEISSING, F. J. 2020. Understanding 'non-genetic' inheritance: insights from molecular-evolutionary crosstalk. *Trends in Ecology and Evolution*, 35, 1078–1089.

ÅGREN, J. A., DAVIES, N. G., & FOSTER, K. R. 2019. Enforcement is central to the evolution of cooperation. *Nature Ecology & Evolution*, 3, 1018–1029.

AGUILAR, C., EICHWALD, C., & EBERL, L. 2015. Multicellularity in bacteria: from division of labor to biofilm formation. *In*: RUIZ-TRILLO, I. & NEDELCU, A. M. (eds.) *Evolutionary Transitions to Multicellular Life*. Dordrecht: Springer Science+Business Media.

AGUILAR, E. G. & AKÇAY, E. 2018. Gene-culture coinheritance of a behavioral trait. *American Naturalist*, 192, 311–320.

AKANUMA, S., NAKAJIMA, Y., YOKOBORI, S., KIMURA, M., NEMOTO, N., MASE, T., MIYAZONO, K., TANOKURA, M., & YAMAGISHI, A. 2013. Experimental evidence for the thermophilicity of ancestral life. *Proceedings of the National Academy of Sciences of the United States of America*, 110, 11067–11072.

AKIL, C. & ROBINSON, R. C. 2018. Genomes of Asgard archaea encode profilins that regulate actin. *Nature*, 562, 439–443.

AKTIPIS, C. A., BODDY, A. M., JANSEN, G., HIBNER, U., HOCHBERG, M. E., MALEY, C. C., & WILKINSON, G. S. 2015. Cancer across the tree of life: cooperation and cheating in multicellularity. *Philosophical Transactions of the Royal Society B*, 370: 20140219.

ALCIATORE, G., UGELVIG, L.V., FRANK, E., BIDAUX, J., GAL, A., SCHMITT, T., KRONAUER, D.J.C., & ULRICH, Y. 2021. Immune challenges increase network centrality in a queenless ant. *Proceedings of the Royal Society B* 288: 20211456.

ALCOCK, J. 2001. *The Triumph of Sociobiology*. Oxford: Oxford University Press.

ALEXANDER, R. D. 1974. The evolution of social behavior. *Annual Review of Ecology and Systematics*, 5, 325–383.

ALEXANDER, R. D. 1979. *Darwinism and Human Affairs*. Seattle, WA: University of Washington Press.

ALEXANDER, R. D. & BORGIA, G. 1978. Group selection, altruism and the levels of the organization of life. *Annual Review of Ecology and Systematics*, 9, 449–474.

ALEXANDER, R. D. & TINKLE, D. W. 1968. A comparative review. *Bioscience*, 18, 245–248.

ALIM, K., ANDREW, N., & PRINGLE, A. 2013. *Physarum*. *Current Biology*, 23, R1082–R1083.

ALIM, K., ANDREW, N., PRINGLE, A., & BRENNER, M. P. 2017. Mechanism of signal propagation in *Physarum polycephalum*. *Proceedings of the National Academy of Sciences of the United States of America*, 114, 5136–5141.

AL-LAWATI, H. & BIENEFELD, K. 2009. Maternal age effects on embryo mortality and juvenile development of offspring in the honey bee (Hymenoptera: Apidae). *Annals of the Entomological Society of America*, 102, 881–888.

ALLEN, G. 1979. The transformation of a science: T.H. Morgan and the emergence of a new American biology. *In*: OLESON, A. & VOSS, J. (eds.) *The Organization of Knowledge in Modern America, 1860–1920*. Baltimore, MD: Johns Hopkins University Press.

AMUNDSON, R. 1996. Historical development of the concept of adaptation. *In*: ROSE, M. R. & LAUDER, G. V. (eds.) *Adaptation*. San Diego, CA: Academic Press.

AMUNDSON, R. 2014. Charles Darwin's reputation: how it changed during the twentieth-century and how it may change again. *Endeavour*, 38, 257–267.

ANDERSEN, S. B., GHOUL, M., MARVIG, R. L., LEE, Z.-B., MOLIN, S., JOHANSEN, H. K., & GRIFFIN, A. S. 2018. Privatisation rescues function following loss of cooperation. *eLife*, 7: e38594.

ANDERSEN, S. B., MARVIG, R. L., MOLIN, S., JOHANSEN, H. K., & GRIFFIN, A. S. 2015. Long-term social dynamics drive loss of function in pathogenic bacteria. *Proceedings of the National Academy of Sciences of the United States of America*, 112, 10756–10761.

ANDERSON, C., BOOMSMA, J. J., & BARTHOLDI, J. J. 2002. Task partitioning in insect societies: bucket brigades. *Insectes Sociaux*, 49, 171–180.

ANDERSON, R. M. & MAY, R. M. 1985. Vaccination and herd immunity to infectious diseases. *Nature*, 318, 323–329.

ANDERSON, T. R. 2013. *The Life of David Lack: Father of Evolutionary Ecology*. Oxford: Oxford University Press.

ANDERSSON, M. 1982. Female choice selects for extreme tail length in a widowbird. *Nature*, 299, 818–820.

AOKI, S. 1982. Soldiers and altruistic dispersal in aphids. *In*: BREED, M. D., MICHENER, C. D., & EVANS, H. E. (eds.) *The Biology of Social Insects*. Boulder, CO: Westview Press.

ARCHER, M. E. 2012. *Vespine Wasps of the World*. Manchester: Siri Scientific Press.

ARCHETTI, M., ÚBEDA, F., FUDENBERG, D., GREEN, J., PIERCE, N. E., & YU, D. W. 2011. Let the right one in: a microeconomic approach to partner choice in mutualisms. *American Naturalist*, 177, 75–85.

ARCHIBALD, J. M. 2012. Lynn Margulis (1938–2011). *Current Biology*, 22, R4–R6.

ARCHIBALD, J. M. 2015. Endosymbiosis and eukaryotic cell evolution. *Current Biology*, 25, R911–R921.

ASCHTGEN, M.-S., WETZEL, K., GOLDMAN, W., MCFALL-NGAI, M., & RUBY, E. 2016. *Vibrio fischeri*-derived outer membrane vesicles trigger host development. *Cellular Microbiology*, 18, 488–499.

ATKINS, P. 2003. *Galileo's Finger*. Oxford: Oxford University Press.

ATKINS, P. W. 1984. *The Second Law*. New York: Freeman.

AXELROD, R. & HAMILTON, W. D. 1981. The evolution of cooperation. *Science*, 211, 1390–1396.

BADCOCK, C. & CRESPI, B. 2008. Battle of the sexes may set the brain. *Nature*, 454, 1054–1055.

BAEDKE, J. 2013. The epigenetic landscape in the course of time: Conrad Hal Waddington's methodological impact on the life sciences. *Studies in History and Philosophy of Biological and Biomedical Sciences*, 44, 756–773.

BAEDKE, J. 2019. O organism, where art thou? Old and new challenges for organism-centered biology. *Journal of the History of Biology*, 52, 293–324.

BAER, B., ARMITAGE, S. A. O., & BOOMSMA, J. J. 2006. Sperm storage induces an immunity cost in ants. *Nature*, 441, 872–875.

BAERENDS, G. P., BARLOW, G. W., BLURTON JONES, N. G., CROOK, J. H., CURIO, E., EISENBERG, J. F., HINDE, R. A., HIRSCH, J., KREBS, J. R., KRUUK, H., MACKINTOSH, N. J., TOBACH, E., ROSENBLATT, J. S., & WICKLER, W. 1976. Multiple review of Wilson's sociobiology - with an author's précis and author's reply. *Animal Behaviour*, 24, 698–718.

BAER-IMHOOF, B., DEN BOER, S. P. A., BOOMSMA, J. J., & BAER, B. 2022. Sperm storage costs determine survival and immunocompetence in newly mated queens of the leaf-cutting ant *Atta colombica*. *Frontiers in Ecology and Evolution*, 9, 759183.

BAKER, J. R. 1938. The evolution of breeding seasons. *In*: DE BEER, G. R. (ed.) *Evolution: Essays on Aspects of Evolutionary Biology*. Oxford: Clarendon Press.

BALDAUF, S. L. 2008. An overview of the phylogeny and diversity of eukaryotes. *Journal of Systematics and Evolution*, 46, 263–273.

BALDWIN, J. M. 1896. A new factor in evolution. *The American Naturalist*, 30, 441–451.

BALDWIN, J. M. 1897. Organic selection. *Nature*, 55, 558.

BANKS, E. M. 1985. Warder Clyde Allee and the Chicago School of animal behavior. *Journal of the History of the Behavioral Sciences*, 21, 345–353.

BARCLAY, P. & KRUPP, D. B. 2016. The burden of proof for a cultural group selection account. Open peer commentary on Richerson et al. (2016) *Behavioral and Brain Sciences*, 39: e33.

BARDEN, P. & GRIMALDI, D. A. 2016. Adaptive radiation in socially advanced stem-group ants from the Cretaceous. *Current Biology*, 26, 515–521.

BARKER, M. G. & WALMSLEY, R. M. 1999. Replicative ageing in the fission yeast *Schizosaccharomyces pombe*. *Yeast*, 15, 1511–1518.

BARRETT, L., BLUMSTEIN, D. T., CLUTTON-BROCK, T. H., & KAPPELER, P. M. 2013. Taking note of Tinbergen, or: the promise of a biology of behaviour. *Philosophical Transactions of the Royal Society B*, 368, 20120352.

BARRETT, S. C. 2008. Major evolutionary transitions in flowering plant reproduction: an overview. *International Journal of Plant Sciences*, 169, 1–5.

BARTON, N. H., BRIGGS, D. E. G., EISEN, J. A., GOLDSTEIN, D. B., & PATEL, N. H. 2007. *Evolution*. Cold Spring Harbor, NY: Cold Spring Harbor Laboratory Press.

BARTON, N. H. & COE, J. B. 2009. On the application of statistical physics to evolutionary biology. *Journal of Theoretical Biology*, 259, 317–324.

BARTON, N. H., NOVAK, S., & PAIXÃO, T. 2014. Diverse forms of selection in evolution and computer science. *Proceedings of the National Academy of Sciences of the United States of America*, 111, 10398–10399.

BARTON, N. H. & TURELLI, M. 2007. Effects of genetic drift on variance components under a general model of epistasis. *Evolution*, 58, 2111–2132.

BASTIAANS, E., DEBETS, A. J. M., & AANEN, D. K. 2016. Experimental evolution reveals that high relatedness protects multicellular cooperation from cheaters. *Nature Communications*, 7: 11435.

BATESON, P. & LALAND, K. N. 2013. On current utility and adaptive significance: a response to Nesse. *Trends in Ecology & Evolution*, 28, 682–683.

BATRA, S. W. T. 1966. Nests and social behavior of halictine bees of India (Hymenoptera: Halictidae). *Indian Journal of Entomology*, 28, 375–393.

BATTY, C. J. K., CREWE, P., GRAFEN, A., & GRATWICK, R. 2014. Foundations of a mathematical theory of darwinism. *Journal of Mathematical Biology*, 69, 295–334.

BAUM, B. & BAUM, D. A. 2020. The merger that made us. *BMC Biology*, 18: 72.

BAUM, D. A. 2015. Selection and the origin of cells. *BioScience*, 65, 678–684.

BEEKMAN, M. & RATNIEKS, F. L. W. 2003. Power over reproduction in social Hymenoptera. *Philosophical Transactions of the Royal Society B*, 358, 1741–1753.

BELL, G. 1982. *The Masterpiece of Nature*. London: Croom Helm Ltd.

BELL, G. 1989. *Sex and Death in Protozoa*. Cambridge: Cambridge University Press.

BELL, G. 1997. *The Basics of Selection*. New York: Chapman & Hall.

BELL, G. & MOOERS, A. O. 1997. Size and complexity among multicellular organisms. *Biological Journal of the Linnaean Society*, 60, 345–363.

BENABENTOS, R., HIROSE, S., SUCGANG, R., CURK, T., KATOH, M., OSTROWSKI, E. A., STRASSMANN, J. E., QUELLER, D. C., ZUPAN, B., SHAULSKY, G., & KUSPA, A. 2009. Polymorphic members of the lag gene family mediate kin discrimination in *Dictyostelium*. *Current Biology*, 19, 567–572.

BENGTSON, S., SALLSTEDT, T., BELIVANOVA, V., & WHITEHOUSE, M. 2017. Three-dimensional preservation of cellular and subcellular structures suggests 1.6 billion-year-old crown-group red algae. *PLoS Biology*, 15: e2000735.

BERENS, A. J., HUNT, J. H., & TOTH, A. L. 2014. Comparative transcriptomics of convergent evolution: different genes but conserved pathways underlie caste phenotypes across lineages of eusocial insects. *Molecular Biology and Evolution*, 32, 690–703.

BERGSON, H. 1911. *Creative Evolution*. New York: The Modern Library.

BERGSTROM, C. T. & DUGATKIN, L. A. 2016. *Evolution*, 2nd edition. New York: W. W. Norton & Co.

BERLIN, I. 1959. *The Crooked Timber of Humanity: Chapters in the History of Ideas*. London: John Murray.

BERNARDES, J. P., JOHN, U., WOLTERMANN, N., VALIADI, M., HERMANN, R. J., & BECKS, L. 2021. The evolution of convex trade-offs enables the transition towards multicellularity. *Nature Communications*, 12: 4222.

BERNSTEIN, H. & BERNSTEIN, C. 2010. Evolutionary origin of recombination during meiosis. *BioScience*, 60, 498–505.

BIEDERMANN, P. H. W., KLEPZIG, K. D., & TABORSKY, M. 2011. Costs of delayed dispersal and alloparental care in the fungus-cultivating ambrosia beetle *Xyleborus affinis* Eichhoff (Scolytinae: Curculionidae). *Behavioral Ecology and Sociobiology*, 65, 1753–1761.

BIEDERMANN, P. H. W. & TABORSKY, M. 2011. Larval helpers and age polyethism in ambrosia beetles. *Proceedings of the National Academy of Sciences of the United States of America*, 108, 17064–17069.

BIJMA, P. 2010. Fisher's fundamental theorem of inclusive fitness and the change in fitness due to natural selection when conspecifics interact. *Journal of Evolutionary Biology*, 23, 194–206.

BIJMA, P. 2014. The quantitative genetics of indirect genetic effects: a selective review of modelling issues. *Heredity*, 112, 61–69.

BIJMA, P. & AANEN, D. K. 2010. Assortment, Hamilton's rule and multilevel selection. *Proceedings of the Royal Society B*, 277, 673–675.

BIRCH, J. 2019. Inclusive fitness as a criterion for improvement. *Studies in History and Philosophy of Science. Part C: Studies in History and Philosophy of Biological and Biomedical Sciences*, 76: 101186.

BIRKHEAD, T., WIMPENNY, J., & MONTGOMERIE, R. 2014. *Ten Thousand Birds: Ornithology since Darwin*. Princeton, NJ: Princeton University Press.

BIRKHEAD, T. R. 1998. Cryptic female choice: criteria for establishing female sperm choice. *Evolution* 52, 1212–1218.

BJÖRKHOLM, P., HARISH, A., HAGSTRÖM, E., ERNST, A. M., & ANDERSSON, S. G. E. 2015. Mitochondrial genomes are retained by selective constraints on protein targeting. *Proceedings of the National Academy of Sciences of the United States of America*, 112, 10154–10161.

BLOOMFIELD, G., PASCHKE, P., OKAMOTO, M., STEVENS, T. J., & URUSHIHARA, H. 2019. Triparental inheritance in *Dictyostelium*. *Proceedings of the National Academy of Sciences of the United States of America*, 116, 2187–2192.

BOEHM, A., ARNOLDINI, M., BERGMILLER, T., RÖÖSLI, T., BIGOSCH, C., & ACKERMANN, M. 2016. Genetic manipulation of glycogen allocation affects replicative lifespan in *E. coli*. *PLOS Genetics*, 12: e1005974.

BOHR, N. 1933. Light and Life. *Nature*, 131, 421–423 and 457–459.

BOISSEAU, R. P., VOGEL, D., & DUSSUTOUR, A. 2016. Habituation in non-neural organisms: evidence from slime moulds. *Proceedings of the Royal Society B*, 283: 20160446.

BOLHUIS, J. J., BECKERS, G. J. L., HUYBREGTS, M. A. C., BERWICK, R. C., & EVERAERT, M. B. H. 2018. Meaningful syntactic structure in songbird vocalizations? *PLoS Biology*, 16, e2005157.

BONABEAU, E., THERAULAZ, G., DENEUBOURG, J.-L., ARON, S., & CAMAZINE, S. 1997. Self-organization in social insects. *Trends in Ecology & Evolution*, 12, 188–193

BONDI, H. 1975. What is progress in science? *In:* HARRÉ, R. (ed.) *Problems of Scientific Revolution: Progress and Obstacles to Progress in the Sciences*. Oxford: Clarendon Press.

BONDURIANSKY, R. & DAY, T. 2018. *Extended Heredity: A New Understanding of Inheritance and Evolution*. Princeton, NJ: Princeton University Press.

BONHOEFFER, S. & SNIEGOWSKI, P. 2002. The importance of being erroneous. *Nature*, 420, 367–369.

BONNER, J. T. 1955. *Cells and Societies*. London: Oxford University Press.

BONNER, J. T. 1967. *The Cellular Slime Molds*. Princeton, NJ: Princeton University Press.

BONNER, J. T. 1974. *On Development: The Biology of Form*. Cambridge, MA: Harvard University Press.

BONNER, J. T. 1988. *The Evolution of Complexity by Means of Natural Selection*. Princeton, NJ: Princeton University Press.

BONNER, J. T. 1993. Dividing the labour in cells and societies. *Current Science*, 64, 459–466.

BONNER, J. T. 1995. The evolution of life's complexity (review of *The Major Transitions in Evolution* by John Maynard Smith and Eörs Szathmáry). *Nature*, 374, 508–509.

BONNER, J. T. 1998. The origins of multicellularity. *Integrative Biology: Issues, News, and Reviews*, 1, 27–36.

BONNER, J. T. 2004. The size-complexity rule. *Evolution*, 58, 1883–1890.

BOOMSMA, J. J. 1996. Split sex ratios and queen-male conflict over sperm allocation. *Proceedings of the Royal Society B*, 263, 697–704.

BOOMSMA, J. J. 2007. Kin selection versus sexual selection: why the ends do not meet. *Current Biology*, 17, R673–R683.

BOOMSMA, J. J. 2009. Lifetime monogamy and the evolution of eusociality. *Philosophical Transactions of the Royal Society B*, 364, 3191–3207.

BOOMSMA, J. J. 2013. Beyond promiscuity: mate-choice commitments in social breeding. *Philosophical Transactions of the Royal Society B*, 368, 20120050.

BOOMSMA, J. J. 2016. Fifty years of illumination about the natural levels of adaptation. *Current Biology*, 26, R1250–R1255.

BOOMSMA, J. J. 2022. Lifetime commitment between social insect families and their fungal cultivars complicates comparisons with human farming. *In*: SCHULTZ, T. R., GAWNE, R., & PEREGRINE, P. N. (eds.) *The Convergent Evolution of Agriculture in Humans and Insects.* Cambridge, MA: MIT Press.

BOOMSMA, J. J., BAER, B., & HEINZE, J. 2005a. The evolution of male traits in social insects. *Annual Review of Entomology*, 50, 395–420.

BOOMSMA, J. J. & GAWNE, R. 2018. Superorganismality and caste differentiation as points of no return: how the major evolutionary transitions were lost in translation. *Biological Reviews*, 93, 28–54.

BOOMSMA, J. J. & GRAFEN, A. 1990. Intraspecific variation in ant sex-ratios and the Trivers-Hare hypothesis. *Evolution*, 44, 1026–1034.

BOOMSMA, J. J., HUSZÁR, D. B., & PEDERSEN, J. S. 2014. The evolution of multiqueen breeding in eusocial lineages with permanent physically differentiated castes. *Animal Behaviour*, 92, 241–252.

BOOMSMA, J. J. & NASH, D. R. 2014. Evolution: sympatric speciation the eusocial way. *Current Biology*, 24, R798–R800.

BOOMSMA, J. J. & RATNIEKS, F. L. W. 1996. Paternity in eusocial Hymenoptera. *Philosophical Transactions of the Royal Society B*, 351, 947–975.

BOOMSMA, J. J., SCHMID-HEMPEL, P., & HUGHES, W. O. H. 2005b. Life histories and parasite pressure across the major groups of social insects. *In*: FELLOWES, M. D. E., HOLLOWAY, G. J., & ROLFF, J. (eds.) *Insect Evolutionary Ecology: Proceedings of the Royal Entomological Society's 22nd Symposium*. Wallingford: CABI Publishing.

BOOTH, A. & DOOLITTLE, W. F. 2015. Eukaryogenesis, how special really? *Proceedings of the National Academy of Sciences of the United States of America*, 112, 10278–10285.

BOUDINOT, B. E., RICHTER, A., KATZKE, J., CHAUL, J. C. M., KELLER, R. A., ECONOMO, E. P., BEUTEL, R. G., & YAMAMOTO, S. 2022. Evidence for the evolution of eusociality in stem ants and a systematic revision of †*Gerontoformica* (Hymenoptera: Formicidae). *Zoological Journal of the Linnean Society*, 195, 1355–1389.

BOULDING, K. E. 1956. General systems theory - the skeleton of science. *Management Science*, 2, 197–208.

BOURGUIGNON, T., LO, N., CAMERON, S. L., SOBOTNÍK, J., HAYASHI, Y., SHIGENOBU, S., WATANABE, D., ROISIN, Y., MIURA, T., & EVANS, T. A. 2014. The evolutionary history of termites as inferred from 66 mitochondrial genomes. *Molecular Biology and Evolution*, 32, 406–421.

BOURKE, A. F. G. 1988. Worker reproduction in the higher eusocial Hymenoptera. *Quarterly Review of Biology*, 63, 291–311.

BOURKE, A. F. G. 1994. Worker matricide in social bees and wasps. *Journal of Theoretical Biology*, 167, 283–292.

BOURKE, A. F. G. 2011a. *Principles of Social Evolution*. Oxford: Oxford University Press.

BOURKE, A. F. G. 2011b. The validity and value of inclusive fitness theory. *Proceedings of the Royal Society B*, 278, 3313–3320.

BOURKE, A. F. G. 2014. Hamilton's rule and the causes of social evolution. *Philosophical Transactions of the Royal Society B*, 369, 20130362.

BOURKE, A.F.G. 2019. Inclusive fitness and the major transitions in evolution. *Current Opinion in Insect Science* 34, 61–67.

BOURKE, A. F. G. 2021. The role and rule of relatedness. *Nature*, 590, 392–394.

BOURKE, A. F. G. & FRANKS, N. R. 1991. Alternative adaptations, sympatric speciation and the evolution of parasitic, inquiline ants. *Biological Journal of the Linnean Society*, 43, 157–178.

BOURKE, A. F. G. & FRANKS, N. R. 1995. *Social Evolution in Ants*. Princeton, NJ: Princeton University Press.

BOURKE, A. F. G., VAN DER HAVE, T. M., & FRANKS, N. R. 1988. Sex-ratio determination and worker reproduction in the slave-making ant *Harpagoxenus sublaevis*. *Behavioral Ecology and Sociobiology*, 23, 233–245.

BOWLER, P. J. 2005. Revisiting the eclipse of Darwinism. *Journal of the History of Biology*, 38, 19–32.

BOWLES, S. 2006. Group competition, reproductive leveling, and the evolution of human altruism. *Science*, 314, 1569–1572.

BOWLES, S. 2009. Did warfare among ancestral hunter-gatherers affect the evolution of human social behaviors? *Science*, 324, 1293–1298.

BOWLES, S. & GINTIS, H. 2011. *A Cooperative Species: Human Reciprocity and Its Evolution.* Princeton, NJ: Princeton University Press.

BOX, G. E. P. 1976. Science and statistics. *Journal of the American Statistical Association*, 71, 791–799.

BOYD, R. & RICHERSON, P. J. 1985. *Culture and the Evolutionary Process.* Chicago, IL: University of Chicago Press.

BOZA, G., SZILÁGYI, A., KUN, A., SANTOS, M., & SZATHMÁRY, E. 2014. Evolution of the division of labor between genes and enzymes in the RNA world. *PLoS Computational Biology*, 10: e1003936.

BRADY, S. G., SCHULTZ, T. R., FISHER, B. L., & WARD, P. S. 2006. Evaluating alternative hypotheses for the early evolution and diversification of ants. *Proceedings of the National Academy of Sciences of the United States of America*, 103, 18172–18177.

BRAHMA, A., LEON, R. G., HERNANDEZ, G. L., & WURM, Y. 2022. Larger, more connected societies of ants have a higher prevalence of viruses. *Molecular Ecology*, 31, 859–865.

BRANDÃO, F. G. S. L., PIANI, M., & HORODECKI, P. 2015. Generic emergence of classical features in quantum Darwinism. *Nature Communications*, 6, 7908.

BRANDON, R. N. 1988. The levels of selection: a hierarchy of interactors. *In*: PLOTKIN, H. C. (ed.) *The Role of Behavior in Evolution.* Cambridge, MA: MIT Press.

BRANDON, R. N. 1990. *Adaptation and Environment.* Princeton, NJ: Princeton University Press.

BRANDVAIN, Y., BARKER, M. S., & WADE, M. J. 2007. Gene co-inheritance and gene transfer. *Science*, 315, 1685.

BRANSTETTER, M. G., DANFORTH, B. N., PITTS, J. P., FAIRCLOTH, B. C., WARD, P. S., BUFFINGTON, M. L., GATES, M. W., KULA, R. R., & BRADY, S. G. 2017. Phylogenomic insights into the evolution of stinging wasps and the origins of ants and bees. *Current Biology*, 27, 1019–1025.

BRAWAND, D., SOUMILLON, M., NECSULEA, A., JULIEN, P., CSÁRDI, G., HARRIGAN, P., WEIER, M., LIECHTI, A., AXIMU-PETRI, A., KIRCHER, M., ALBERT, F. W., ZELLER, U., KHAITOVICH, P., GRÜTZNER, F., BERGMANN, S., NIELSEN, R., PÄÄBO, S., & KAESSMANN, H. 2011. The evolution of gene expression levels in mammalian organs. *Nature*, 478, 343–348.

BRAWLEY, S. H., BLOUIN, N. A., FICKO-BLEAN, E., WHEELER, G. L., LOHR, M., GOODSON, H. V., JENKINS, J. W., BLABY-HAAS, C. E., HELLIWELL, K. E., CHAN, C. X., MARRIAGE, T. N., BHATTACHARYA, D., KLEIN, A. S., BADIS, Y., BRODIE, J., CAO, Y., COLLÉN, J., DITTAMI, S. M., GACHON, C., GREEN, B. R., & PROCHNIK, S. E. 2017. Insights into the red algae and eukaryote evolution from the genome of *Porphyra umbilicalis* (Bangiophyceae, Rhodophyta). *Proceedings of the National Academy of Sciences of the United States of America*, 114, E6361–E6370.

BRAZELTON, W. 2017. Hydrothermal vents. *Current Biology*, 27, R450–R452.

BRETL, D. J. & KIRBY, J. R. 2016. Molecular mechanisms of signalling in *Myxococcus xanthus* development. *Journal of Molecular Biology*, 428, 3805–3830.

BRIAN, M. V. 1965. *Social Insect Populations.* London: Academic Press.

BRIGANDT, I. 2006. Homology and heterochrony: the evolutionary embryologist Gavin Rylands de Beer (1899–1972). *Journal of Experimental Zoology. Molecular and Developmental Evolution*, 306B, 317–328.

BRIGGS, D. E. G. 2015. The Cambrian explosion. *Current Biology*, 25, R864–R868.

BRIGGS, J. A., WEINREB, C., WAGNER, D. E., MEGASON, S., PESHKIN, L., KIRSCHNER, M. W., & KLEIN, A. M. 2018. The dynamics of gene expression in vertebrate embryogenesis at single-cell resolution. *Science*, 360, eaar5780.

BROCK, D. A., CALLISON, W. É., STRASSMANN, J. E., & QUELLER, D. C. 2016. Sentinel cells, symbiotic bacteria and toxin resistance in the social amoeba *Dictyostelium discoideum*. *Proceedings of the Royal Society B*, 283, 20152727.

BROCK, D. A., DOUGLAS, T. E., QUELLER, D. C., & STRASSMANN, J. E. 2011. Primitive agriculture in a social amoeba. *Nature*, 469, 393–396.

BROCK, D. A., READ, S., BOZHCHENKO, A., QUELLER, D. C., & STRASSMANN, J. E. 2013.

Social amoeba farmers carry defensive symbionts to protect and privatize their crops. *Nature Communications*, 4: 2385.

BRONSTEIN, J. L. (ed.) 2015. *Mutualism*. Oxford: Oxford University Press.

BROOKS, D. S. 2017. In defense of levels: layer cakes and guilt by association. *Biological Theory*, 12, 142–156.

BROWN, J. L. 1975. *The Evolution of Behavior*. New York: W. W. Norton & Co.

BROWN, J. L. 1980. Fitness in complex avian social systems. *In*: MARKL, H. (ed.) *Evolution of Social Behavior: Hypotheses and Empirical Tests*. Weinheim: Verlag Chemie GmbH.

BRUCE, J. B., COOPER, G. A., CHABAS, H., WEST, S. A., & GRIFFIN, A. S. 2017. Cheating and resistance to cheating in natural populations of the bacterium *Pseudomonas fluorescens*. *Evolution*, 71, 2484–2495.

BRÜCKNER, S. & MÖSCH, H.-U. 2012. Choosing the right lifestyle: adhesion and development in *Saccharomyces cerevisiae*. *FEMS Microbiology Reviews*, 36, 25–58.

BUCEK, A., SOBOTNÍK, J., HE, S., SHI, M., MCMAHON, D. P., HOLMES, E. C., ROISIN, Y., LO, N., & BOURGUIGNON, T. 2019. Evolution of termite symbiosis informed by transcriptome-based phylogenies. *Current Biology*, 29, 3728–3734.

BUCHNER, P. 1953. *Endosymbiose der Tiere mit Pflanzlichen Mikroorganismen*, Basel/Stuttgart, Verlag Birkhäuser.

BUDIN, I. & SZOSTAK, J. W. 2011. Physical effects underlying the transition from primitive to modern cell membranes. *Proceedings of the National Academy of Sciences of the United States of America*, 108, 5249–5254.

BULL, J. J. & RICE, W. R. 1991. Distinguishing mechanisms for the evolution of cooperation. *Journal of Theoretical Biology*, 149, 63–74.

BULL, J. J. & WANG, I.-N. 2010. Optimality models in the age of experimental evolution and genomics. *Journal of Evolutionary Biology*, 23, 1820–1838.

BURKE, T. 1989. DNA fingerprinting and other methods for the study of mating success. *Trends Ecology and Evolution*, 4, 139–144.

BURKHARDT, R. W. 1992. Huxley and the rise of ethology. *In*: WATERS, C. K. & VAN HELDEN, A. (eds.) *Julian Huxley: Biologist and Statesman of Science*. Houston, TX: Rice University Press.

BURKHARDT, R. W. 2005. *Patterns of Behavior: Konrad Lorenz, Niko Tinbergen, and the Founding of Ethology*. Chicago, IL: University of Chicago Press.

BURT, A. & TRIVERS, R. 2006. *Genes in Conflict*. Cambridge, MA: Harvard University Press.

BURTON-CHELLEW, M. N. & DUNBAR, R. I. M. 2015. Hamilton's rule predicts anticipated social support in humans. *Behavioral Ecology*, 26, 130–137.

BURTON-CHELLEW, M. N., EL MOUDEN, C., & WEST, S. A. 2016. Conditional cooperation and confusion in public good experiments. *Proceedings of the National Academy of Sciences of the United States of America*, 113, 1291–1296.

BURTON-CHELLEW, M. N., EL MOUDEN, C., & WEST, S. A. 2017. Social learning and the demise of costly cooperation in humans. *Proceedings of the Royal Society B*, 284, 20170067.

BURTON-CHELLEW, M. N. & WEST, S. A. 2013. Prosocial preferences do not explain human cooperation in public good games. *Proceedings of the National Academy of Sciences of the United States of America*, 110, 216–221.

BUSCHINGER, A. 2009. Social parasitism among ants: a review (Hymenoptera: Formicidae). *Myrmecological News*, 12, 219–235.

BUSS, L. W. 1982. Somatic cell parasitism and the evolution of somatic tissue compatibility. *Proceedings of the National Academy of Sciences of the United States of America*, 79, 5337–5341.

BUSS, L. W. 1987. *The Evolution of Individuality*. Princeton, NJ: Princeton University Press.

BUSS, L. W. 1988. Diversification and germ-line determination. *Paleobiology*, 14, 313–321.

BUTAITE, E., BAUMGARTNER, M., WYDER, S., & KÜMMERLI, R. 2017. Siderophore cheating and cheating resistance shape competition for iron in soil and freshwater *Pseudomonas* communities. *Nature Communications*, 8: 414.

BUTTERY, N. J., ROZEN, D. E., WOLF, J. B., & THOMPSON, C. R. L. 2009. Quantification of social behavior in *D. discoideum* reveals complex fixed and facultative strategies. *Current Biology*, 19, 1373–1377.

BYARS, S. G., STEARNS, S. C., & BOOMSMA, J. J. 2014. Opposite risk patterns for autism and schizophrenia are associated with normal variation in birth size: phenotypic support for hypothesized diametric gene-dosage effects. *Proceedings of the Royal Society B*, 281, 20140604.

CAIN, A. J. 1964. The perfection of animals. *In*: CARTHY, J. D. & DUDDINGTON, C. L. (eds.) *Viewpoints in Biology*. London: Butterworth.

CAIN, J. 2010. Julian Huxley, general biology and the London Zoo, 1935–42. *Notes and Records of the Royal Society of London*, 64, 359–378.

CALCOTT, B. & STERELNY, K. (eds.) 2011. *The Major Transitions in Evolution Revisited*. Cambridge MA: MIT Press.

CAMAZINE, S., DENEUBOURG, J.-L., FRANKS, N. R., SNEYD, J., THERAULAZ, G., & BONABEAU, E. 2001. *Self-Organization in Biological Systems*. Princeton, NJ: Princeton University Press.

CANIGLIA, G. 2017. "How complex and even perverse the real world can be" W.D. Hamilton's early work on social wasps (1964–1968). *Studies in History and Philosophy of Biological and Biomedical Sciences*, 64, 41–52.

CANT, M. A. & CROFT, D. P. 2019. Life-history evolution: grandmothering in space and time. *Current Biology*, 29, R215–R218.

CARDINAL, S. & DANFORTH, B. N. 2011. The antiquity and evolutionary history of social behavior in bees. *PLoS One*, 6: e21086.

CARDINAL, S. & DANFORTH, B. N. 2013. Bees diversified in the age of eudicots. *Proceedings of the Royal Society B*, 280, 20122686.

CARLSON, E. A. 2004. *Mendel's Legacy: The Origin of Classical Genetics*. Cold Spring Harbor, NY: Cold Spring Harbor Laboratory Press.

CASTILLO, D. I., SWITZ, G. T., FOSTER, K. R., QUELLER, D. C., & STRASSMANN, J. E. 2005. A cost to chimerism in *Dictyostelium discoideum* on natural substrates. *Evolutionary Ecology Research*, 7, 263–271.

CHALKER, D. L., MEYER, E., & MOCHIZUKI, K. 2013. Epigenetics of Ciliates. *Cold Spring Harbor Perspectives in Biology*, 5: a017764.

CHAPMAN, T., LIDDLE, L. F., KALB, J. M., WOLFNER, M. F., & PARTRIDGE, L. 1995. Cost of mating in *Drosophila melanogaster* females is mediated by male accessory gland products. *Nature*, 373, 241–244.

CHAPUISAT, M. 2010. Social evolution: sick ants face death alone. *Current Biology*, 20, R104–R105.

CHARLESWORTH, B. 1978. Some models of the evolution of altruistic behaviour between siblings. *Journal of Theoretical Biology*, 72, 297–319.

CHARLESWORTH, B. 1990. Optimization models, quantitative genetics, and mutation. *Evolution*, 44, 520–538.

CHARLESWORTH, B. 2000. Fisher, Medawar, Hamilton and the evolution of aging. *Genetics*, 156, 927–931.

CHARLESWORTH, B. & CHARLESWORTH, D. 2009. Darwin and genetics. *Genetics*, 183, 757–766.

CHARLESWORTH, D., BARTON, N. H., & CHARLESWORTH, B. 2017. The sources of adaptive variation. *Proceedings of the Royal Society B*, 284, 20162864.

CHARNOV, E. L. 1976. Optimal foraging, the marginal value theorem. *Theoretical Population Biology*, 9, 129–136.

CHARNOV, E. L. 1977. An elementary treatment of the genetical theory of kin selection. *Journal of Theoretical Biology*, 66, 541–550.

CHARNOV, E. L. 1982. *The Theory of Sex Allocation*. Princeton, NJ: Princeton University Press.

CHARNOV, E. L. 1984. Behavioural ecology of plants. *In*: KREBS, J. R. & DAVIES, N. B. (eds.) *Behavioural Ecology: An Evolutionary Approach*, 2nd edition. Oxford: Blackwell Scientific Publications.

CHARNOV, E. L. 1993. *Life History Invariants*. Oxford, England: Oxford University Press.

CHARNOV, E.L. & KREBS, J.R. 1974. On clutch-size and fitness. *Ibis*, 116, 217–219.

CHARNOV, E. L. & SCHAFFER, W. M. 1973. Life-history consequences of natural selection: Cole's result revisited. *American Naturalist*, 107, 791–793.

CHASTAIN, E., LIVNAT, A., PAPADIMITRIOU, C., & VAZIRANI, U. 2014. Algorithms, games, and evolution. *Proceedings of the National Academy of Sciences of the United States of America*, 111, 10620–10623.

CHEN, H., JOLLY, C., BUBLYS, K., MARCU, D., & IMMLER, S. 2020. Trade-off between somatic and germline repair in a vertebrate supports the expensive germ line hypothesis. *Proceedings of the National Academy of Sciences of the United States of America*, 117, 8973–8979.

CHEN, I. A., ROBERTS, R. W., & SZOSTAK, J. W. 2004. The emergence of competition between model protocells. *Science*, 305, 1474–1476.

CHOI, J. & KIM, S.-H. 2017. A genome tree of life for the Fungi kingdom. *Proceedings of the National Academy of Sciences of the United States of America*, 114, 9391–9396.

CHOI, J.-K. & BOWLES, S. 2007. The coevolution of parochial altruism and war. *Science*, 318, 636–640.

CHOMICKI, G., WERNER, G. D. A., WEST, S. A., & KIERS, E. T. 2020. Compartmentalization drives the evolution of symbiotic cooperation. *Philosophical Transactions of the Royal Society B*, 375, 20190602.

CHRISTIANSEN, F. B. 1991. On conditions for evolutionary stability for a continuously varying character. *American Naturalist*, 138, 37–50.

CHURCHILL, F. B. 2015. *August Weismann: Development, Heredity and Evolution*. Cambridge, MA: Harvard University Press.

CINI, A., SUMNER, S., & CERVO, R. 2019. Inquiline social parasites as tools to unlock the secrets of insect sociality. *Philosophical Transactions of the Royal Society B*, 374, 20180193.

CLAIDIÈRE, N. & ANDRÉ, J.-B. 2012. The transmission of genes and culture: a questionable analogy. *Evolutionary Biology*, 39, 12–24.

CLAIDIÈRE, N., SCOTT-PHILLIPS, T. C., & SPERBER, D. 2014. How Darwinian is cultural evolution? *Philosophical Transactions of the Royal Society B*, 369, 20130368.

CLAIDIÈRE, N. & SPERBER, D. 2010. Imitation explains the propagation, not the stability of animal culture. *Proceedings of the Royal Society B*, 277, 651–659.

CLARKE, E. 2010. The problem of biological individuality. *Biological Theory*, 5, 312–325.

CLARKE, E. 2013. The multiple realizability of biological individuals. *Journal of Philosophy*, 110, 413–435.

CLARKE, E. 2014. Origins of evolutionary transitions. *Journal of Biosciences*, 39, 303–317.

CLUTTON-BROCK, T. H. 1982. The function of antlers. *Behaviour*, 79, 108–125.

CLUTTON-BROCK, T. H. 1988. *Reproductive Success: Studies of Individual Variation in Contrasting Breeding Systems*. Chicago, IL: University of Chicago Press.

CLUTTON-BROCK, T. H., ALBON, S. D., & HARVEY, P. H. 1980. Antlers, body-size and breeding group size in the Cervidae. *Nature*, 285, 565–567.

CLUTTON-BROCK, T. H., GUINNESS, F. E., & ALBON, S. D. 1982. *Red Deer: Behavior and Ecology of Two Sexes*. Chicago, IL: University of Chicago Press.

CLUTTON-BROCK, T. H. & HARVEY, P. H. 1979. Comparison and adaptation. *Proceedings of the Royal Society B*, 205, 547–565.

COCK, J. M., STERCK, L., ROUZÉ, P., SCORNET, D., ALLEN, A. E., AMOUTZIAS, G., ANTHOUARD, V., ARTIGUENAVE, F., AURY, J. M., BADGER, J. H., BESZTERI, B., BILLIAU, K., BONNET, E., BOTHWELL, J. H., BOWLER, C., BOYEN, C., BROWNLEE, C., CARRANO, C. J., CHARRIER, B., CHO, G. Y., ..., & WINCKLER, P. 2010. The *Ectocarpus* genome and the independent evolution of multicellularity in brown algae. *Nature*, 465, 617–621.

COCKBURN, A. 1991. *An Introduction to Evolutionary Ecology*. Oxford: Blackwell Scientific Publications.

COELHO, M., DERELI, A., HAESE, A., KÜHN, S., MALINOVSKA, L., DESANTIS, M. E., SHORTER, J., ALBERTI, S., GROSS, T., & TOLIC-NØRRELYKKE, I. M. 2013. Fission yeast does not age under favorable conditions, but does so after stress. *Current Biology*, 23, 1844–1852.

COFNAS, N. 2018. Power in cultural evolution and the spread of prosocial norms. *Quarterly Review of Biology*, 93, 297–318.

COLLINS, J. P. 1986. Evolutionary ecology and the use of natural selection in ecological theory. *Journal of the History of Biology*, 19, 257–288.

CONROY, T. E. & HOLMAN, L. 2022. Social immunity in the honey bee: do immune-challenged workers enter enforced or self-imposed exile? *Behavioral Ecology and Sociobiology*, 76: 32.

COOPER, G. A., LEVIN, S. R., WILD, G., & WEST, S. A. 2018. Modelling relatedness and demography in social evolution. *Evolution Letters*, 2, 260–271.

COOPER, G. A. & WEST, S. A. 2018. Division of labour and the evolution of extreme specialization. *Nature Ecology & Evolution*, 2, 1161–1167.

CORNING, P. A. & KLINE, S. J. 1998a. Thermodynamics, information and life revisited, Part I: 'to be or entropy'. *Systems Research and Behavioral Science*, 15, 273–295.

CORNING, P. A. & KLINE, S. J. 1998b. Thermodynamics, information and life revisited, Part II: 'thermoeconomics' and 'control information'. *Systems Research and Behavioral Science*, 15, 453–482.

CORNWALLIS, C. K., BOTERO, C. A., RUBENSTEIN, D. R., DOWNING, P. A., WEST, S. A., & GRIFFIN, A. S. 2017. Cooperation facilitates the colonization of harsh environments. *Nature Ecology and Evolution*, 1: 0057.

CORNWALLIS, C. K., WEST, S. A., DAVIS, K. E., & GRIFFIN, A. S. 2010. Promiscuity and the evolutionary transition to complex societies. *Nature*, 466, 969–972.

CORNWALLIS, C. K., WEST, S. A., & GRIFFIN, A. S. 2009. Routes to indirect fitness in cooperatively breeding vertebrates: kin discrimination and limited dispersal. *Journal of Evolutionary Biology*, 22, 2445–2457.

COSMIDES, L. & TOOBY, J. 1992. Cognitive adaptations for social exchange. *In*: BARKOW, J. H., COSMIDES, L., & TOOBY, J. (eds.) *The Adapted Mind*. New York: Oxford University Press.

COSMIDES, L. M. & TOOBY, J. 1981. Cytoplasmic inheritance and intragenomic conflict. *Journal of Theoretical Biology*, 89, 83–129.

COSTA, J. T. 2006. *The Other Insect Societies*. Cambridge, MA: Belknap Press of Harvard University Press.

COSTA, J. T. 2013. Hamiltonian inclusive fitness: a fitter fitness concept. *Biology Letters*, 9: 20130335.

COUVET, D., RONCE, O., & GLIDDON, C. 1998. The maintenance of nucleocytoplasmic polymorphism in a metapopulation: the case of gynodioecy. *American Naturalist*, 152, 59–70.

COYNE, J. A. 2009. *Why Evolution is True*. New York: Viking.

COYNE, J. A., BARTON, N. H., & TURELLI, M. 2000. Is Wright's shifting balance process important in evolution? *Evolution*, 54, 306–317.

COYTE, K. Z., SCHLUTER, J., & FOSTER, K. R. 2015. The ecology of the microbiome: networks, competition, and stability. *Science*, 350, 663–666.

CREMER, S. 2019. Social immunity in insects. *Current Biology*, 29, R458–R463.

CREMER, S., ARMITAGE, S. A. O., & SCHMID-HEMPEL, P. 2007. Social immunity. *Current Biology*, 17, R693–R702.

CREMER, S., PULL, C.D., & FÜRST, M.A. 2017. Social immunity: emergence and evolution of colony-level disease protection. *Annual Review of Entomology*, 63, 105–123.

CREMER, S. & SIXT, M. 2009. Analogies in the evolution of individual and social immunity. *Philosophical Transactions of the Royal Society B*, 364, 129–142.

CRESPI, B. J. 2001. The evolution of social behavior in microorganisms. *Trends in Ecology & Evolution*, 16, 178–183.

CRESPI, B. & SPRINGER, S. 2003. Social slime molds meet their match. *Science*, 299, 56–57.

CRESPI, B. & SUMMERS, K. 2005. Evolutionary biology of cancer. *Trends in Ecology and Evolution*, 20, 545–552.

CRESPI, B. J. 1992. Eusociality in Australian gall thrips. *Nature*, 359, 724–726.

CREWE, P., GRATWICK, R., & GRAFEN, A. 2018. Defining fitness in an uncertain world. *Journal of Mathematical Biology*, 76, 1059–1099.

CRONIN, H. 1991. *The Ant and the Peacock*. Cambridge: Cambridge University Press.

CRONIN, H. 2005. Adaptation: "a critique of some current evolutionary thought". *Quarterly Review of Biology*, 80, 19–27.

CRONIN, K.A. 2016. Unnatural paradigm calls into question whether macaques' social decisions represent empathy. *Proceedings of the National Academy of Sciences of the United States of America* 113, E1331.

CROW, J. F. 1991. Some optimality principles in evolution. Open peer commentary on Schoemaker (1991). *Behavioral and Brain Sciences*, 14, 218–219.

CROW, J. F. 2002. Here's to Fisher, additive genetic variance, and the fundamental theorem of natural selection. *Evolution*, 56, 1313–1316.

CROW, J. F. 2008. Commentary: Haldane and beanbag genetics. *International Journal of Epidemiology*, 37, 442–445.

CROZIER, R.H. 1992. The genetic evolution of flexible strategies. *American Naturalist*, 139, 218–223.

CROZIER, R. H. & PAMILO, P. 1996. *Evolution of Social Insects Colonies: Sex Allocation and Kin Selection*. Oxford: Oxford University Press.

CUMMINS, J. M. 2000. Fertilization and elimination of the paternal mitochondrial genome. *Human Reproduction*, 15, 92–101.

CURRIE, C. R., POULSEN, M., MENDENHALL, J., BOOMSMA, J. J., & BILLEN, J. 2006. Coevolved crypts and exocrine glands support mutualistic bacteria in fungus-growing ants. *Science*, 311, 81–83.

DAAN, S., DIJKSTRA, C., & TINBERGEN, J. M. 1990. Family planning in the kestrel (*Falco tinnunculus*): the ultimate control of covariation of laying date and clutch size. *Behavior*, 114, 83–116.

DALLAI, R., THIPAKSORN, A., GOTTARDO, M., MERCATI, D., MACHIDA, R., & BEUTEL, R. G. 2014. The sperm structure of *Cryptocercus punctulatus* Scudder (Blattoida) and sperm evolution in Dictyoptera. *Journal of Morphology*, 276, 361–369.

DALY, M. 1991. Natural selection doesn't have goals, but it's the reason organisms do. Open peer commentary on Schoemaker (1991). *Behavioral and Brain Sciences*, 14, 219–220.

DAMUTH, J. & HEISLER, I. L. 1988. Alternative formulations of multilevel selection. *Biology and Philosophy*, 3, 407–430.

DANDEKAR, A. A., CHUGANI, S., & GREENBERG, E. P. 2012. Bacterial quorum sensing and metabolic incentives to cooperate. *Science*, 338, 264–266.

DANFORTH, B. N., CARDINAL, S., PRAZ, C., ALMEIDA, E. A. B., & MICHEZ, D. 2013. The impact of molecular data on our understanding of bee phylogeny and evolution. *Annual Review of Entomology*, 58, 57–78.

DARCH, S. E., WEST, S. A., WINZER, K., & DIGGLE, S. P. 2012. Density-dependent fitness benefits in quorum-sensing bacterial populations. *Proceedings of the National Academy of Sciences of the United States of America*, 109, 8259–8263.

DARWIN, C. 1859. *On the Origin of Species by Means of Natural Selection*. London: John Murray.

DAVIES, N. B. 1992. *Dunnock Behaviour and Social Evolution*. Oxford: Oxford University Press.

DAVIES, N. B., KREBS, J. R., & WEST, S. A. 2012. *An Introduction to Behavioural Ecology*, 4th edition. Oxford: Wiley.

DAVIES, N. G. & GARDNER, A. 2018. Monogamy promotes altruistic sterility in insect societies. *Royal Society Open Science*, 5, 172190.

DAWKINS, R. 1976. *The Selfish Gene*. Oxford: Oxford University Press.

DAWKINS, R. 1982a. *The Extended Phenotype*. Oxford: Oxford University Press.

DAWKINS, R. 1982b. Replicators and vehicles. *In*: KING'S COLLEGE SOCIOBIOLOGY GROUP (ed.) *Current Problems in Sociobiology*. Cambridge: Cambridge University Press.

DAWKINS, R. 2006. *Climbing Mount Improbable*. London: Penguin.

DAWKINS, R. 2013. *An Appetite for Wonder*. London: Black Swan.

DAWKINS, R. & KREBS, J. R. 1979. Arms races between and within species. *Proceedings of the Royal Society B*, 205, 489–511.

DAY, T., PARSONS, T., LAMBERT, A., & GANDON, S. 2020. The Price equation and evolutionary epidemiology. *Philosophical Transactions of the Royal Society B*, 375, 20190357.

DE BEER, G. R. E. 1938. *Evolution: Essays on Aspects of Evolutionary Biology*. Oxford: Clarendon Press.

DE BEKKER, C., WILL, I., DAS, B., & ADAMS, R. M. M. 2018. The ants (Hymenoptera: Formicidae) and their parasites: effects of parasitic manipulations and host responses on ant behavioral ecology. *Myrmecological News*, 28, 1–24.

DE DUVE, C. 2005a. The onset of selection. *Nature*, 433, 581–582.

DE DUVE, C. 2005b. *Singularities: Landmarks on the Pathways of Life*. New York: Cambridge University Press.

DE DUVE, C. 2011. Life as a cosmic imperative? *Philosophical Transactions of the Royal Society B*, 369, 620–623.

DE JONG, T. & KLINKHAMER, J. 2005. *Evolutionary Ecology of Plant Reproductive Strategies*. Cambridge: Cambridge University Press.

DE LA MORA, A., SANKOVITZ, M., & PURCELL, J. 2020. Ants (Hymenoptera: Formicidae) as host and intruder: recent advances and future directions in the study of exploitative strategies. *Myrmecological News*, 30, 53–71.

DE MENDOZA, A., SEBÉ-PEDRÓS, A., SESTAK, M. S., MATEJCIC, M., TORRUELLA, G., DOMAZET-LOSO, T., & RUIZ-TRILLO, I. 2013. Transcription factor evolution in eukaryotes and the assembly of the regulatory toolkit in multicellular lineages. *Proceedings of the National Academy of Sciences of the United States of America*, 110, E4858–E4866.

DECAESTECKER, E., GABA, S., RAEYMAEKERS, J. A. M., STOKS, R., VAN KERCKHOVEN, L., EBERT, D., & DE MEESTER, L. 2007. Host–parasite 'Red Queen' dynamics archived in pond sediment. *Nature*, 450, 870–873.

DELGADO, M. D. M., ROSLIN, T., MEYKE, E., LO, C., GURARIE, E., ABADONOVA, M., ABDURAIMOV, O., ADRIANOVA, O., AKIMOVA, T., AKKIEV, M., ANANIN, A., ANDREEVA, E., ANDRIYCHUK, N., ANTIPIN, M., ARZAMASCEV, K., BABINA, S., BABUSHKIN, M., BAKIN, O., BARABANCOVA, A., . . . , & OVASKAINEN, O. 2020. Differences in spatial versus temporal reaction norms for spring and autumn phenological events. *Proceedings of the National Academy of Sciences of the United States of America*, 117, 31249–31258.

DEN BOER, S. P. A., BAER, B., & BOOMSMA, J. J. 2010. Seminal fluid mediates ejaculate competition in social insects. *Science*, 327, 1506–1509.

DEN BOER, S. P. A., BAER, B., DREIER, S., ARON, S., NASH, D. R., & BOOMSMA, J. J. 2009. Prudent sperm use by leaf-cutter ant queens. *Proceedings of the Royal Society B*, 276, 3945–3953.

DENNETT, D. C. 1991. *Consciousness Explained*. London: Penguin.

DENNETT, D. C. 1995. *Darwin's Dangerous Idea: Evolution and the Meanings of Life*. New York: Simon & Schuster.

DENNETT, D. C. 2017. *From Bacteria to Bach and Back: The Evolution of Minds*. New York: W. W. Norton.

DÍAZ-MUÑOZ, S. L., SANJUÁN, R., & WEST, S. 2017. Sociovirology: conflict, cooperation and communication among viruses. *Cell Host & Microbe*, 22, 437–441.

DICKINSON, H. G. & GRANT-DOWNTON, R. 2009. Bridging the generation gap: flowering plant gametophytes and animal germlines reveal unexpected similarities. *Biological Reviews*, 84, 589–615.

DIGGLE, S. P., GARDNER, A., WEST, S. A., & GRIFFIN, A. S. 2007a. Evolutionary theory of bacterial quorum sensing: when is a signal not a signal? *Philosophical Transactions of the Royal Society B*, 362, 1241–1249.

DIGGLE, S. P., GRIFFIN, A. S., CAMPBELL, G. S., & WEST, S. A. 2007b. Cooperation and conflict in quorum-sensing bacterial populations. *Nature*, 450, 411–414.

DIJKSTRA, M. B. & BOOMSMA, J. J. 2006. Are workers of *Atta* leafcutter ants capable of reproduction? *Insectes Sociaux*, 53, 136–140.

DIJKSTRA, M. B. & BOOMSMA, J. J. 2007. The economy of worker reproduction in *Acromyrmex?* leafcutter ants. *Animal Behaviour*, 74, 519–529.

DISALVO, S., HASELKORN, T. S., BASHIR, U., JIMENEZ, D., BROCK, D. A., QUELLER, D. C., & STRASSMANN, J. E. 2015. *Burkholderia* bacteria infectiously induce the proto-farming symbiosis of *Dictyostelium* amoebae and food bacteria. *Proceedings of the National Academy of Sciences of the United States of America*, 112, E5029–E5037.

DIVALL, C. 1992. From a Victorian to a modern: Julian Huxley and the English intellectual climate. *In*: WATERS, C. K. & VAN HELDEN, A. (eds.) *Julian Huxley: Biologist and Statesman of Science*. Houston, TX: Rice University Press.

DOBZHANSKY, T. 1937. *Genetics and the Origin of Species*. New York: Columbia University Press.

DOBZHANSKY, T. 1958. Species after Darwin. *In*: BARNETT, S. A. (ed.) *A Century of Darwin*. London: Heinemann.

DOBZHANSKY, T. 1973. Nothing in biology makes sense except in the light of evolution. *The American Biology Teacher*, 35, 125–129.

DODD, M. S., PAPINEAU, D., GRENNE, T., SLACK, J. F., RITTNER, M., PIRAJNO, F., O'NEIL, J., & LITTLE, C. T. S. 2017. Evidence for early life in Earth's oldest hydrothermal vent precipitates. *Nature*, 543, 60–64.

DOLLO, L. 1893. Les lois de l'évolution. *Bulletin de la Société belge de géologie, de paléontologie et d'hydrologie*, 7, 164–166.

DOOLITTLE, W. F. 2020. Evolution: two domains of life or three? *Current Biology*, 30, R177–R179.

DOSSELLI, R., GRASSL, J., DEN BOER, S. P. A., KRATZ, M., MORAN, J. M., BOOMSMA, J. J., & BAER, B. 2019. Protein-level interactions as mediators of sexual conflict in ants. *Molecular & Cellular Proteomics*, 18, S34–S45.

DOUGLAS, A. E. 1983. Uric acid utilization in *Platymonas convolutae* and symbiotic *Convoluta roscoffensis*. *Journal of the Marine Biological Association of the United Kingdom*, 63, 435–447.

DOUGLAS, A. E. 2010. *The Symbiotic Habit*. Princeton, NJ: Princeton University Press.

DOUGLAS, T. E., QUELLER, D. C., & STRASSMANN, J. E. 2017. Social amoebae mating types do not invest unequally in sexual offspring. *Journal of Evolutionary Biology*, 30, 926–937.

DOWNING, P. A., GRIFFIN, A. S., & CORNWALLIS, C. K. 2020a. The benefits of help in cooperative birds: non-existent or difficult to detect? *American Naturalist*, 195, 1085–1091.

DOWNING, P. A., GRIFFIN, A. S., & CORNWALLIS, C. K. 2020b. Group formation and the evolutionary pathway to complex sociality in birds. *Nature Ecology & Evolution*, 4, 479–486.

DRONAMRAYU, K. R. 1993. *If I Am to Be Remembered: The Life and Work of Julian Huxley with Selected Correspondence*. Singapore: World Scientific Publishing Co. Pte. Ltd.

DUARTE, A., COTTER, S. C., REAVEY, C. E., WARD, R. J. S., DE GASPERIN, O., & KILNER, R. M. 2016. Social immunity of the family: parental contributions to a public good modulated by brood size. *Evolutionary Ecology*, 30, 123–135.

DUARTE, A., WEISSING, F. J., PEN, I., & KELLER, L. 2011. An evolutionary perspective on self-organized division of labor in social insects. *Annual Review of Ecology, Evolution, and Systematics*, 42, 91–110.

DUGATKIN, L. A. 2006. *The Altruism Equation*. Princeton, NJ: Princeton University Press.

DUMAS, Z. & KUMMERLI, R. 2012. Cost of cooperation rules selection for cheats in bacterial metapopulations. *Journal of Evolutionary Biology*, 25, 473–484.

DUNN, C. 2009. Siphonophores. *Current Biology*, 19, R233–R234.

DUNN, C. W., PUGH, P. R., & HADDOCK, S. H. D. 2005. Molecular phylogenetics of the Siphonophora (Cnidaria), with implications for the evolution of functional specialization. *Systematic Biology*, 54, 916–935.

DUPUIS, M. É., VILLION, M., MAGADÁN, A. H., & MOINEAU, S. 2013. CRISPR-Cas and restriction–modification systems are compatible and increase phage resistance. *Nature Communications*, 4: 2087.

DURANT, J. R. 1992. The tension at the heart of Huxley's evolutionary ethology. *In*: WATERS, C. K. & VAN HELDEN, A. (eds.) *Julian Huxley: Biologist and Statesman of Science*. Houston, TX: Rice University Press.

DUSSUTOUR, A., LATTY, T., BEEKMAN, M., & SIMPSON, S. J. 2010. Amoeboid organism solves complex nutritional challenges. *Proceedings of the National Academy of Sciences of the United States of America*, 107, 4607–4611.

DWORKIN, M. & BONNER, J. T. 1972. The Myxobacteria: new directions in studies of prokaryote development. *CRC Critical Reviews in Microbiology*, 1, 435–452.

DYSON, C. J., CROSSLEY, H. G., RAY, C. H. & GOODISMAN, M. A. D. 2022. Social structure of perennial *Vespula squamosa* wasp colonies. *Ecology and Evolution*, 12:e8569.

EBERHARD, W. G. 1985. *Sexual Selection and Animal Genitalia*. Cambridge, MA: Harvard University Press.

EBERL, G. & PRADEU, T. 2018. Towards a general theory of immunity? *Trends in Immunology*, 39, 261–263.

EDWARDS, A. W. F. 1994. The fundamental theorem of natural selection. *Biological Reviews*, 69, 443–474.

EGERTON, F. N. 2013. History of ecological sciences, part 45: ecological aspects of entomology during the 1800s. *Bulletin of the Ecological Society of America*, 94, 36–88.

EL MOUDEN, C., ANDRÉ, J.-B., MORIN, O., & NETTLE, D. 2014. Cultural transmission and the evolution of human behaviour: a general approach based on the Price equation. *Journal of Evolutionary Biology*, 27, 231–241.

EL MOUDEN, C., WEST, S. A., & GARDNER, A. 2010. The enforcement of cooperation by policing. *Evolution*, 64, 2139–2152.

ELTON, C. 1927. *Animal Ecology*. New York: The MacMillan Company.

ELWICK, J. 2003. Herbert Spencer and the disunity of the social organism. *History of Science*, 41, 35–72.

EMERSON, A. E. 1939. Social coordination and the superorganism. *American Midland Naturalist*, 21, 182–209.

EMERSON, A. E. 1960. The evolution of adaptation in population systems. *In*: TAX, S. (ed.) *Evolution After Darwin, Volume 1: The Evolution of Life*. Chicago, IL: University of Chicago Press.

ENGEL, M. S., BARDEN, P., RICCIO, M. L., & GRIMALDI, D. A. 2016. Morphologically specialized termite castes and advanced sociality in the early Cretaceous. *Current Biology*, 26, 522–530.

ENNIS, H. L., DAO, D. N., PUKATZKI, S. U., & KESSIN, R. H. 2000. *Dictyostelium* amoebae lacking an F-box protein form spores rather than stalk in chimeras with wild type. *Proceedings of the National Academy of Sciences of the United States of America*, 97, 3292–3297.

EREZ, Z., STEINBERGER-LEVY, I., SHAMIR, M., DORON, S., STOKAR-AVIHAIL, A., PELEG, Y., MELAMED, S., LEAVITT, A., SAVIDOR, A., ALBECK, S., AMITAI, G., & SOREK, R. 2017. Communication between viruses guides lysis-lysogeny decisions. *Nature*, 541, 488–493.

ESHEL, I. & FELDMAN, M. W. 2001. Optimality and evolutionary stability under short-term and long-term selection. *In*: ORZACK, S. H. & SOBER, E. (eds.) *Adaptationism and Optimality*. Cambridge: Cambridge University Press.

EVANGELISTA, D. A., WIPFLER, B., BÉTHOUX, O., DONATH, A., FUJITA, M., KOHLI, M. K., LEGENDRE, F., LIU, S., MACHIDA, R., MISOF, B., PETERS, R. S., PODSIADLOWSKI, L., RUST, J., SCHUETTE, K., TOLLENAAR, W., WARE, J. L., WAPPLER, T., ZHOU, X., MEUSEMANN, K., & SIMON, S. 2019. An integrative phylogenomic approach illuminates the evolutionary history of cockroaches and termites (Blattodea). *Proceedings of the Royal Society B*, 286, 20182076.

EVANS, M. A. & EVANS, H. E. 1970. *William Morton Wheeler, Biologist*. Cambridge, MA: Harvard University Press.

EWENS, W. J. 1992. An optimizing principle of natural selection in evolutionary population genetics. *Theoretical Population Biology*, 42, 333–346.

FALCONER, D. S. 1960. *Introduction to Quantitative Genetics*. Edinburgh: Oliver & Boyd.

FALCONER, D. S. & MACKAY, T. F. C. 1996. *Introduction to Quantitative Genetics*, 4th edition. Harlow, Essex: Longmans Green.

FARRELL, J. A., WANG, Y., RIESENFELD, S., SHEKHAR, K., REGEV, A., & SCHIER, A. 2018. Single-cell reconstruction of developmental trajectories during zebrafish embryogenesis. *Science*, 360, 979 and eaar3131.

FELDMAN, M. W. 1992. Heritability: some theoretical ambiguities. *In*: FOX KELLER, E. & LLOYD, E. A. (eds.) *Keywords in Evolutionary Biology*. Cambridge, MA: Harvard University Press.

FELSENSTEIN, J. 1985. Phylogenies and the comparative method. *American Naturalist*, 125, 1–15.

FELSENSTEIN, J. 1988. Phylogenies and quantitative characters. *Annual Review of Ecology and Systematics*, 19, 445–471.

FENCHEL, T. & FINLAY, B. J. 1994. The evolution of life without oxygen. *American Scientist*, 82, 22–29.

FERRIERE, R., BRONSTEIN, J. L., RINALDI, S., LAW, R., & GAUDUCHON, M. 2002. Cheating and the evolutionary stability of mutualisms. *Proceedings of the Royal Society B*, 269, 773–780.

FIEGNA, F. & VELICER, G. J. 2005. Exploitative and hierarchical antagonism in a cooperative bacterium. *PLoS Biology*, 3, 1980–1987.

FIEGNA, F., YU, Y.-T. N., KADAM, S. V., & VELICER, G. J. 2006. Evolution of an obligate social cheater to a superior cooperator. *Nature*, 441, 310–314.

FIELD, J. 2008. The ecology and evolution of helping in hover wasps (Hymenoptera: Stenogastrinae). *In*: KORB, J. & HEINZE, J. (eds.) *Ecology of Social Evolution*. Berlin: Springer-Verlag.

FIGUEIREDO, A. R. T. & KÜMMERLI, R. 2020. Microbial Mutualism: Will You Still Need Me, Will You Still Feed Me? *Current Biology*, 30, R1041–R1043.

FINET, C., TIMME, R. E., DELWICHE, C. F., & MARLÉTAZ, F. 2010. Multigene phylogeny of the green lineage reveals the origin and diversification of land plants. *Current Biology*, 20, 2217–2222.

FINNEGAN, D. J. 2012. Retrotransposons. *Current Biology*, 22, R432–R437.

FISHER, R. A. 1918. The correlation between relatives on the supposition of mendelian inheritance. *Transactions of the Royal Society of Edinburgh*, 52, 399–433.

FISHER, R. A. 1930. The Genetical Theory of Natural Selection. Oxford: Oxford University Press.

FISHER, R. M., CORNWALLIS, C. K., & WEST, S. A. 2013. Group formation, relatedness and the evolution of multicellularity. *Current Biology*, 23, 1120–1125.

FISHER, R. M., HENRY, L. M., CORNWALLIS, C. K., KIERS, E. T., & WEST, S. A. 2017. The evolution of host-symbiont dependence. *Nature Communications*, 8: 15973.

FISHER, R. M. & REGENBERG, B. 2019. Multicellular group formation in *Saccharomyces cerevisiae*. *Proceedings of the Royal Society B*, 286, 20191098.

FISHER, R. M., SHIK, J. Z., & BOOMSMA, J. J. 2020. The evolution of multicellular complexity: the role of relatedness and environmental constraints. *Proceedings of the Royal Society B*, 287, 20192963.

FJERDINGSTAD, E. J. & BOOMSMA, J. J. 2000. Queen mating frequency and relatedness in young *Atta sexdens* colonies. *Insectes Sociaux*, 47, 354–356.

FLATT, T. 2005. The evolutionary genetics of canalization. *Quarterly Review of Biology*, 80, 287–316.

FLETCHER, D. J. C. & MICHENER, C. D. E. (eds.) 1987. *Kin Recognition in Animals*. New York: Wiley & Sons.

FLETCHER, J. A. & DOEBELI, M. 2006. How altruism evolves: assortment and synergy. *Journal of Evolutionary Biology*, 19, 1389–1393.

FLETCHER, J. A. & DOEBELI, M. 2009. A simple and general explanation for the evolution of altruism. *Proceedings of the Royal Society B*, 276, 13–19.

FLETCHER, J. A. & DOEBELI, M. 2010. Assortment is a more fundamental explanation for the evolution of altruism than inclusive fitness or multilevel selection: reply to Bijma and Aanen. *Proceedings of the Royal Society B*, 277, 677–678.

FLETCHER, J. A. & ZWICK, M. 2006. Unifying the theories of inclusive fitness and reciprocal altruism. *American Naturalist*, 168, 252–262.

FLOREA, M. 2017. Aging and immortality in unicellular species. *Mechanisms of Ageing and Development*, 167, 5–15.

FOLSE, H. J. & ROUGHGARDEN, J. 2010. What is an individual organism? A multilevel selection perspective. *Quarterly Review of Biology*, 85, 447–472.

FORD, E. B. 1938. The genetic basis of adaptation. *In*: DE BEER, G. R. (ed.) *Evolution: Essays on Aspects of Evolutionary Biology*. Oxford: Clarendon Press.

FORD, E. B. 1971. *Ecological Genetics and Evolution*. Oxford: Blackwell Scientific Publications.

FORTUNATO, A., QUELLER, D. C., & STRASSMANN, J. E. 2003a. A linear dominance hierarchy among clones in chimeras of the social amoeba *Dictyostelium discoideum*. *Journal of Evolutionary Biology*, 16, 438–445.

FORTUNATO, A., STRASSMANN, J. E., SANTORELLI, L., & QUELLER, D. C. 2003b. Co-occurrence in nature of different clones of the social amoeba, *Dictyostelium discoideum*. *Molecular Ecology*, 12, 1031–1038.

FORTUNATO, L. & ARCHETTI, M. 2010. Evolution of monogamous marriage by maximization of inclusive fitness. *Journal of Evolutionary Biology*, 23, 149–156.

FOSTER, K. R. 2004. Diminishing returns in social evolution: the not-so-tragic commons. *Journal of Evolutionary Biology*, 17, 1058–1072.

FOSTER, K. R. 2009. A defense of sociobiology. *Cold Spring Harbor Symposia on Quantitative Biology*, 74, 403–418.

FOSTER, K. R., FORTUNATO, A., STRASSMANN, J. E., & QUELLER, D. C. 2002. The costs and benefits of being a chimera. *Proceedings of the Royal Society B*, 269, 2357–2362.

FOSTER, K. R., SCHLUTER, J., COYTE, K. Z., & RAKOFF-NAHOUM, S. 2017. The evolution of the host microbiome as an ecosystem on a leash. *Nature*, 548, 43–51.

FOSTER, K. R. & WENSELEERS, T. 2006. A general model for the evolution of mutualisms. *Journal of Evolutionary Biology*, 19, 1283–1293.

FRANK, E.T., WEHRHAHN, M. & LINSENMAIR, K.E. 2018. Wound treatment and selective help in a termite-hunting ant. Proceedings of the Royal Society B 285: 20172457.

FRANK, S. A. 1994. Genetics of mutualism: the evolution of altruism between species. *Journal of Theoretical Biology*, 170, 393–400.

FRANK, S. A. 1995a. The origin of synergistic symbiosis. *Journal of Theoretical Biology*, 176, 403–410.

FRANK, S. A. 1995b. George Price's contributions to evolutionary genetics. *Journal of Theoretical Biology*, 175, 373–388.

FRANK, S. A. 1996a. Host-symbiont conflict over the mixing of symbiotic lineages. *Proceedings of the Royal Society B*, 263, 339–344.

FRANK, S. A. 1996b. Models of parasite virulence. *Quarterly Review of Biology*, 71, 37–78.

FRANK, S. A. 1997a. Models of symbiosis. *American Naturalist*, 150, S80–S99.

FRANK, S. A. 1997b. Multivariate analysis of correlated selection and kin selection, with an ESS maximization method. *Journal of Theoretical Biology*, 189, 307–316.

FRANK, S. A. 1997c. The Price equation, Fisher's fundamental theorem, kin selection, and causal analysis. *Evolution*, 51, 1712–1729.

FRANK, S. A. 1998. *Foundations of Social Evolution*. Princeton, NJ: Princeton University Press.

FRANK, S. A. 2003. Repression of competition and the evolution of cooperation. *Evolution*, 57, 693–705.

FRANK, S. A. 2007. *Dynamics of Cancer: Incidence, Inheritance, and Evolution*. Princeton, NJ: Princeton University Press.

FRANK, S. A. 2009. Natural selection maximizes Fisher information. *Journal of Evolutionary Biology*, 22, 231–244.

FRANK, S. A. 2010. Somatic evolutionary genomics: mutations during development cause highly variable genetic mosaicism with risk of cancer and neurodegeneration. *Proceedings of the National Academy of Sciences of the United States of America*, 107, 1725–1730.

FRANK, S. A. 2011a. Natural selection. I. Variable environments and uncertain returns on investment. *Journal of Evolutionary Biology*, 24, 2299–2309.

FRANK, S. A. 2011b. Natural selection. II. Developmental variability and evolutionary rate. *Journal of Evolutionary Biology*, 24, 2310–2320.

FRANK, S. A. 2012a. Natural selection. III. Selection versus transmission and the levels of selection. *Journal of Evolutionary Biology*, 25, 227–243.

FRANK, S. A. 2012b. Natural selection. IV. The Price equation. *Journal of Evolutionary Biology*, 25, 1002–1019.

FRANK, S. A. 2012c. Natural selection. V. How to read the fundamental equations of evolutionary change in terms of information theory. *Journal of Evolutionary Biology*, 25, 2377–2396.

FRANK, S. A. 2012d. Wright's adaptive landscape versus Fisher's fundamental theorem. *In*: SVENSSON, E. & CALSBEEK, R. (eds.) *The Adaptive Landscape in Evolutionary Biology*. Oxford: Oxford University Press.

FRANK, S. A. 2013a. Natural selection. VI. Partitioning the information in fitness and characters by path analysis. *Journal of Evolutionary Biology*, 26, 457–471.

FRANK, S. A. 2013b. Natural selection. VII. History and interpretation of kin selection theory. *Journal of Evolutionary Biology*, 26, 1151–1184.

FRANK, S. A. 2017. Universal expressions of population change by the Price equation: natural selection, information, and maximum entropy production. *Ecology and Evolution*, 7, 3381–2296.

FRANK, S. A. 2018. The Price equation program: simple invariances unify population dynamics, thermodynamics, probability, information and inference. *Entropy*, 20, 978.

FRANK, S. A. 2020. Simple unity among the fundamental equations of science. *Philosophical Transactions of the Royal Society B*, 375, 20190351.

FRANK, S. A. & FOX, G. A. 2020. The inductive theory of natural selection. *In*: SCHEINER, S. M. & MINDELL, D. P. (eds.) *The Theory of Evolution*. Chicago, IL: University of Chicago Press.

FROMHAGE, L. & JENNIONS, M. D. 2016. Coevolution of parental investment and sexually selected traits drives sex-role divergence. *Nature Communications*, 7: 12517.

FROMHAGE, L. & JENNIONS, M. D. 2019. The strategic reference gene: an organismal theory of inclusive fitness. *Proceedings of the Royal Society B*, 286, 20190459.

FUTUYMA, D. J. 2013. *Evolution*, 3rd edition. Sunderland, MA: Sinauer Associates.

FUTUYMA, D. J. 2017. Evolutionary biology today and the call for an extended synthesis. *Interface Focus*, 7: 20160145.

GADAGKAR, R. 2001. *The Social Biology of Ropalidia Marginata: Toward Understanding the Evolution of Eusociality*. Cambridge, MA: Harvard University Press.

GAGO, S., ELENA, S. F., FLORES, R., & SANJUÁN, R. 2009. Extremely high mutation rate of a hammerhead viroid. *Science*, 323, 1308.

GANTI, T. 1975. Organization of chemical reactions into dividing and metabolizing units: the chemotons. *BioSystems*, 7, 15–21.

GAO, C., LIU, C., SCHENZ, D., LI, X., ZHANG, Z., JUSUP, M., WANG, Z., BEEKMAN, M., & NAKAGAKI, T. 2019. Does being multi-headed make you better at solving problems? A survey of *Physarum*-based models and computations. *Physics of Life Reviews*, 29, 1–26.

GARCIA-COSTOYA, G. & FROMHAGE, L. 2021. Realistic genetic architecture enables organismal adaptation as predicted under the folk definition of inclusive fitness. *Journal of Evolutionary Biology*, 34, 1087–1094.

GARDNER, A. 2009. Adaptation as organism design. *Biology Letters*, 5, 861–864.

GARDNER, A. 2015a. The genetical theory of multi-level selection. *Journal of Evolutionary Biology*, 28, 305–319.

GARDNER, A. 2015b. More on the genetical theory of multilevel selection. *Journal of Evolutionary Biology*, 28, 1747–1751.

GARDNER, A. 2015c. Group selection versus group adaptation. *Nature*, 524, E3–E4.

GARDNER, A. 2015d. Kin selection. *In*: WRIGHT, J. D. (ed.) *International Encyclopedia of the Social and Behavioral Sciences*, 2nd edition, volume 13. New York: Elsevier.

GARDNER, A. 2019. The greenbeard effect. *Current Biology*, 29, R430–R431.

GARDNER, A., ARCE, A., & ALPEDRINHA, J. 2009. Budding dispersal and the sex ratio. *Journal of Evolutionary Biology*, 22, 1036–1045.

GARDNER, A. & GRAFEN, A. 2009. Capturing the superorganism: a formal theory of group adaptation. *Journal of Evolutionary Biology*, 22, 659–671.

GARDNER, A. & UBEDA, F. 2017. The meaning of intragenomic conflict. *Nature Ecology & Evolution*, 1, 1807–1815.

GARDNER, A. & WELCH, J. J. 2011. A formal theory of the selfish gene. *Journal of Evolutionary Biology*, 24, 1801–1813.

GARDNER, A. & WEST, S. A. 2006. Demography, altruism, and the benefits of budding. *Journal of Evolutionary Biology*, 19, 1707–1716.

GARDNER, A. & WEST, S. A. 2009. Greenbeards. *Evolution*, 64, 25–38.

GARDNER, A., WEST, S. A., & BARTON, N. H. 2007. The relation between multilocus population genetics and social evolution theory. *American Naturalist*, 169, 207–226.

GARDNER, A., WEST, S. A., & BUCKLING, A. 2004. Bacteriocins, spite and virulence. *Proceedings of the Royal Society B*, 271, 1529–1535.

GARDNER, A., WEST, S. A., & WILD, G. 2011. The genetical theory of kin selection. *Journal of Evolutionary Biology*, 24, 1020–1043.

GAUCHER, E. A., GOVINDARAJAN, S., & GANESH, O. K. 2008. Paleotemperature trend for Precambrian life inferred from resurrected proteins. *Nature*, 451, 704–707.

GAWNE, R., MCKENNA, K. Z., & LEVIN, M. 2020. Competitive and coordinative interactions between body parts produce adaptive developmental outcomes. *BioEssays*, 42, 1900245.

GAWNE, R., MCKENNA, K. Z., & NIJHOUT, H. F. 2018. Unmodern synthesis: developmental hierarchies and the origin of phenotypes. *BioEssays*, 40, 1600265.

GAYON, J. 2000. History of the concept of allometry. *American Zoologist*, 40, 748–758.

GE, P., SCHOLL, D., PROKHOROV, N. S., AVAYLON, J., SHNEIDER, M. M., BROWNING, C., BUTH, S. A., PLATTNER, M., CHAKRABORTY, U., DING, K., LEIMAN, P. G., MILLER, J. F., & ZHOU, Z. H. 2020. Action of a minimal contractile bactericidal nanomachine. *Nature*, 580, 658–662.

GHISELIN, M. T. 1974. *The Economy of Nature and the Evolution of Sex*. Berkeley, CA: University of California Press.

GHISELIN, M. T. 2011. A consumer's guide to superorganisms. *Perspectives in Biology and Medicine*, 54, 152–167.

GHOUL, M., GRIFFIN, A. S., & WEST, S. A. 2013. Towards an evolutionary definition of cheating. *Evolution*, 68, 318–331.

GIBSON, A. H. 2013. Edward O. Wilson and the organicist tradition. *Journal of the History of Biology*, 46, 599–630.

GIEHR, J. & HEINZE, J. 2018. Queens stay, workers leave: caste-specific responses to fatal infections in an ant. *BMC Evolutionary Biology*, 18, 202.

GILBERT, O. M., FOSTER, K. R., MEHDIABADI, N. J., STRASSMANN, J. E., & QUELLER, D. C.

2007. High relatedness maintains multicellular cooperation in a social amoeba by controlling cheater mutants. *Proceedings of the National Academy of Sciences of the United States of America*, 104, 8913–8917.

GILBERT, W. 1986. The RNA World. *Nature*, 319, 618.

GLASS, D. J. & HALL, N. 2008. A brief history of the hypothesis. *Cell*, 134, 378–381.

GLÖCKNER, G., LAWAL, H. M., FELDER, M., SINGH, R., SINGER, G., WEIJER, C. J., & SCHAAP, P. 2016. The multicellularity genes of dictyostelid social amoebas. *Nature Communications*, 7: 12085.

GOBIN, B., ITO, F., BILLEN, J., & PEETERS, C. 2008. Degeneration of sperm reservoir and the loss of mating ability in worker ants. *Naturwissenschaften*, 95, 1041–1048.

GODFREY, H. C. J., PARTRIDGE, L., & HARVEY, P. H. 1991. Clutch size. *Annual Review of Ecology and Systematics*, 22, 409–429.

GODFREY, R. K. & GRONENBERG, W. 2019. Brain evolution in social insects: advocating for the comparative approach. *Journal of Comparative Physiology A*, 205, 13–32.

GODFREY-SMITH, P. 2001. Three kinds of adaptationism. *In*: ORZACK, S. H. & SOBER, E. (eds.) *Adaptationism and Optimality*. Cambridge: Cambridge University Press.

GODFREY-SMITH, P. 2009. *Darwinian Populations and Natural Selection*. Oxford: Oxford University Press.

GONCALVES, A. P., HELLER, J., SPAN, E. A., ROSENFIELD, G., DO, H. P., PALMA-GUERRERO, J., REQUENA, N., MARLETTA, M. A., & GLASS, N. L. 2019. Allorecognition upon fungal cell-cell contact determines social cooperation and impacts the acquisition of multicellularity. *Current Biology*, 29, 3006–3017.

GOODNIGHT, C. J. 2015. Multilevel selection theory and evidence: a critique of Gardner, 2015. *Journal of Evolutionary Biology*, 28, 1734–1746.

GOODNIGHT, C. J. & WADE, M. J. 2000. The ongoing synthesis: a reply to Coyne, Barton, and Turelli. *Evolution*, 54, 317–324.

GOTOH, A., BILLEN, J., HASHIM, R., & ITO, F. 2008. Comparison of spermatheca morphology between reproductive and non-reproductive females in social wasps. *Arthropod Structure & Development*, 37, 199–209.

GOULD, S. B. 2018. Membranes and evolution. *Current Biology*, 28, R381–385.

GOULD, S. J. 1983. The hardening of the modern synthesis. *In*: GREEN, M. (ed.) *Dimensions of Darwinism*. New York: Cambridge University Press.

GOULD, S. J. 1988. On replacing the idea of progress with an operational notion of directionality. *In*: NITECKI, M. H. (ed.) *Evolutionary Progress*. Chicago, IL: University of Chicago Press.

GOULD, S. J. & LEWONTIN, R. C. 1979. The spandrels of San Marco and the Panglossian Paradigm: a critique of the adaptationist programme. *Proceedings of the Royal Society B*, 205, 581–598.

GOULD, S. J. & VRBA, E. S. 1982. Exaptation—a missing term in the science of form. *Paleobiology*, 8, 4–15.

GRAFEN, A. 1984. Natural selection, kin selection and group selection. *In*: KREBS, J. R. & DAVIES, N. B. (eds.) *Behavioural Ecology: An Evolutionary Approach*, 2nd edition, Oxford: Blackwell.

GRAFEN, A. 1985a. A geometric view of relatedness. *In*: DAWKINS, R. & RIDLEY, M. (eds.) *Oxford Surveys in Evolutionary Biology*. Oxford: Oxford University Press.

GRAFEN, A. 1985b. Hamilton's rule OK. *Nature*, 318, 310–311.

GRAFEN, A. 1989. The phylogenetic regression. *Philosophical Transactions of the Royal Society B*, 326, 119–157.

GRAFEN, A. 1990a. Do animals really recognize kin? *Animal Behaviour*, 39, 42–54.

GRAFEN, A. 1990b. Biological signals as handicaps. *Journal of Theoretical Biology*, 144, 517–546.

GRAFEN, A. 1991. Modelling in behavioural ecology. *In*: KREBS, J. R. & DAVIES, N. B. (eds.) *Behavioural Ecology: An Evolutionary Approach*, 3rd edition. Oxford: Blackwell Scientific Publications.

GRAFEN, A. 2002. A first formal link between the Price equation and an optimization program. *Journal of Theoretical Biology*, 217, 75–91.

GRAFEN, A. 2006. Optimization of inclusive fitness. *Journal of Theoretical Biology*, 238, 541–563.

GRAFEN, A. 2007. The formal Darwinism project: a mid-term report. *Journal of Evolutionary Biology*, 20, 1243–1254.

GRAFEN, A. 2009. Formalizing Darwinism and inclusive fitness theory. *Philosophical Transactions of the Royal Society B*, 364, 3135–3141.

GRAFEN, A. 2014a. The formal darwinism project in outline. *Biology & Philosophy*, 29, 155–174.

GRAFEN, A. 2014b. The formal darwinism project in outline: response to commentaries. *Biology & Philosophy*, 29, 281–292.

GRAFEN, A. 2015. Biological fitness and the fundamental theorem of natural selection. *American Naturalist*, 186, 1–14.

GRAFEN, A. 2018. The left hand side of the fundamental theorem of natural selection. *Journal of Theoretical Biology*, 456, 175–189.

GRAFEN, A. 2019. Should we ask for more than consistency of Darwinism with Mendelism? *Studies in History and Philosophy of Biological and Biomedical Sciences*, 78: 101224.

GRAFEN, A. 2020. The Price equation and reproductive value. *Philosophical Transactions of the Royal Society B*, 375: 20190356.

GRANATO, E. T., MEILLER-LEGRAND, T. A., & FOSTER, K. R. 2019. The evolution and ecology of bacterial warfare. *Current Biology*, 29, R521–R537.

GREY, D., HUTSON, V., & SZATHMÁRY, E. 1995. A re-examination of the stochastic corrector model. *Proceedings of the Royal Society B*, 262, 29–35.

GRIESSER, M., DROBNIAK, S. M., NAKAGAWA, S., & BOTERO, C. A. 2017. Family living sets the stage for cooperative breeding and ecological resilience in birds. *PLoS Biology*, 15: e2000483.

GRIFFIN, A. S. & WEST, S. A. 2002. Kin selection: fact and fiction. *Trends in Ecology & Evolution*, 17, 15–21.

GRIFFIN, A. S., WEST, S. A., & BUCKLING, A. 2004. Cooperation and competition in pathogenic bacteria. *Nature*, 430, 1024–1027.

GRIFFITHS, P. E. 2009. In what sense does 'nothing make sense except in the light of evolution'? *Acta Biotheoretica*, 57, 11–32.

GROSBERG, R. K. & STRATHMANN, R. R. 2007. The evolution of multicellularity: a minor major transition? *Annual Review of Ecology, Evolution and Systematics*, 38, 621–654.

GROSHOLZ, E. R. 1991. *Cartesian Method and the Problem of Reduction*. Oxford: Oxford University Press.

GRUM-GRZHIMAYLO, A. A., BASTIAANS, E., VAN DEN HEUVEL, J., BERENGUER MILLANES, C., DEBETS, A. J. M., & AANEN, D. K. 2021. Somatic deficiency causes reproductive parasitism in a fungus. *Nature Communications*, 12: 783.

GUERRIERI, F. J., NEHRING, V., JØRGENSEN, C. G., NIELSEN, J., GALIZIA, C. G., & D'ETTORRE, P. 2009. Ants recognize foes and not friends. *Proceedings of the Royal Society B*, 276, 2461–2468.

GUPTA, M., PRASAD, N.G., DEY, S., JOSHI, A., & VIDYA, T.N.C. 2017. Niche construction in evolutionary theory: the construction of an academic niche? *Journal of Genetics* 96, 491–504.

HAECKEL, E. 1879. *The Evolution of Man*. London: C. Kegan Paul and Co.

HAIG, D. 1997. Parental antagonism, relatedness asymmetries, and genomic imprinting. *Proceedings of the Royal Society B*, 264, 1657–1662.

HAIG, D. 2004. Genomic imprinting and kinship: how good is the evidence? *Annual Review of Genetics*, 38, 553–585.

HAIG, D. 2007. Weismann rules! OK? Epigenetics and the Lamarckian temptation. *Biology and Philosophy*, 22, 415–428.

HAIG, D. 2010. Transfers and transitions: parent-offspring conflict, genomic imprinting, and the evolution of human life history. *Proceedings of the National Academy of Sciences of the United States of America*, 107, 1731–1735.

HAIG, D. 2011. Genomic imprinting and the evolutionary psychology of human kinship. *Proceedings of the National Academy of Sciences of the United States of America*, 108, 10878–10885.

HAIG, D. 2012. The epidemiology of epigenetics. *International Journal of Epidemiology*, 41, 13–16.

HAIG, D. 2016. Transposable elements: self seekers of the germline, team-players of the soma. *Bioessays*, 38, 1158–1166.

HAIG, D. & WESTOBY, M. 1989. Parent-specific gene expression and the triploid endosperm. *The American Naturalist*, 134, 147–155.

HALDANE, J. B. S. 1932. *The Causes of Evolution*, 1990 edition. London: Longmans, Green.

HAMILTON, W. D. 1963. Evolution of altruistic behavior. *American Naturalist*, 97, 354–356.

HAMILTON, W. D. 1964a. The genetical evolution of social behaviour I. *Journal of Theoretical Biology*, 7, 1–16.

HAMILTON, W. D. 1964b. The genetical evolution of social behaviour II. *Journal of Theoretical Biology*, 7, 17–52.

HAMILTON, W. D. 1966. The moulding of senescence by natural selection. *Journal of Theoretical Biology*, 12, 12–45.

HAMILTON, W. D. 1967a. Extraordinary sex ratios. *Science*, 156, 477–488.

HAMILTON, W. D. 1967b. Social insect populations. *Science Progress*, 55, 496–501.

HAMILTON, W. D. 1970. Selfish and spiteful behaviour in an evolutionary model. *Nature*, 228, 1218–1220.

HAMILTON, W. D. 1972. Altruism and related phenomena, mainly in social insects. *Annual Review of Ecology and Systematics*, 3, 193–232.

HAMILTON, W. D. 1975. Innate social aptitudes of man: an approach from evolutionary genetics. *In*: FOX, R. (ed.) *Biosocial Anthropology*. London: Malaby Press.

HAMILTON, W. D. 1980. Sex versus non-sex versus parasite *Oikos*, 35, 282–290.

HAMILTON, W.D. 1987a. Kinship, recognition, disease, and intelligence: constraints of social evolution. *In*: ITO, Y., BROWN, J.L., & KIKKAWA, J. (eds.) *Animal Societies: Theories and Facts*. Tokyo: Japan Scientific Societies Press.

HAMILTON, W. D. 1987b. Discriminating nepotism: expectable, common, overlooked. *In*: FLETCHER, D. J. C. & MICHENER, C. D. (eds.) *Kin Recognition in Animals*. New York: Wiley.

HAMILTON, W. D. 1996. *Narrow Roads of Gene Land: Evolution of Social Behaviour*. Oxford, New York, Heidelberg: Freeman.

HAMILTON, W. D. 2000. My intended burial and why. *Ethology, Ecology & Evolution*, 12, 111–122.

HAMILTON, W. D., AXELROD, R., & TANESE, R. 1990. Sexual reproduction as an adaptation to resist parasites (a review). *Proceedings of the National Academy of Sciences of the United States of America*, 87, 3566–3573.

HAMILTON, W. D. & ZUK, M. 1982. Heritable true fitness and bright birds: a role for parasites? *Science*, 218, 384–387.

HÄMMERLING, J. 1953. Nucleo-cytoplasmic relationships in the development of *Acetabularia*. *International Review of Cytology*, 2, 475–498.

HAMMERSTEIN, P. 1996. Darwinian adaptation, population genetics and the streetcar theory of evolution. *Journal of Mathematical Biology*, 34, 511–532.

HAMMOND, R. L., BRUFORD, M. W., & BOURKE, A. F. G. 2003. Male parentage does not vary with colony kin structure in a multiple-queen ant. *Journal of Evolutionary Biology*, 16, 446–455.

HAMMOND, R. L. & KELLER, L. 2004. Conflict over male parentage in social insects. *PLoS Biology*, 2: e248.

HARDIN, G. 1968. The tragedy of the commons. *Science*, 162, 1243–1248.

HARDY, I. C. W. (ed.) 2002. *Sex ratios: concepts and research methods*: Cambridge University Press.

HÄRING, M., VESTERGAARD, G., RACHEL, R., CHEN, L., GARRETT, R. A. & PRANGISHVILI, D. 2005. Independent virus development outside a host. *Nature*, 436, 1101–1102.

HARMAN, O. 2010. *The Price of Altruism*. London: Bodley Head.

HAROLD, F. M. 2014. *In Search of Cell History*. Chicago, IL: University of Chicago Press.

HART, A. G. & RATNIEKS, F. L. W. 2002. Waste management in the leaf-cutting ant *Atta colombica*. *Behavioral Ecology*, 13, 224–231.

HARTKE, T. R. & BAER, B. 2011. The mating biology of termites: a comparative review. *Animal Behaviour*, 82, 927–936.

HARTMANN, A., WANTIA, J., TORRES, J. A., & HEINZE, J. 2003. Worker policing without genetic conflicts in a clonal ant. *Proceedings of the National Academy of Sciences of the United States of America*, 100, 12836–12840.

HARVEY, P. H. & CLUTTON-BROCK, T. H. 1983. The survival of the theory. *New Scientist*, 98, 312–315.

HARVEY, P. H. & MACE, G. M. 1982. Comparisons between taxa and adaptive trends: problems of methodology. *In*: KING'S COLLEGE SOCIOBIOLOGY GROUP (ed.) *Current Problems in Sociobiology*. Cambridge: Cambridge University Press.

HARVEY, P. H. & PAGEL, M. 1991. *The Comparative Method in Evolutionary Biology*. Oxford: Oxford University Press.

HEINZE, J. 1990. Dominance behavior among ant females. *Naturwissenschaften*, 77, 41–43.

HEINZE, J. 2016. Life-history evolution in ants: the case of *Cardiocondyla*. *Proceedings of the Royal Society B*, 284, 20161406.

HEINZE, J. & WALTER, B. 2010. Moribund ants leave their nests to die in isolation. *Current Biology*, 20, 249–252.

HELANTERÄ, H. & SUNDSTRÖM, L. 2007. Worker reproduction in *Formica* ants. *American Naturalist* 170, E14–E25.

HELANTERÄ, H. & ULLER, T. 2010. The Price equation and extended inheritance. *Philosophy, Theory, and Practice in Biology*, 2: e101.

HEMMINGS, N. & BIRKHEAD, T. R. 2015. Polyspermy in birds: sperm numbers and embryo survival. *Proceedings of the Royal Society B*, 282, 20151682.

HENIKOFF, S. 2018. Darwin meets Waddington. *Current Biology*, 28, R682–R684.

HENRICH, J. 2016. *The Secret of Our Success: How Culture is Driving Human Evolution, Domesticating our Species, and Making us Smarter*. Princeton, NJ: Princeton University Press.

HERRE, E. A. 1985. Sex ratio adjustment in fig wasps. *Science*, 228, 896–898.

HERRE, E. A., JANDER, K. C., & MACHADO, C. A. 2008. Evolutionary ecology of figs and their associates: recent progress and outstanding puzzles. *Annual Review of Ecology, Evolution and Systematics*, 39, 439–458.

HERRE, E. A., KNOWLTON, N., MUELLER, U. G., & REHNER, S. A. 1999. The evolution of mutualisms: exploring the paths between conflict and cooperation. *Trends in Ecology and Evolution*, 14, 49–53.

HERRING, E. 2018. 'Great is Darwin and Bergson his poet': Julian Huxley's other evolutionary synthesis. *Annals of Science*, 75, 40–54.

HERRON, M. D. 2016. Origins of multicellular complexity: *Volvox* and the volvocine algae. *Molecular Ecology*, 25, 1213–1223.

HERRON, M. D. & MICHOD, R. E. 2008. Evolution of complexity in the volvocine algae: transitions in individuality through Darwin's eye. *Evolution*, 62, 436–451.

HERRON, M. D. & NEDELCU, A. M. 2015. Volvocine algae: from simple to complex multicellularity. *In*: RUIZ-TRILLO, I. & NEDELCU, A. M. (eds.) *Evolutionary Transitions to Multicellular Life*. Dordrecht: Springer.

HEYES, C. 2020. Culture. *Current Biology*, 30, R1246–R1250.

HILL, W. G. & MACKAY, T. F. C. 2004. D.S. Falconer and *Introduction to Quantitative Genetics*. *Genetics*, 167, 1529–1536.

HILLMANN, F., FORBES, G., NOVOHRADSKÁ, S., FERLING, I., RIEGE, K., GROTH, M., WESTERMANN, M., MARZ, M., SPALLER, T., WINCKLER, T., SCHAAP, P., & GLOCKNER, G. 2018. Multiple roots of fruiting body formation in Amoebozoa. *Genome Biology and Evolution*, 10, 591–606.

HIMMELFARB, G. 2004. *The Roads to Modernity: The British, French and American Enlightenments*. London: Vintage Books.

HINES, H. M., HUNT, J. H., O'CONNOR, T. K., GILLESPIE, J. J., & CAMERON, S. A. 2007. Multigene phylogeny reveals eusociality evolved twice in vespid wasps. *Proceedings of the National Academy of Sciences of the United States of America*, 104, 3295–3299.

HIROSE, S., BENABENTOS, R., HO, H.-I., KUSPA, A., & SHAULSKY, G. 2011. Self-recognition in social amoebae is mediated by allelic pairs of tiger genes. *Science*, 333, 467–470.

HO, H.-I., HIROSE, S., KUSPA, A., & SHAULSKY, G. 2013. Kin recognition protects cooperators against cheaters. *Current Biology*, 23, 1590–1595.

HÖLLDOBLER, B. & WILSON, E. O. 2009. *The Superorganism*. New York: W. W. Norton & Co.

HOLM, E. A. 2019. In defense of the black box. *Science*, 364, 26–27.

HOLMAN, L. 2014. Caste load and the evolution of reproductive skew. *American Naturalist*, 183, 84–95.

HOLMES, W. G. & SHERMAN, P. W. 1982. The ontogeny of kin recognition in two species of ground squirrels. *American Zoologist*, 22, 491–517.

HOWARD, J. C. 2009. Why didn't Darwin discover Mendel's laws? *Journal of Biology*, 8: 15.

HOWARD, R. S. & LIVELY, C. M. 1994. Parasitism, mutation accumulation and the maintenance of sex. *Nature*, 367, 554–557.

HOWE, J., SCHIØTT, M., LI, Q., WANG, Z., ZHANG, G. & BOOMSMA, J. J. 2020. A novel method for using RNA-seq data to identify imprinted genes in social Hymenoptera with multiply mated queens. *Journal of Evolutionary Biology*. 33, 1770–1782.

HRDY, S. B. 1980. *The Langurs of Abu*. Cambridge, MA: Harvard University Press.

HRDY, S. B. 2009. *Mothers and Others*. Cambridge, MA: Belknap Press of Harvard University Press.

HUEY, R. B., GARLAND, T., & TURELLI, M. 2019. Revisiting a key innovation in evolutionary biology: Felsenstein's "Phylogenies and the Comparative Method". *American Naturalist*, 193, 755–772.

HUGHES, C. R., QUELLER, D. C., STRASSMAN, J. E., & DAVIS, S. K. 1993. Relatedness and altruism in *Polistes* wasps. *Behavioral Ecology*, 4, 128–137.

HUGHES, W. O. H., OLDROYD, B. P., BEEKMAN, M., & RATNIEKS, F. L. W. 2008. Ancestral monogamy shows kin selection is key to the evolution of eusociality. *Science*, 320, 1213–1216.

HULL, D.L. 1988a. Interactors versus vehicles. *In*: PLOTKIN, H.C. (ed.) *The Role of Behavior in Evolution*. Cambridge, MA: MIT Press.

HULL, D. L. 1988b. Progress in ideas of progress. *In*: NITECKI, M. H. (ed.) *Evolutionary Progress*. Chicago, IL: University of Chicago Press.

HUNT, J. H. & AMDAM, G. V. 2005. Bivoltinism as an antecedent to eusociality in the paper wasp genus *Polistes*. *Science*, 308, 264–267.

HURST, L. D. 1991. The incidences and evolution of cytoplasmic male killers. *Proceedings of the Royal Society of London B*, 244, 91–99.

HUTCHINSON, G. E. 1957. Concluding remarks. *Cold Spring Harbor Symposium on Quantitative Biology*, 22, 415–427.

HUTSON, V., LAW, R., & LEWIS, D. 1985. Dynamics of ecologically obligate mutualisms - effects of spatial diffusion on resilience of the interacting species. *American Naturalist*, 126, 445–449.

HUXLEY, A. 1932. *Brave New World*. London: Chatto & Windus.

HUXLEY, J. S. 1923a. Biology and sociology. *In: Essays of a Biologist*. London: Chatto & Windus. pp. 67–98.

HUXLEY, J. S. 1923b. Progress, biological and other. *In: Essays of a Biologist*. London: Chatto & Windus. pp. 3–64.

HUXLEY, J. S. 1912. *The Individual in the Animal Kingdom*. Cambridge: Cambridge University Press.

HUXLEY, J. S. 1916. Bird-watching and biological science. *The Auk*, 33, 142–161, 256–270.

HUXLEY, J. S. 1926. The biological basis of individuality. *Journal of Philosophical Studies*, 1, 305–319.

HUXLEY, J. S. 1930. *Ants*. London: Arrow Books.

HUXLEY, J. S. 1931. The relative size of antlers in deer. *Proceedings of the Zoological Society of London*, 101, 819–864.

HUXLEY, J. S. 1936. Natural selection and evolutionary progress. *Nature*, 138, 603–605.

HUXLEY, J. S. 1940. Science, natural and social. *The Scientific Monthly*, 50, 5–16.

HUXLEY, J. S. 1942. *Evolution: The Modern Synthesis*. Cambridge, MA: MIT Press.

HUXLEY, J. S. 1960a. The emergence of Darwinism. *In*: TAX, S. (ed.) *Evolution After Darwin, Volume 1: The Evolution of Life*. Chicago, IL: University of Chicago Press.

HUXLEY, J. S. 1960b. The evolutionary vision. *In*: TAX, S. & CALLENDER, C. (eds.) *Evolution after Darwin, Volume 3: Issues in Evolution*. Chicago, IL: University of Chicago Press.

HUXLEY, J. S. 2010. *Evolution, The Modern Synthesis: The Definitive Edition with a New Foreword by Massimo Pigliucci and Gerd B. Müller*. Cambridge, MA: MIT Press.

HUXLEY, J. S. 2022. *The Individual in the Animal Kingdom with a New Foreword by Richard Gawne and Jacobus J. Boomsma*. Cambridge MA: MIT Press.

IMACHI, H., NOBU, M. K., NAKAHARA, N., MORONO, Y., OGAWARA, M., TAKAKI, Y., TAKANO, Y., UEMATSU, K., IKUTA, T., ITO, M., MATSUI, Y., MIYAZAKI, M., MURATA, K., SAITO, Y., SAKAI, S., SONG, C., TASUMI, E., YAMANAKA, Y., YAMAGUCHI, T., KAMAGATA, Y., TAMAKI, H., & TAKAI, K. 2020. Isolation of an archaeon at the prokaryote-eukaryote interface. *Nature*, 577, 519–525.

INABA, M. & YAMASHITA, Y. M. 2012. Asymmetric stem cell division: precision for robustness. *Cell Stem Cell*, 11, 461–469.

INWARD, D. J. G., VOGLER, A. P., & EGGLETON, P. 2007. A comprehensive phylogenetic analysis of termites (Isoptera) illuminates key aspects of their evolutionary biology. *Molecular Phylogenetics and Evolution*, 44, 953–967.

IRASTORTZA-OLAZIREGI, M. & AMSTER-CHODER, O. 2021. Coupled transcription-translation in prokaryotes: an old couple with new surprises. *Frontiers in Microbiology*, 11: 624830.

ITO, F. & OHKAWARA, K. 1994. Spermatheca size differentiation between queens and workers in primitive ants. *Naturwissenschaften*, 81, 138–140.

IWASA, Y. 1988. Free fitness that always increases in evolution. *Journal of Theoretical Biology*, 135, 265–281.

IWASA, Y., POMIANKOWSKI, A., & NEE, S. 1991. The evolution of costly mate preferences II. The "handicap" principle. *Evolution*, 45, 1431–1442.

JABLONKA, E. & LAMB, M. J. 1995. *Epigenetic Inheritance and Evolution: The Lamarckian Dimension*. Oxford: Oxford University Press.

JABLONKA, E. & LAMB, M. J. 2002. The changing concept of epigenetics. *Annals of the New York Academy of Sciences*, 981, 82–96.

JABLONKA, E. & LAMB, M. J. 2005. *Evolution in Four Dimensions*. Cambridge, MA: MIT Press.

JACOB, F. 1974. *The Logic of Living Systems*. London: Allen Lane.

JAGERS OP AKKERHUIS, G. A. J. M. 2008. Analysing hierarchy in the organization of biological and physical systems. *Biological Reviews*, 83, 1–12.

JANDÉR, K. C. & HERRE, E. A. 2010. Host sanctions and pollinator cheating in the fig tree–fig wasp mutualism. *Proceedings of the Royal Society B*, 277, 1481–1488.

JANDÉR, K. C., HERRE, E. A., & SIMMS, E. L. 2012. Precision of host sanctions in the fig tree–fig wasp mutualism: consequences for uncooperative symbionts. *Ecology Letters*, 15, 1362–1369.

JANOUSKOVEC, J., TIKHONENKOV, D. V., BURKI, F., HOWE, A. T., ROHWER, F. L., MYLNIKOV, A. P., & KEELING, P. J. 2017. A new lineage of eukaryotes illuminates early mitochondrial genome reduction. *Current Biology*, 27, 3717–3724.

JAVEAUX, E. J. 2019. Challenges in evidencing the earliest traces of life. *Nature*, 572, 451–460.

JEANNE, R. L. 1991. The swarm-founding Polistinae. *In*: ROSS, K. G. & MATTHEWS, R. W. (eds.) *The Social Biology of Wasps*. Ithaca, NY: Cornell University Press.

JENNINGS, H. S. & LYNCH, R. S. 1928. Age, mortality, fertility, and individual diversities in the rotifer *Proales sordida* Gosse. I. Effect of the age of the parent on characteristics of the offspring. *Journal of Experimental Zoology*, 50, 345–407.

JETZ, W. & RUBENSTEIN, D. R. 2011. Environmental uncertainty and the global biogeography of cooperative breeding in birds. *Current Biology*, 21, 72–78.

JIRICNY, N., DIGGLE, S. P., WEST, S. A., EVANS, B. A., BALLANTYNES, G., ROSS-GILLESPIE, A., & GRIFFIN, A. S. 2010. Fitness correlates with the extent of cheating in a bacterium. *Journal of Evolutionary Biology*, 23, 738–747.

JOHNSON, B. R., BOROWIEC, M. L., CHIU, J. C., LEE, E. K., ATALLAH, J., & WARD, P. S. 2013. Phylogenomics resolves evolutionary relationships among ants, bees, and wasps. *Current Biology*, 23, 2058–2062.

JOHNSON, B. R. & LAM, S. K. 2010. Self-organization, natural selection, and evolution: cellular hardware and genetic software. *BioScience*, 60, 879–885.

JOHNSTON, T. D. 1995. The influence of Weismann's germ-plasm theory on the distinction between learned and innate behavior. *Journal of the History of the Behavioral Sciences*, 31, 115–128.

JONES, E. I., AFKHAMI, M. E., AKÇAY, E., BRONSTEIN, J. L., BSHARY, R., M.E., F., HEATH, K. D., HOEKSEMA, J. D., NESS, J. H., PANKEY, M. S., PORTER, S. S., SACHS, J. L., SCHARNAGL, K., & FRIESEN, M. L. 2015. Cheaters must prosper: reconciling theoretical and empirical perspectives on cheating in mutualism. *Ecology Letters*, 18, 1270–1284.

JONES, J. D. G., VANCE, R. E., & DANGL, J. L. 2016. Intracellular innate immune surveillance devices in plants and animals. *Science*, 354, 1117 and aaf6395.

JUTHE, A. 2005. Argument by analogy. *Argumentation*, 19, 1–27.

KAHNEMAN, D. 2012. *Thinking, Fast and Slow*. London: Penguin Books.

KAISER, D. 1979. Social gliding is correlated with the presence of pili in *Myxococcus xanthus*. *Proceedings of the National Academy of Sciences of the United States of America*, 76, 5952–5956.

KAPSETAKI, S. E. & WEST, S. A. 2019. The costs and benefits of multicellular group formation in algae. *Evolution*, 73, 1296–1308.

KATAJISTO, P., DÖHLA, J., CHAFFER, C.L., PENTINMIKKO, N., MARJANOVIC, N., IQBAL, S., ZONCU, R., CHEN, W., WEINBERG, R.A., & SABATINI, D.M. 2015. Asymmetric apportioning of aged mitochondria between daughter cells is required for stemness. *Science*, 348, 340–343.

KAUFFMAN, S. A. 1993. *Origins of Order: Self-Organization and Selection in Evolution*. Oxford: Oxford University Press.

KAUFFMAN, S. A. 2000. *Investigations*. Oxford: Oxford University Press.

KAY, T., KELLER, L., & LEHMANN, L. 2020. The evolution of altruism and the serial rediscovery of the role of relatedness. *Proceedings of the National Academy of Sciences of the United States of America*, 117, 28894–28898.

KEELING, P. J. 2004. Diversity and evolutionary history of plastids and their hosts. *American Journal of Botany*, 91, 1481–1493.

KEELING, P. J. 2013. The number, speed, and impact of plastid endosymbioses in eukaryotic evolution. *Annual Review of Plant Biology*, 64, 583–607.

KEELING, P. J. 2016. Genomics: evolution of the genetic code. *Current Biology*, 26, R851–R853.

KEELING, P. J. & MCCUTCHEON, J. P. 2017. Endosymbiosis: the feeling is not mutual. *Journal of Theoretical Biology*, 434, 75–79.

KELLER, L. (ed.) 1999. *Levels of Selection in Evolution*. Princeton, NJ: Princeton University Press.

KELLER, L. & GENOUD, M. 1997. Extraordinary lifespans in ants: a test of evolutionary theories of ageing. *Nature*, 389, 958–960.

KELLER, L. & SURETTE, M. G. 2006. Communication in bacteria: an ecological and evolutionary perspective. *Nature Reviews Microbiology*, 4, 249–258.

KENNEDY, P., HIGGINSON, A. D., RADFORD, A. N., & SUMNER, S. 2018. Altruism in a volatile world. *Nature*, 555, 359–362.

KENT, D. S. & SIMPSON, J. A. 1992. Eusociality in the beetle *Austroplatypus incompertus* (Coleoptera, Curculionidae). *Naturwissenschaften*, 79, 86–87.

KERR, B., GODFREY-SMITH, P., & FELDMAN, M. W. 2004. What is altruism? *Trends in Ecology & Evolution*, 19, 135–140.

KERR, B., NEUHAUSER, C., BOHANNAN, B. J. M., & DEAN, A. M. 2006. Local migration promotes competitive restraint in a host-pathogen 'tragedy of the commons'. *Nature*, 442, 75–78.

KESÄNIEMI, J., KOSKIMÄKI, J. J., & JURVANSUU, J. 2019. Corpse management of the invasive Argentine ant inhibits growth of pathogenic fungi. *Scientific Reports*, 9: 7593.

KHARE, A., SANTORELLI, L. A., STRASSMANN, J. E., QUELLER, D. C., KUSPA, A., & SHAULSKY, G. 2009. Cheater-resistance is not futile. *Nature*, 461, 980–982.

KIERS, E. T. & DENISON, R. F. 2008. Sanctions, cooperation, and the stability of plant-rhizosphere mutualisms. *Annual Review of Ecology, Evolution, and Systematics*, 39, 215–236.

KIERS, E. T., DUHAMEL, M., BEESETTY, Y., MENSAH, J. A., FRANKEN, O., VERBRUGGEN, E., FELLBAUM, C. R., KOWALCHUK, G. A., HART, M. M., BAGO, A., PALMER, T. M., WEST, S. A., VANDENKOORNHUYSE, P., JANSA, J., & BÜCKING, H. 2011. Reciprocal rewards stabilize cooperation in the mycorrhizal symbiosis. *Science*, 333, 880–882.

KIERS, E. T., ROUSSEAU, R. A., WEST, S. A., & DENISON, R. F. 2003. Host sanctions and the legume-rhizobium mutualism. *Nature*, 425, 78–81.

KIERS, E. T. & WEST, S. A. 2016. Evolution: Welcome to symbiont prison. *Current Biology*, 26, R66–R68.

KIERS, E. T., WEST, S. A., WYATT, G. A. K., GARDNER, A., BÜCKING, H., & WERNER, G. D. A. 2016. Misconceptions on the application of biological market theory to the mycorrhizal symbiosis. *Nature Plants*, 2: 16063.

KIMLER, W. C. 1983. Mimicry: views of naturalists and ecologists before the modern synthesis.

In: GREENE, M. (ed.) *Dimensions of Darwinism*. Cambridge: Cambridge University Press.

KIMLER, W. C. 1986. Advantage, adaptiveness, and evolutionary ecology. *Journal of the History of Biology*, 19, 215–233.

KIMURA, M. 1961. Natural selection as a process of accumulating genetic information in adaptive evolution. *Genetic Research*, 2, 127–140.

KIRBY, W. & SPENCE, W. 1818. *An Introduction to Entomology*, 2nd edition, volume II. London: Longman, Hurst, Rees, Orme, and Brown.

KIRKENDALL, L. R., BIEDERMANN, P. H. W., & JORDAL, B. H. 2015. Evolution and diversity of bark and ambrosia beetles. *In*: F.E. VEGA and R.W. HOFSTETTER (eds.) *Bark Beetles: Biology and Ecology of Native and Invasive Species*. New York: Academic Press.

KIRKWOOD, T. B. L. 1977. Evolution of ageing. *Nature*, 270, 301–304.

KIRKWOOD, T. B. L. 1985. Comparative and evolutionary aspects of longevity. *In*: FINCH, C. E. & SCHNEIDER, E. L. (eds.) *Handbook of the Biology of Aging*, 2nd edition. New York: Van Nostrand Reinhold.

KOASHI, M. & WINTER, A. 2004. Monogamy of quantum entanglement and other correlations. *Physical Review A*, 69, 022309.

KOCHER, S. D., TSURUDA, J. M., GIBSON, J. D., EMORE, C. M., ARECHAVALETA-VELASCO, M. E., QUELLER, D. C., STRASSMANN, J. E., GROZINGER, C. M., GRIBSKOV, M. R., SAN MIGUEL, P., WESTERMAN, R., & HUNT, G. J. 2015. A search for parent-of-origin effects on honey bee gene expression. *G3. Genes, Genomics, Genetics*, 5, 1657–1662.

KOEHLER, S., GAEDEKE, R., THOMPSON, C., BONGRAND, C., VISICK, K. L., RUBY, E., & MCFALL-NGAI, M. J. 2019. The model squid-vibrio symbiosis provides a window into the impact of strain- and species-level differences during the initial stages of symbiont engagement. *Environmental Microbiology*, 21, 3269–3283.

KOHN, M. 2004. *A Reason for Everything: Natural Selection and the British Imagination*. London: Faber & Faber Ltd.

KOKKO, H. & JENNIONS, M. D. 2008. Parental investment, sexual selection and sex ratios. *Journal of Evolutionary Biology*, 21, 919–948.

KONRAD, M., PULL, C. D., METZLER, S., SEIF, K., NADERLINGER, E., GRASSE, A. V., & CREMER, S. 2018. Ants avoid superinfections by performing risk-adjusted sanitary care. *Proceedings of the National Academy of Sciences of the United States of America*, 115, 2782–2787.

KOOIJ, P. W., AANEN, D. K., SCHIØTT, M. & BOOMSMA, J. J. 2015. Evolutionarily advanced ant farmers rear polyploid fungal crops. *Journal of Evolutionary Biology*, 28, 1911–1924.

KOONIN, E. V. 2015a. Energetics and population genetics at the root of eukaryotic cellular and genomic complexity. *Proceedings of the National Academy of Sciences of the United States of America*, 112, 15777–15778.

KOONIN, E. V. 2015b. Archaeal ancestors of eukaryotes: not so elusive any more. *BMC Biology*, 13: 84.

KOONIN, E. V. 2017. Evolution of RNA- and DNA-guided antivirus defense systems in prokaryotes and eukaryotes: common ancestry vs convergence. *Biology Direct*, 12: 5.

KOONIN, E. V., DOLJA, V. V., & KRUPOVIC, M. 2015. Origins and evolution of viruses of eukaryotes: the ultimate modularity. *Virology*, 479–480, 2–25.

KOONIN, E. V. & MAKAROVA, K. S. 2018. Anti-CRISPRs on the march. *Science*, 362, 156–157.

KOONIN, E. V. & MARTIN, W. 2005. On the origin of genomes and cells within organic compartments. *Trends in Genetics*, 21, 647–654.

KOONIN, E. V., SENKEVICH, T. G., & DOLJA, V. V. 2006. The ancient virus world and evolution of cells. *Biology Direct*, 1: 29.

KORB, J. 2010. Social insects, major evolutionary transitions and multilevel selection. *In*: KAPPELER, P. (ed.) *Animal Behaviour: Evolution & Mechanisms*. Berlin: Springer.

KORB, J., BUSCHMANN, M., SCHAFBERG, S., LIEBIG, J., & BAGNÈRES, A.-G. 2012. Brood care and social evolution in termites. *Proceedings of the Royal Society B*, 279, 2662–2671.

KORB, J. & HEINZE, J. 2016. Major hurdles for the evolution of sociality. *Annual Review of Entomology*, 61, 297–316.

KORB, J. & HEINZE, J. 2021. Ageing and sociality: why, when and how does sociality change ageing patterns? *Philosophical Transactions of the Royal Society B*, 376, 20190727.

KRAEMER, S. A. & VELICER, G. J. 2014. Social complementation and growth advantages promote socially defective bacterial isolates. *Proceedings of the Royal Society B*, 281, 20140036.

KRAEMER, S. A., WIELGOSS, S., FIEGNA, F., & VELICER, G. J. 2016. The biogeography of kin discrimination across microbial neighbourhoods. *Molecular Ecology*, 25, 4875–4888.

KRAMER, B. H., NEHRING, V., BUTTSTEDT, A., HEINZE, J., KORB, J., LIBBRECHT, R., MEUSEMANN, K., PAXTON, R. J., SÉGURET, A., SCHAUB, F. & BERNADOU, A. 2021. Oxidative stress and senescence in social insects: a significant but inconsistent link? *Philosophical Transactions of the Royal Society B*, 376: 20190732.

KRAMER, K. L. & RUSSELL, A. F. 2015. Was monogamy a key step on the Hominin road? Reevaluating the monogamy hypothesis in the evolution of cooperative breeding. *Evolutionary Anthropology*, 24, 73–83.

KRASNOW, M. M. & DELTON, A. W. 2016. The sketch is blank: no evidence for an explanatory role for cultural group selection. Open peer commentary on Richerson et al. (2016). *Behavioral and Brain Sciences*, 39: e43.

KRAŠOVEC, R., BELAVKIN, R. V., ASTON, J. A. D., CHANNON, A., ASTON, E., RASH, B. M., KADIRVEL, M., FORBES, S., & KNIGHT, C. G. 2014. Mutation rate plasticity in rifampicin resistance depends on *Escherichia coli* cell-cell interactions. *Nature Communications*, 5: 3742.

KRAŠOVEC, R., RICHARDS, H., GIFFORD, D. R., HATCHER, C., FAULKNER, K. J., BELAVKIN, R. V., CHANNON, A., ASTON, E., MCBAIN, A. J., & KNIGHT, C. G. 2017. Spontaneous mutation rate is a plastic trait associated with population density across domains of life. *PLoS Biology*, 15: e2002731.

KREBS, J. R. & DAVIES, N. B. (eds.) 1978. *Behavioural Ecology: An Evolutionary Approach*, 1st edition. Oxford: Blackwell Scientific Publications.

KREBS, J. R. & DAVIES, N. B. 1981. *An Introduction to Behavioural Ecology*, 1st edition. Oxford: Blackwell Scientific Publications.

KREBS, J. R. & DAVIES, N. B. (eds.) 1984. *Behavioural Ecology: An Evolutionary Approach*, 2nd edition. Oxford: Blackwell Scientific Publications.

KREBS, J. R. & DAVIES, N. B. (eds.) 1991. *Behavioural Ecology: An Evolutionary Approach*, 3rd edition. Oxford: Blackwell Scientific Publications.

KREBS, J. R. & DAVIES, N. B. (eds.) 1997. *Behavioural Ecology: An Evolutionary Approach*, 4th edition. Oxford: Blackwell Science Ltd.

KREMER, N., PHILIPP, E. E. R., CARPENTIER, M.-C., BRENNAN, C. A., KRAEMER, L., ALTURA, M. A., AUGUSTIN, R., HÄSLER, R., HEATH-HECKMAN, E. A. C., PEYER, S. M., SCHWARTZMAN, J., RADER, B. A., RUBY, E. G., ROSENSTIEL, P., & MCFALL-NGAI, M. J. 2013. Initial symbiont contact orchestrates host-organ-wide transcriptional changes that prime tissue colonization. *Cell Host & Microbe*, 14, 183–194.

KRONAUER, D. J. C. 2008. Genomic imprinting and kinship in the social Hymenoptera: what are the predictions? *Journal of Theoretical Biology*, 254, 737–740.

KROPOTKIN, P. 1902. *Mutual Aid*, 3rd edition. Montreal: Black Rose Books.

KRUPOVIC, M. 2012. Recombination between RNA viruses and plasmids might have played a central role in the origin and evolution of small DNA viruses. *BioEssays*, 34, 867–870.

KRUPOVIC, M., CVIRKAITE-KRUPOVIC, V., IRANZO, J., PRANGISHVILI, D., & KOONIN, E. V. 2018. Viruses of archaea: structural, functional, environmental and evolutionary genomics. *Virus Research*, 244, 181–193.

KRUPOVIC, M. & KOONIN, E. V. 2017. Multiple origins of viral capsid proteins from cellular ancestors. *Proceedings of the National Academy of Sciences of the United States of America*, 114, E2401–E2410.

KRUUK, H. 2003. *Niko's Nature*. Oxford: Oxford University Press.

KU, C. & MARTIN, W. F. 2016. A natural barrier to lateral gene transfer from prokaryotes to eukaryotes revealed from genomes: the 70% rule. *BMC Biology*, 14: 89.

KU, C., NELSON-SATHI, S., ROETTGER, M., SOUSA, F. L., LOCKHART, P. J., BRYANT, D., HAZKANI-COVO, E., MCINERNEY, J. O., LANDAN, G., & MARTIN, W. F. 2015. Endosymbiotic origin and differential loss of eukaryotic genes. *Nature*, 524, 427–432.

KUHN, T. S. 1962. *The Structure of Scientific Revolutions*. Chicago, IL: University of Chicago Press.

KÜMMERLI, R., JIRICNY, N., CLARKE, L. S., WEST, S. A., & GRIFFIN, A. S. 2009a. Phenotypic plasticity of a cooperative behaviour in bacteria. *Journal of Evolutionary Biology*, 22, 589–598.

KÜMMERLI, R., GRIFFIN, A. S., WEST, S. A., BUCKLING, A., & HARRISON, F. 2009b. Viscous medium promotes cooperation in the pathogenic bacterium *Pseudomonas aeruginosa*. *Proceedings of the Royal Society B*, 276, 3531–3538.

KÜMMERLI, R., SCHIESSL, K. T., WALDVOGEL, T., MCNEILL, K., & ACKERMANN, M. 2014. Habitat structure and the evolution of diffusible siderophores in bacteria. *Ecology Letters*, 17, 1536–1544.

KUMRU, C. S. & VESTERLUND, L. 2010. The effect of status on charitable giving. *Journal of Public Economic Theory*, 12, 709–735.

KURZBAN, R. & AKTIPIS, C. A. 2006. Modular minds, multiple motives. *In*: SCHALLER, M., SIMPSON, J. A., & KENRICK, D. T. (eds.) *Evolution and Social Psychology*. New York: Psychology Press.

KURZBAN, R., BURTON-CHELLEW, M. N., & WEST, S. A. 2014. The evolution of altruism in humans. *Annual Review of Psychology*, 66, 575–599.

KUTSCHERA, U. & HOPPE, T. 2019. Plasmodial slime molds and the evolution of microbial husbandry. *Theory in Biosciences*, 138, 127–132.

KUZDZAL-FICK, J. J., FOX, S. A., STRASSMANN, J. E., & QUELLER, D. C. 2011. High relatedness is necessary and sufficient to maintain multicellularity in *Dictyostelium*. *Science*, 334, 1548–1551.

LACK, D. 1948. The significance of clutch-size – part III - some interspecific comparisons. *Ibis*, 90, 25–45.

LACK, D. 1949. The significance of ecological isolation. *In*: JEPSEN, G. L., MAYR, E., & SIMPSON, G. G. (eds.) *Genetics, Paleontology, and Evolution*. Princeton, NJ: Princeton University Press.

LACK, D. 1954a. The evolution of reproductive rates. *In*: HUXLEY, J. S., HARDY, A. C., & FORD, E. B. (eds.) *Evolution as a Process*. London: Allen & Unwin.

LACK, D. 1954b. *The Natural Regulation of Animal Numbers*. Oxford: Oxford University Press.

LACK, D. 1966. *Population Studies of Birds*. Oxford: Clarendon Press.

LAHTI, D. C., JOHNSON, N. A., AJIE, B. C., OTTO, S. P., HENDRY, A. P., BLUMSTEIN, D. T., COSS, R. G., DONOHUE, K., & FOSTER, S. A. 2009. Relaxed selection in the wild. *Trends in Ecology and Evolution*, 24, 487–496.

LALAND, K., ULLER, T., FELDMAN, M., STERELNY, K., MULLER, G. B., MOCZEK, A., JABLONKA, E., & ODLING-SMEE, J. 2014. Does evolutionary theory need a rethink? POINT: Yes, urgently. *Nature*, 514, 161–164.

LALAND, K. N., ULLER, T., FELDMAN, M. W., STERELNY, K., MÜLLER, G. B., MOCZEK, A., JABLONKA, E., & ODLING-SMEE, J. 2015. The extended evolutionary synthesis: its structure, assumptions and predictions. *Proceedings of the Royal Society B*, 282, 20151019.

LANDE, R. & ARNOLD, S. J. 1983. The measurement of selection on correlated characters. *Evolution*, 37, 1210–1226.

LANDENMARK, H. K. E., FORGAN, D. H., & COCKELL, C. S. 2015. An estimate of the total DNA in the biosphere. *PLoS Biology*, 13: e1002168.

LANE, N. & MARTIN, W. 2010. The energetics of genome complexity. *Nature*, 467, 929–934.

LANE, N. & MARTIN, W. F. 2015. Eukaryotes really are special, and mitochondria are why. *Proceedings of the National Academy of Sciences of the United States of America*, 112, E4823.

LANE, N. & MARTIN, W. F. 2016. Mitochondria, complexity, and evolutionary deficit spending. *Proceedings of the National Academy of Sciences of the United States of America*, 113, E666.

LANFEAR, R. 2018. Do plants have a segregated germline? *PLoS Biology*, 16: e2005439.

LANG, D. & RENSING, S. A. 2015. The evolution of transcriptional regulation in the Viridiplantae and its correlation with morphological complexity. *In*: RUIZ-TRILLO, I. & NEDELCU, A. M. (eds.) *Evolutionary Transitions to Multicellular Life*. Dordrecht: Springer.

LANGEFORS, Å., HASSELQUIST, D., & VON SCHANTZ, T. 1998. Extra-pair fertilizations in the Sedge Warbler. *Journal of Avian Biology*, 29, 134–144.

LANSING, A. I. 1947. A transmissible, cumulative, and reversible factor in aging. *Journal of Gerontology*, 2, 228–239.

LAPLANE, L., MANTOVANI, P., ADOLPHS, R., CHANG, H., MANTOVANI, A., MCFALL-NGAI, M., ROVELLI, C., SOBER, E., & PRADEU, T. 2019. Why science needs philosophy. *Proceedings of the National Academy of Sciences of the United States of America*, 116, 3948–3952.

LAPOLLA, J. S., DLUSSKY, G. M. & PERRICHOT, V. 2013. Ants and the Fossil Record. *Annual Review of Entomology*, 58, 609–630.

LARSEN, B. B., MILLER, E. C., RHODES, M. K., & WIENS, J. J. 2017. Inordinate fondness multiplied and redistributed: the number of species on earth and the new pie of life. *Quarterly Review of Biology*, 92, 229–265.

LEA, A. M. & RYAN, M. J. 2015. Irrationality in mate choice revealed by túngara frogs. *Science*, 349, 964–966.

LEADBEATER, E., CARRUTHERS, J. M., GREEN, J. P., ROSSER, N. S., & FIELD, J. 2011. Nest inheritance is the missing source of direct fitness in a primitively eusocial insect. *Science*, 333, 874–876.

LEEKS, A., DOS SANTOS, M., & WEST, S. A. 2019. Transmission, relatedness, and the evolution of cooperative symbionts. *Journal of Evolutionary Biology*, 32, 1036–1045.

LEEKS, A., WEST, S. A., & GHOUL, M. 2021. The evolution of cheating in viruses. *Nature Communications*, 12: 6928.

LEHMANN, L., BARGUM, K., & REUTER, M. 2006. An evolutionary analysis of the relationship between spite and altruism. *Journal of Evolutionary Biology*, 19, 1507–1516.

LEHMANN, L., FELDMAN, M. W., & FOSTER, K. R. 2008. Cultural transmission can inhibit the evolution of altruistic helping. *American Naturalist*, 172, 12–24.

LEHMANN, L. & KELLER, L. 2006. The evolution of cooperation and altruism – a general framework and a classification of models. *Journal of Evolutionary Biology*, 19, 1365–1376.

LEHMANN, L., KELLER, L., WEST, S., & ROZE, D. 2007. Group selection and kin selection: two concepts but one process. *Proceedings of the National*

Academy of Sciences of the United States of America, 104, 6736–6739.

LEHMANN, L. & ROUSSET, F. 2014. Fitness, inclusive fitness, and optimization. *Biology and Philosophy*, 29, 181–195.

LEHMANN, L. & ROUSSET, F. 2020. When do individuals maximize their inclusive fitness? *American Naturalist*, 195, 717–732.

LEHMANN, L., ROUSSET, F., ROZE, D., & KELLER, L. 2007. Strong reciprocity or strong ferocity? A population genetic view on the evolution of altruistic punishment. *American Naturalist*, 170, 21–36.

LEHTONEN, J. 2016. Multilevel selection in kin selection language. *Trends in Ecology & Evolution*, 31, 752–762.

LEHTONEN, J. 2020. The Price equation and the unity of social evolution theory. *Philosophical Transactions of the Royal Society B*, 375: 20190362.

LEHTONEN, J. & HELANTERÄ, H. 2020. Superorganismal anisogamy: queen-male dimorphism in eusocial insects. *Proceedings of the Royal Society B* 287: 20200635.

LEIGH, E. G. 1977. How does selection reconcile individual advantage with the good of the group? *Proceedings of the National Academy of Sciences of the United States of America*, 74, 4542–4546.

LEIGH, E. G. 1999. The modern synthesis, Ronald Fisher and creationism. *Trends in Ecology and Evolution*, 14, 495–498.

LEIGH, E. G. 2001. Adaptation, adaptationism, and optimality. *In*: ORZACK, S. H. & SOBER, E. (eds.) *Adaptationism and Optimality*. Cambridge: Cambridge University Press.

LEIGH, E. G. 2010a. The evolution of mutualism. *Journal of Evolutionary Biology*, 23, 2507–2528.

LEIGH, E. G. 2010b. The group selection controversy. *Journal of Evolutionary Biology*, 23, 6–19.

LELIAERT, F., SMITH, D. R., MOREAU, H., HERRON, M. D., VERBRUGGEN, H., DELWICHE, C. F., & DECLERCK, O. 2012. Phylogeny and molecular evolution of the green algae. *Critical Reviews in Plant Sciences*, 31, 1–46.

LERAT, E. & MORAN, N. A. 2004. The evolutionary history of quorum-sensing systems in bacteria. *Molecular Biology and Evolution*, 21, 903–913.

LERNER, I. M. 1950. *Population Genetics and Animal Improvement*. Cambridge: Cambridge University Press.

LEROI, A. M. 2014. *The Lagoon: How Aristotle Invented Science*. New York: Bloomsbury/Viking.

LESSARD, S. & EWENS, W. J. 2019. The left-hand side of the fundamental theorem of natural selection: a reply. *Journal of Theoretical Biology*, 472, 77–83.

LEVALLOIS, C. 2018. The development of sociobiology in relation to animal behavior studies, 1946–1975. *Journal of the History of Biology*, 51, 419–444.

LEVIN, S. R., GANDON, S., & WEST, S. A. 2020. The social coevolution hypothesis for the origin of enzymatic cooperation. *Nature Ecology & Evolution*, 4, 132–137.

LEVIN, S. R. & GRAFEN, A. 2019. Inclusive fitness is an indispensable approximation for understanding organismal design. *Evolution*, 73, 1066–1076.

LEVIN, S. R. & GRAFEN, A. 2021. Extending the range of additivity in using inclusive fitness. *Ecology and Evolution*, 11, 1970–1983.

LEVIN, S. R., SCOTT, T. W., COOPER, H. S., & WEST, S. A. 2017. Darwin's aliens. *International Journal of Astrobiology*, 18, 1–9.

LEVINS, R. 1968. *Evolution in Changing Environments*. Princeton, NJ: Princeton University Press.

LEVIS, N. A. & PFENNIG, D. W. 2016. Evaluating 'plasticity-first' evolution in nature: key criteria and empirical approaches. *Trends in Ecology and Evolution*, 31, 563–574.

LEVIS, N. A. & PFENNIG, D. W. 2019. Plasticity-led evolution: evaluating the key prediction of frequency-dependent adaptation. *Proceedings of the Royal Society B*, 286, 20182754.

LIDGARD, S. & NYHART, L. K. 2017. The work of biological individuality: concepts and contexts. *In*: LIDGARD, S. & NYHART, L. K. (eds.) *Biological Individuality*. Chicago, IL: University of Chicago Press.

LIEBERMAN, D., TOOBY, J., & COSMIDES, L. 2007. The architecture of human kin detection. *Nature*, 445, 727–731.

LINDNER, A. B., MADDEN, R., DEMAREZ, A., STEWART, E. J., & TADDEI, F. 2008. Asymmetric segregation of protein aggregates is associated with cellular aging and rejuvenation. *Proceedings of the National Academy of Sciences of the United States of America*, 105, 3076–3081.

LINKSVAYER, T. A. 2006. Direct, maternal, and sib-social genetic effects on individual and colony traits in an ant. *Evolution*, 60, 2552–2561.

LINKSVAYER, T. A. & WADE, M. J. 2005. The evolutionary origin and elaboration of sociality in the aculeate Hymenoptera: maternal effects, sib-social effects, and heterochrony. *Quarterly Review of Biology*, 80, 317–336.

LIONAKI, E., MARKAKI, M., PALIKARAS, K., & TAVERNARAKIS, N. 2015. Mitochondria, autophagy and age-associated neurodegenerative diseases: new insights into a complex interplay. *Biochimica et Biophysica Acta*, 1847, 1412–1423.

LONG, H., DOAK, T. G., & LYNCH, M. 2018. Limited mutation-rate variation within the *Paramecium aurelia* species complex. *G3. Genes, Genomes, Genetics*, 8, 2523–2526.

LOOPE, K. J. 2015. Queen killing is linked to high worker-worker relatedness in a social wasp. *Current Biology*, 25, 2976–2979.

LOOPE, K. J. & WILSON RANKIN, E. E. 2021. Viral load, not food availability or temperature, predicts colony longevity in an invasive eusocial wasp with plastic life history. *Scientific Reports*, 11: 10087.

LÓPEZ-RIQUELME, G. O. & FANJUL-MOLES, M. L. 2013. The funeral ways of social insects. Social strategies for corpse disposal. *Trends in Entomology*, 9, 73–129.

LORENZ, K. 1966. *On Aggression*. New York: Harcourt, Brace & World.

LORIES, B., ROBERFROID, S., DIELTJENS, L., DE COSTER, D., FOSTER, K. R., & STEENACKERS, H. P. 2020. Biofilm bacteria use stress responses to detect and respond to competitors. *Current Biology*, 30, 1231–1244.

LOVELOCK, J. 2000. *Gaia: A New Look at Life on Earth*. Oxford: Oxford Paperbacks.

LOWE, C. D., MINTER, E. J., CAMERON, D. D., & BROCKHURST, M. A. 2016. Shining a light on exploitative host control in a photosynthetic endosymbiosis. *Current Biology*, 26, 207–211.

LUHMANN, N. 1992. Operational closure and structural coupling: the differentiation of the legal system. *Cardoso Law Review*, 13, 1419–1441.

LUJÁN, A. M., GÓMEZ, P., & BUCKLING, A. 2015. Siderophore cooperation of the bacterium *Pseudomonas fluorescens* in soil. *Biology Letters*, 11: 20140934.

LUKAS, D. & CLUTTON-BROCK, T. 2012. Cooperative breeding and monogamy in mammalian societies. *Proceedings of the Royal Society B*, 279, 2151–2156.

LUKAS, D. & CLUTTON-BROCK, T. 2018. Social complexity and kinship in animal societies. *Ecology Letters*, 21, 1129–1134.

LUO, D., XU, H., LIU, Z., GUO, J., LI, H., CHEN, L., FANG, C., ZHANG, Q., BAI, M., YAO, N., WU, H., WU, H., JI, C., ZHENG, H., CHEN, Y., YE, S., LI, X., ZHAO, X., LI, R., & LIU, Y.-G. 2013. A detrimental mitochondrial-nuclear interaction causes cytoplasmic male sterility in rice. *Nature Genetics*, 45, 573–577.

LÜSCHER, M. 1953. The termite and the cell. *Scientific American*, 188, 74–78.

LUTZONI, F., NOWAK, M. D., ALFARO, M. E., REEB, V., MIADLIKOWSKA, J., KRUG, M., ARNOLD, A. E., LEWIS, L. A., SWOFFORD, D. L., HIBBETT, D., HILU, K., JAMES, T. Y., QUANDT, D., & MAGALLÓN, S. 2018. Contemporaneous radiations of fungi and plants linked to symbiosis. *Nature Communications*, 9: 5451.

LYNCH, M. & MARINOV, G. K. 2015. The bioenergetic costs of a gene. *Proceedings of the National Academy of Sciences of the United States of America*, 112, 15690–15695.

LYNE, J. & HOWE, H. F. 1990. The rhetoric of expertise: E.O. Wilson and sociobiology. *Quarterly Journal of Speech*, 76, 134–151.

LYONS, N. A., KRAIGHER, B., STEFANIC, P., MANDIC-MULEC, I., & KOLTER, R. 2016. A combinatorial kin discrimination system in *Bacillus subtilis*. *Current Biology*, 26, 733–742.

MACARTHUR, R. H. & WILSON, E. O. 1967. *The Theory of Island Biogeography*. Princeton, NJ: Princeton University Press.

MACE, R. & SILVA, A. S. 2016. The role of cultural group selection in explaining human cooperation is a hard case to prove. Open peer commentary on Richerson et al. (2016). *Behavioral and Brain Sciences*, 39: e45.

MACORANO, L. & NOWACK, E. C. M. 2021. *Paulinella chromatophora*. *Current Biology*, 31, R1017–R1034.

MADGWICK, P. G., BELCHER, L. J., & WOLF, J. B. 2019. Greenbeard genes: theory and reality. *Trends in Ecology & Evolution*, 34, 1092–1103.

MAKLAKOV, A. A. & CHAPMAN, T. 2019. Evolution of ageing as a tangle of trade-offs: energy versus function. *Proceedings of the Royal Society B*, 286, 20191604.

MAKLAKOV, A. A. & IMMLER, S. 2016. The expensive germline and the evolution of ageing. *Current Biology*, 26, R577–R586.

MALIK, S. S., AZEM-E-ZAHRA, S., KIM, K. M., CAETANO-ANOLLÉS, G., & NASIR, A. 2017. Do viruses exchange genes across superkingdoms of life? *Frontiers in Microbiology*, 8: 2110.

MARINO, N. D., ZHANG, J. Y., BORGES, A. L., SOUSA, A. A., LEON, L. M., RAUCH, B. J., WALTON, R. T., BERRY, J. D., JOUNG, J. K., KLEINSTIVER, B. P., & BONDY-DENOMY, J. 2018. Discovery of widespread type I and type V CRISPR-Cas inhibitors. *Science*, 362, 240–242.

MARKL, H. E. 1980. *Evolution of Social Behavior: Hypotheses and Empirical Tests*. Weinheim: Verlag Chemie.

MARSHALL, J. A. R. 2015. *Social Evolution and Inclusive Fitness Theory: An Introduction*. Princeton, NJ: Princeton University Press.

MARSHALL, J. A. R. 2016. What is inclusive fitness theory, and what is it for? *Current Opinion in Behavioral Sciences*, 12, 103–108.

MARTIN, W. & KOONIN, E. V. 2006. Introns and the origin of nucleus-cytosol compartmentalization. *Nature*, 440, 41–45.

MARTIN, W. F., SOUSA, F. L., & LANE, N. 2014. Energy at life's origin. *Science*, 344, 1092–1093.

MATSUMURA, S., KUN, A., RYCKELYNCK, M., COLDREN, F., SZILÁGYI, A., JOSSINET, F., RICK, C., NGHE, P., SZATHMÁRY, E., & GRIFFITHS, A. D. 2016. Transient compartmentalization of RNA replicators prevents extinction due to parasites. *Science*, 354, 1293–1296.

MATSUURA, K., VARGO, E. L., KAWATSU, K., LABADIE, P. E., NAKANO, H., YASHIRO, T., & TSUJI, K. 2009. Queen succession through asexual reproduction in termites. *Science*, 323, 1687.

MATURANA, H. & VARELA, F. 1980. *Autopoiesis and Cognition: The Realization of the Living*. Dordrecht: Reidel Publishing.

MAYNARD SMITH, J. 1958. *The Theory of Evolution*. Harmondsworth, Middlesex: Penguin Books Ltd.

MAYNARD SMITH, J. 1964. Group selection and kin selection. *Nature*, 201, 1145–1147.

MAYNARD SMITH, J. 1974. The theory of games and the evolution of animal conflict. *Journal of Theoretical Biology*, 47, 209–221.

MAYNARD SMITH, J. 1976a. Evolution and the theory of games. *American Scientist*, 64, 41–45.

MAYNARD SMITH, J. 1976b. *The Evolution of Sex*. Cambridge: Cambridge University Press.

MAYNARD SMITH, J. 1976c. Group selection. *Quarterly Review of Biology*, 51, 277–283.

MAYNARD SMITH, J. 1978a. Optimization theory in evolution. *Annual Review of Ecology and Systematics*, 9, 31–56.

MAYNARD SMITH, J. 1978b. The evolution of behavior. *Scientific American*, 239, 176–192.

MAYNARD SMITH, J. 1978c. The ecology of sex. *In*: KREBS, J. R. & DAVIES, N. B. (eds.) *Behavioural Ecology: An Evolutionary Approach*, 1st edition. Oxford: Blackwell Scientific Publications.

MAYNARD SMITH, J. 1979. Hypercycles and the origin of life. *Nature*, 280, 445–446.

MAYNARD SMITH, J. 1982a. *Evolution and the Theory of Games*, Cambridge, England, Cambridge University Press.

MAYNARD SMITH, J. 1982b. The evolution of social behaviour – a classification of models. *In*: KING'S COLLEGE SOCIOBIOLOGY GROUP (ed.) *Current Problems in Sociobiology*. Cambridge: Cambridge University Press.

MAYNARD SMITH, J. 1984. The ecology of sex. *In*: KREBS, J. R. & DAVIES, N. B. (eds.) *Behavioural Ecology: An Evolutionary Approach*, 2nd edition. Oxford: Blackwell Scientific Publications.

MAYNARD SMITH, J. 1987. When learning guides evolution. *Nature*, 329, 761–762.

MAYNARD SMITH, J. 1988. Evolutionary progress and levels of selection. *In*: NITECKI, M. H. (ed.) *Evolutionary Progress*. Chicago, IL: Chicago University Press.

MAYNARD SMITH, J. 1991. A Darwinian view of symbiosis. *In*: MARGULIS, L. & FESTER, R. (eds.) *Symbiosis as a Source of Evolutionary Innovation*. Cambridge, MA: MIT Press.

MAYNARD SMITH, J. 2000. The concept of information in biology. *Philosophy of Science*, 67, 177–194.

MAYNARD SMITH, J. & PARKER, G. A. 1976. The logic of asymmetric contests. *Animal Behaviour*, 24, 159–175.

MAYNARD SMITH, J. & PRICE, G. R. 1973. The logic of animal conflict *Nature*, 246, 15–18.

MAYNARD SMITH, J. & SZATHMÁRY, E. 1993. The origin of chromosomes I. Selection for linkage. *Journal of Theoretical Biology*, 164, 437–446.

MAYNARD SMITH, J. & SZATHMÁRY, E. 1995. *The Major Transitions in Evolution*. Oxford: W. H. Freeman Spektrum.

MAYR, E. 1942. *Systematics and the Origin of Species*. Cambridge: Harvard University Press.

MAYR, E. 1960. The emergence of evolutionary novelties. *In*: TAX, S. (ed.) *Evolution After Darwin, Volume 1: The Evolution of Life*. Chicago, IL: University of Chicago Press.

MAYR, E. 1961. Cause and effect in biology – kinds of causes, predictability, and teleology are viewed by a practicing biologist. *Science*, 134, 1501–1506.

MAYR, E. 1974. Teleological and teleonomic, a new analysis. *In*: COHEN, R. S. & WARTOFSKY, M. W. (eds.) *Methodological and Historical Essays in the Natural and Social Sciences*. Dordrecht: D. Reidel Publishing Company.

MAYR, E. 1982. *The Growth of Biological Thought: Diversity, Evolution, and Inheritance*. Cambridge, MA: Belknap Press of Harvard University Press.

MAYR, E. 1983a. How to carry out the adaptationist program. *American Naturalist*, 121, 324–334.

MAYR, E. 1983b. Darwin, intellectual revolutionary. *In*: BENDALL, D. S. (ed.) *Evolution from Molecules to Men*. Cambridge: Cambridge University Press.

MAYR, E. 2001. *What Evolution Is*. New York: Basic Books.

MCFALL-NGAI, M. J. 2014a. The importance of microbes in animal development: lessons from the Squid-Vibrio symbiosis. *Annual Review of Microbiology*, 68, 177–194.

MCFALL-NGAI, M. J. 2014b. Divining the essence of symbiosis: insights from the squid–vibrio model. *PLoS Biology*, 12: e1001783.

MCGLOTHLIN, J. W. & BRODIE III, E. D. 2009. How to measure indirect genetic effects: the congruence of trait-based and variance partitioning approaches. *Evolution*, 63, 1785–1795.

MCGLOTHLIN, J. W., WOLF, J. B., BRODIE III, E. D., & MOORE, A. J. 2014. Quantitative genetic versions of Hamilton's rule with empirical applications. *Philosophical Transactions of the Royal Society B*, 369, 20130358.

MCKENZIE, S. K., FETTER-PRUNEDA, I., RUTA, V., & KRONAUER, D. J. C. 2016. Transcriptomics and neuroanatomy of the clonal raider ant implicate an expanded clade of odorant receptors in chemical communication. *Proceedings of the National Academy of Sciences of the United States of America*, 113, 14091–14096.

MCMAHON, D. P., FÜRST, M. A., CASPAR, J., THEODOROU, P., BROWN, M. J. F., & PAXTON, R. J. 2015. A sting in the spit: widespread cross-infection of multiple RNA viruses across wild and managed bees. *Journal of Animal Ecology*, 84, 615–624.

MCNAMARA, J. M. & LEIMAR, O. 2020. *Game Theory in Biology: Concepts and Frontiers*. Oxford: Oxford University Press.

MCSHEA, D. W. 2001a. The hierarchical structure of organisms: a scale and documentation of a trend in the maximum. *Paleobiology*, 27, 405–423.

MCSHEA, D. W. 2001b. The minor transitions in hierarchical evolution and the question of a directional bias. *Journal of Evolutionary Biology*, 14, 502–518.

MCSHEA, D. W. 2016. Three trends in the history of life: an evolutionary syndrome. *Evolutionary Biology*, 43, 531–542.

MCSHEA, D. W. & BRANDON, R. N. 2010. *Biology's First Law: The Tendency for Diversity & Complexity to Increase in Evolutionary Systems*. Chicago, IL: University of Chicago Press.

MCSHEA, D. W. & SIMPSON, C. 2011. The miscellaneous transitions in evolution. *In*: CALCOTT, B., & STERELNY, K. (eds.) *The Major Transitions in Evolution Revisited*. Cambridge, MA: MIT Press.

MEDAWAR, P. B. 1952. *An Unsolved Problem of Biology*. London: H. K. Lewis and Co.

MEDINA, J. D., SCHREENIDHI, P. M., LARSEN, T. J., QUELLER, D. C., & STRASSMANN, J. E. 2019. Cooperation and conflict in the social amoeba *Dictyostelium discoideum*. *International Journal of Developmental Biology*, 63, 371–382.

MEHDIABADI, N. J., JACK, C. N., FARNHAM, T. T., PLATT, T. G., KALLA, S. E., SHAULSKY, G., QUELLER, D. C., & STRASSMANN, J. E. 2006. Kin preference in a social amoeba. *Nature*, 442, 881–882.

MELANDER, A. L. & CARPENTER, F. M. 1937. William Morton Wheeler. *Annals of the Entomological Society of America*, 30, 433–437.

MENAKER, M. 1996. Colin S. Pittendrigh (1918–96). *Nature*, 381, 24.

MESOUDI, A. & THORNTON, A. 2018. What is cumulative cultural evolution? *Proceedings of the Royal Society B*, 285, 20180712.

MEUNIER, J., WEST, S. A., & CHAPUISAT, M. 2008. Split sex ratios in the social Hymenoptera: a meta-analysis. *Behavioral Ecology*, 19, 382–390.

MEYER, S. N., AMOYEL, M., BERGANTIÑOS, C., DE LA COVA, C., SCHERTEL, C., BASLER, K. & JOHNSTON, L. A. 2014. An ancient defense system eliminates unfit cells from developing tissues during cell competition. *Science*, 346, 1199 and 1258236.

MICHENER, C. D. 2000. *The Bees of the World*. Baltimore, MD: Johns Hopkins University Press.

MICHENER, C. D. 1985. From solitary to eusocial: need there be a series of intervening species? *In*: HÖLLDOBLER, B. & LINDAUER, M. (eds.) *Experimental Behavioral Ecology and Sociobiology*. Stuttgart: Gustav Fisher Verlag.

MICHOD, R. E. 1982. The theory of kin selection. *Annual Review of Ecology and Systematics*, 13, 23–55.

MICHOD, R. E. 2000. *Darwinian Dynamics: Evolutionary Transitions in Fitness and Individuality*. Princeton, NJ: Princeton University Press.

MICHOD, R. E., NEDELCU, A. M., & ROZE, D. 2003. Cooperation and conflict in the evolution of individuality IV. Conflict mediation and evolvability in *Volvox carteri*. *BioSystems*, 69, 95–114.

MICHOD, R. E. & ROZE, D. 2000. Some aspects of reproductive mode and the origin of multicellularity. *Selection*, 1, 97–109.

MICHOD, R. E. & SANDERSON, M. J. 1985. Behavioural structure and the evolution of cooperation. *In*: GREENWOOD, J. J. & SLATKIN, M. (eds.) *Evolution - Essays in Honour of John Maynard*

Smith. Cambridge: Cambridge University Press.

MILLER, M. R., WHITE, A., & BOOTS, M. 2006. The evolution of parasites in response to tolerance in their hosts: the good, the bad, and apparent commensalism. *Evolution*, 60, 945–956.

MILUTINOVÍC, B., STOCK, M., GRASSE, A. V., NADERLINGER, E., HILBE, C., & CREMER, S. 2020. Social immunity modulates competition between coinfecting pathogens. *Ecology Letters*, 23, 565–574.

MISOF, B., LIU, S., MEUSEMANN, K., PETERS, R. S., DONATH, A., MAYER, C., FRANDSEN, P. B., WARE, J., FLOURI, T., BEUTEL, R. G., NIEHUIS, O., PETERSEN, M., IZQUIERDO-CARRASCO, F., WAPPLER, T., RUST, J., ABERER, A. J., ASPÖCK, U., ASPÖCK, H., BARTEL, D., BLANKE, A., . . ., & ZHOU, X. 2014. Phylogenomics resolves the timing and pattern of insect evolution. *Science*, 346, 763–767.

MITCHELL, S. D. 1995. The superorganism metaphor: then and now. *In*: MAASEN, S., MENDELSOHN, E., & WEINGART, P. (eds.) *Biology as Society, Society as Biology: Metaphors.* Dordrecht: Kluwer Academic Publishers.

MITMAN, G. 1988. From the population to society: the cooperative metaphors of W.C. Allee and A.E. Emerson. *Journal of the History of Biology*, 21, 173–194.

MIZUUCHI, R., FURUBAYASHI, T. & ICHIHASHI, N. 2022. Evolutionary transition from a single RNA replicator to a multiple replicator network. *Nature Communications*, 13:1460.

MOCK, D. W. 2004. *More Than Kin and Less Than Kind: The Evolution of Family Conflict.* Cambridge, MA: Harvard University Press.

MOCK, D. W. & PARKER, G. A. 1998. Siblicide, family conflict and the evolutionary limits of selfishness. *Animal Behaviour*, 56, 1–10.

MONAGHAN, P., MAKLAKOV, A. A., & METCALFE, N. B. 2020. Intergenerational transfer of aging: parental age and offspring lifespan. *Trends in Ecology and Evolution*, 35, 927–937.

MONAGHAN, P. & METCALFE, N. B. 2019. The deteriorating soma and the indispensable germline: gamete senescence and offspring fitness. *Proceedings of the Royal Society B*, 286, 20192187.

MONOD, J. L. 1970. *Chance and Necessity.* New York: Knopf.

MONOD, J. L. 1975. On the molecular theory of evolution. *In*: HARRÉ, R. (ed.) *Problems of Scientific Revolution: Progress and Obstacles to Progress in the Sciences.* Oxford: Clarendon Press.

MOORAD, J., PROMISLOW, D. & SILVERTOWN, J. 2019. Evolutionary ecology of senescence and a reassessment of Williams' 'extrinsic mortality' hypothesis. *Trends in Ecology & Evolution*, 34, 519–530.

MOORE, T. & HAIG, D. 1991. Genomic imprinting in mammalian development: a parental tug-of-war. *Trends in Genetics*, 7, 45–49.

MORAN, N. A. 1996. Accelerated evolution and Muller's rachet in endosymbiotic bacteria. *Proceedings of the National Academy of Sciences*, 93, 2873–2878.

MOREIRA, D. & LÓPEZ-GARCÍA, P. 2009. Ten reasons to exclude viruses from the tree of life. *Nature Reviews Microbiology*, 7, 306–311.

MORELOS-JUÁREZ, C., WALKER, T. N., LOPES, J. F. S., & HUGHES, W. O. H. 2010. Ant farmers practice proactive personal hygiene to protect their fungus crops. *Current Biology*, 20, R553–R554.

MORENO, A. & MOSSIO, M. 2015. *Biological Autonomy: A Philosophical and Theoretical Enquiry.* Dordrecht: Springer.

MORITZ, R. F. A. & SOUTHWICK, E. E. 1992. *Bees as Superorganisms: An Evolutionary Reality.* Berlin: Springer.

MORRELL, J. 1997. *Science at Oxford 1914–1939: Transforming an Arts University.* Oxford: Oxford University Press.

MULLER, H. J. 1949. Redintegration of the symposium on genetics, paleontology, and evolution. In: Jepsen, G.L., Mayr, E., & Simpson, G.G. (eds.) *Genetics, Paleontology and Evolution.* Princeton, NJ: Princeton University Press.

MULLER, H. J. 1960. The guidance of human evolution. *In*: TAX, S. (ed.) *Evolution After Darwin, Volume 2: The Evolution of Man.* Chicago, IL: University of Chicago Press.

MUNRO, C., SIEBERT, S., ZAPATA, F., HOWISON, M., DAMIAN-SERRANO, A., CHURCH, S. H., GOETZ, F. E., PUGH, P. R., HADDOCK, S. H. D., & DUNN, C. W. 2018. Improved phylogenetic resolution within Siphonophora (Cnidaria) with implications for trait evolution. *Molecular Phylogenetics and Evolution*, 127, 823–833.

MURRAY, E. A., BOSSERT, S., & DANFORTH, B. N. 2018. Pollinivory and the diversification of bees. *Biology Letters*, 14: 20180530.

NAGEL, M., QIU, B., BRANDENBORG, L. E., LARSEN, R. S., NING, D., BOOMSMA, J. J., & ZHANG, G. 2020. The gene expression network regulating queen brain remodeling after insemination and its parallel use in ants with reproductive workers. *Science Advances*, 6: eaaz5772.

NAGY, L. G., KOVÁCS, G. M., & KRIZSÁN, K. 2018. Complex multicellularity in fungi: evolutionary convergence, single origin, or both? *Biological Reviews*, 93, 1778–1794.

NAKAGAKI, T., YAMADA, H. & TÓTH, Á. 2000. Maze-solving by an amoeboid organism. *Nature*, 407, 470.

NALEPA, C. & BELL, W. J. 1997. Postovulation parental investment and parental care in cockroaches. *In*: CHOE, J. C. & CRESPI, B. J. (eds.) *Social Behavior in Insects and Arachnids*. Cambridge: Cambridge University Press.

NARANJO-ORTIZ, M. A. & GABALDÓN, T. 2019. Fungal evolution: major ecological adaptations and evolutionary transitions. *Biological Reviews*, 94, 1443–1476.

NASIR, A. & CAETANO-ANOLLÉS, G. 2015. A phylogenetic data-driven exploration of viral origins and evolution. *Science Advances*, 1: e1500527.

NEE, S. 1989. Does Hamilton's rule describe the evolution of reciprocal altruism? *Journal of Theoretical Biology*, 141, 81–91.

NESSE, R. M. 2008. Why a lot of people with selfish genes are pretty nice—except for their hatred of *the selfish gene*. *In*: GRAFEN, A. & RIDLEY, M. (eds.) *Richard Dawkins: How a Scientist Changed the Way We Think*. Oxford: Oxford University Press.

NESSE, R.M. 2009. Social selection and the origins of culture. *In*: SCHALLER, M., NORENZAYAN, A., HEINE, S. J., YAMAGISHI, T., & KAMEDA, T. (eds.) *Evolution, Culture, and the Human Mind*. Philadelphia, PA: Lawrence Erlbaum Associates.

NESSE, R. M. 2013. Tinbergen's four questions, organized: a response to Bateson and Laland. *Trends in Ecology & Evolution*, 28, 681–682.

NESSE, R. M. 2016. Social selection is a powerful explanation for prosociality. Open peer commentary on Richerson et al. (2016). *Behavioral and Brain Sciences*, 39: e47.

NEWMAN, S. A. 2022. Self-organization in embryonic development: myth and reality. *In*: DAMBRINCOURT MALASSÉ, A. (ed.) *Self-Organization as a New Paradigm in Evolutionary Biology*. Cham: Springer.

NICHOLSON, A. J. 1960. The role of population dynamics in natural selection. *In*: TAX, S. (ed.) *Evolution After Darwin, Volume I: The Evolution of Life*. Chicago, IL: Chicago University Press.

NICHOLSON, D. J. 2010. Biological atomism and cell theory. *Studies in History and Philosophy of Biological and Biomedical Sciences*, 41, 202–211.

NICHOLSON, D. J. 2013. Organisms ≠ machines. *Studies in History and Philosophy of Biological and Biomedical Sciences*, 44, 669–678.

NICHOLSON, D. J. 2014. The machine conception of the organism in development and evolution: a critical analysis. *Studies in History and Philosophy of Biological and Biomedical Sciences*, 48, 162–174.

NICHOLSON, D. J. 2019. Is the cell *really* a machine? *Journal of Theoretical Biology*, 477, 108–126.

NICHOLSON, D. J. & GAWNE, R. 2014. Rethinking Woodger's legacy in the philosophy of biology. *Journal of the History of Biology*, 47, 243–292.

NICHOLSON, D. J. & GAWNE, R. 2015. Neither logical empiricism nor vitalism, but organicism: what the philosophy of biology was. *History and Philosophy of the Life Sciences*, 37, 345–381.

NIEUWENHUIS, B. P. S., DEBETS, A. J. M., & AANEN, D. K. 2013. Fungal fidelity: nuclear divorce from a dikaryon by mating or monokaryon regeneration. *Fungal Biology*, 117, 261–267.

NIJHOUT, H. F. 1990. Metaphors and the role of genes in development. *BioEssays*, 12, 441–446.

NISHIMURA, N. J. & MANDOLI, D. F. 1992. Vegetative growth of *Acetabularia acetabulum* (Chlorophyta): structural evidence for juvenile and adult phases in development. *Journal of Phycology*, 28, 669–677.

NOË, R. & HAMMERSTEIN, P. 1995. Biological markets. *Trends in Ecology and Evolution*, 10, 336–339.

NOË, R. & KIERS, E. T. 2018. Mycorrhizal markets, firms, and co-ops. *Trends in Ecology and Evolution*, 33, 777–789.

NOH, S., GEIST, K. S., TIAN, X., STRASSMANN, J. E., & QUELLER, D. C. 2018. Genetic signatures of microbial altruism and cheating in social amoebas in the wild. *Proceedings of the National Academy of Sciences of the United States of America*, 115, 3096–3101.

NOLL, F. B., DA SILVA, M., SOLEMAN, R. A., LOPES, R. B., GRANDINETE, Y. C., ALMEIDA, E. A. B., WENZEL, J. W., & CARPENTER, J. M. 2021. Marimbondos: systematics, biogeography, and evolution of social behaviour of neotropical swarm-founding wasps (Hymenoptera: Vespidae: Epiponini). *Cladistics*, 37, 423–441.

NOLL, F. B. & WENZEL, J. W. 2008. Caste in the swarming wasps: 'queenless' societies in highly social insects. *Biological Journal of the Linnean Society*, 93, 509–522.

NONACS, P. 2019. Hamilton's rule is essential but insufficient for understanding monogamy's role in social evolution. *Royal Society Open Science*, 6: 180913.

NORMAN, A., HANSEN, L. H. & SØRENSEN, S. J. 2009. Conjugative plasmids: vessels of the communal gene pool. *Philosophical Transactions of the Royal Society B*, 364, 2275–2289.

NOWAK, M. A., TARNITA, C. E., & WILSON, E. O. 2010. The evolution of eusociality. *Nature*, 466, 1057–1062.

NUOTCLÀ, J. A., BIEDERMANN, P. H. W., & TABORSKY, M. 2019. Pathogen defence is a potential driver of social evolution in ambrosia beetles. *Proceedings of the Royal Society B*, 286, 20192332.

NURSE, P. 2021. Biology must generate ideas as well as data. *Nature*, 597, 305.

NYGAARD, S., HU, H., LI, C., SCHIØTT, M., CHEN, Z., YANG, Z., XIE, Q., MA, C., DENG, Y., DIKOW, R. B., RABELING, C., NASH, D. R., WCISLO, W. T., BRADY, S. G., SCHULTZ, T. R., ZHANG, G., & BOOMSMA, J. J. 2016. Reciprocal genomic evolution in the ant-fungus agricultural symbiosis. *Nature Communications*, 7, 12233.

ODENBAUGH, J. 2013. Searching for patterns, hunting for causes: Robert MacArthur, the mathematical naturalist. *In*: HARMAN, O. & DIETRICH, M. R. (eds.) *Outsider Scientists: Routes to Innovation in Biology*. Chicago, IL, Chicago University Press.

ODLING SMEE, J., LALAND, K. N. & FELDMAN, M. W. 2003. *Niche Construction: The Neglected Process in Evolution*, Princeton, NJ, Princeton University Press.

ODUM, E. P. 1971. *Fundamentals of Ecology*, 3rd edition. Philadelphia, PA: W. B. Saunders Company.

OKASHA, S. 2006. *Evolution and the Levels of Selection*. Oxford: Oxford University Press.

OLDROYD, B. P. 2002. The Cape honeybee: an example of a social cancer. *Trends in Ecology and Evolution*, 17, 249–251.

OLDROYD, B. P. & YAGOUND, B. 2021. Parent-of-origin effects, allele-specific expression, genomic imprinting and paternal manipulation in social insects. *Philosophical Transactions of the Royal Society B*, 376, 20200425.

OLDROYD, G. E. D. 2013. Speak, friend, and enter: signalling systems that promote beneficial symbiotic associations in plants. *Nature Reviews Microbiology*, 11, 252–263.

OLESON, A. & VOSS, J. (eds.) 1979. *The Organization of Knowledge in Modern America*. Baltimore, MD: Johns Hopkins University Press.

OSTER, G. F. & WILSON, E. O. 1978. *Caste and Ecology in the Social Insects*. Princeton, NJ: Princeton University Press.

OSTROM, E. 1990. *Governing the Commons: The Evolution of Institutions for Collective Action*. Cambridge: Cambridge University Press.

OSTROM, E., BURGER, J., FIELD, C. B., NORGAARD, R. B., & POLICANSKY, D. 1999. Revisiting the commons: local lessons, global challenges. *Science*, 284, 278–282.

OSTROWSKI, E. A. 2019. Enforcing cooperation in the social amoebae. *Current Biology*, 29, R474–R484.

OSTROWSKI, E. A., SHEN, Y., TIAN, X., SUCGANG, R., JIANG, H., QU, J., KATOH-KURASAWA, M., BROCK, D. A., DINH, C., LARA-GARDUNO, F., LEE, S. L., KOVAR, C. L., DINH, H. H., KORCHINA, V., JACKSON, L., PATIL, S., HAN, Y., CHABOUB, L., SHAULSKY, G., MUZNY, D. M., WORLEY, K. C., GIBBS, R. A., RICHARDS, S., KUSPA,

A., STRASSMANN, J. E., & QUELLER, D. C. 2015. Genomic signatures of cooperation and conflict in the social amoeba. *Current Biology*, 25, 1661–1665.

OTANI, S., CHALLINOR, V. L., KREUZENBECK, N. B., KILDGAARD, S., CHRISTENSEN, S. K., MUNK LARSEN, L. L., AANEN, D. K., RASMUSSEN, S. A., BEEMELMANNS, C., & POULSEN, M. 2019. Disease-free monoculture farming by fungus-growing termites. *Scientific Reports*, 9: 8819.

OTTO, S. P. & ROSALES, A. 2020. Theory in service of narratives in evolution and ecology. *American Naturalist*, 195, 290–299.

ÖZKAYA, Ö., BALBONTíN, R., GORDO, I., & XAVIER, K. B. 2018. Cheating on cheaters stabilizes cooperation in *Pseudomonas aeruginosa*. *Current Biology*, 28, 2070–2080.

PADIAN, K. 2009. Ten myths about Charles Darwin. *BioScience*, 59, 800–804.

PAGE, R. E. & AMDAM, G. V. 2007. The making of a social insect: developmental architectures of social design. *BioEssays*, 29, 334–343.

PAGE, R. E. & METCALF, R. A. 1982. Multiple mating, sperm utilization, and social evolution. *American Naturalist*, 119, 263–281.

PAGEL, M. 2012. *Wired for culture: origins of the human social mind*. New York: W. W. Norton & Company.

PAIS, A. 1997. *A Tale of Two Continents*. Princeton, NJ: Princeton University Press.

PÁL, C. & PAPP, B. 2000. Selfish cells threaten multicellular life. *Trends in Ecology and Evolution*, 15, 351–352.

PALEY, W. 1809. *Natural Theology*. London: J. Faulder.

PANDE, S. & VELICER, G. J. 2018. Chimeric synergy in natural social groups of a cooperative microbe. *Current Biology*, 28, 262–267.

PARKER, G. A. 1970. Sperm competition and its evolutionary consequences in the insects. *Biological Reviews*, 45, 525–567.

PARKER, G. A., BAKER, R. R., & SMITH, V. G. F. 1972. The origin and evolution of gamete dimorphism and the male-female phenomenon. *Journal of Theoretical Biology*, 36, 529–553.

PARKER, G. A. & MAYNARD SMITH, J. 1990. Optimality theory in evolutionary biology. *Nature*, 348, 27–33.

PARTRIDGE, L. & BARTON, N. H. 1993. Optimally, mutation and the evolution of ageing. *Nature*, 362, 305–311.

PARTRIDGE, L. & GEMS, D. 2002. Mechanisms of aging: public or private? *Nature Reviews Genetics*, 3, 165–175.

PATALANO, S., HORE, T. A., REIK, W., & SUMNER, S. 2012. Shifting behaviour: epigenetic reprogramming in eusocial insects. *Current Opinion in Cell Biology*, 24, 367–373.

PATALANO, S., VLASOVA, A., WYATT, C., EWELS, P., CAMARA, F., FERREIRA, P. G., ASHER, C. L., JURKOWSKI, T. P., SEGONDS-PICHON, A., BACHMAN, M., GONZÁLEZ-NAVARRETE, I., MINOCHE, A. E., KRUEGER, F., LOWY, E., MARCET-HOUBEN, M., RODRIGUEZ-ALES, J. L., NASCIMENTO, F. S., BALASUBRAMANIAN, S., GABALDON, T., TARVER, J. E., ANDREWS, S., HIMMELBAUER, H., HUGHES, W. O. H., GUIGÓ, R., REIK, W., & SUMNER, S. 2015. Molecular signatures of plastic phenotypes in two eusocial insect species with simple societies. *Proceedings of the National Academy of Sciences of the United States of America*, 112, 13970–13975.

PAULY, P. J. 1984. The appearance of academic biology in late nineteenth-century America. *Journal of the History of Biology*, 17, 369–397.

PEETERS, C. 2012. Convergent evolution of wingless reproductives across all subfamilies of ants, and sporadic loss of winged queens (Hymenoptera: Formicidae). *Myrmecological News*, 16, 75–91.

PEETERS, C. & ITO, F. 2015. Wingless and dwarf workers underlie the ecological success of ants (Hymenoptera: Formicidae). *Myrmecological News*, 21, 117–130.

PEMBERTON, J. M., ALBON, S. D., GUINNESS, F. E., CLUTTON-BROCK, T. H., & DOVER, G. A. 1992. Behavioral estimates of male mating success tested by DNA fingerprinting in a polygynous mammal. *Behavioral Ecology*, 3, 66–75.

PENN, D. C. & POVINELLI, D. J. 2007. On the lack of evidence that non-human animals possess anything remotely resembling a 'theory of mind'. *Philosophical Transactions of the Royal Society B*, 362, 731–744.

PEPPER, J. W. 2000. Relatedness in trait group models of social evolution. *Journal of Theoretical Biology*, 206, 355–368.

PERRARD, A., LOPEZ-OSORIO, F., & CARPENTER, J. M. 2016. Phylogeny, landmark analysis and the use of wing venation to study the evolution of social wasps (Hymenoptera: Vespidae: Vespinae). *Cladistics*, 32, 406–425.

PETERS, R. S., KROGMANN, L., MAYER, C., DONATH, A., GUNKEL, S., MEUSEMANN, K., KOZLOV, A., PODSIADLOWSKI, L., PETERSEN, M., LANFEAR, R., DIEZ, P. A., HERATY, J., KJER, K. M., KLOPFSTEIN, S., MEIER, R., POLIDORI, C., SCHMITT, T., LIU, S., ZHOU, X., WAPPLER, T., RUST, J., MISOF, B., & NIEHUIS, O. 2017. Evolutionary history of the Hymenoptera. *Current Biology*, 27, 1013–1018.

PETERSON, E. 2016. *The Life Organic*. Pittsburgh, PA: University of Pittsburgh Press.

PFAFF, D., TABANSKY, I., & HAUBENSAK, W. 2019. Tinbergen's challenge for the neuroscience of behavior. *Proceedings of the National Academy of Sciences of the United States of America*, 116, 9704–9710.

PHILIPPI, T. & SEGER, J. 1989. Hedging one's evolutionary bets, revisited. *Trends in Ecology and Evolution*, 4, 41–44.

PHILLIPS, J. A. 2009. Reproductive ecology of *Caulerpa taxifolia* (Caulerpaceae, Bryopsidales) in subtropical eastern Australia. *European Journal of Phycology*, 44, 81–88.

PIANKA, E. R. 1970. On *r*-and *K*-selection. *American Naturalist*, 104, 592–597.

PIEKARSKI, P. K., CARPENTER, J. M., LEMMON, A. R., LEMMON, E. M., & SHARANOWSKI, B. J. 2018. Phylogenomic evidence overturns current conceptions of social evolution in wasps (Vespidae). *Molecular Biology and Evolution*, 35, 2097–2109.

PIGLIUCCI, M. 2007. Do we need an extended evolutionary synthesis? *Evolution*, 61 2743–2749.

PIGLIUCCI, M. & FINKELMAN, L. 2014. The extended (evolutionary) synthesis debate: where science meets philosophy. *BioScience* 64, 511–516.

PINKER, S. 2002. *The Blank Slate*. London: Penguin Books.

PITTENDRIGH, C. S. 1958. Adaptation, natural selection, and behavior. *In*: ROE, A. & SIMPSON, G. G. (eds.) *Behavior and Evolution*. New Haven, CT: Yale University Press.

PLATT, J. R. 1964. Strong inference. *Science*, 146, 347–353.

POMIANKOWSKI, A. 1988. The evolution of female mating preferences for male genetic quality. *In*: HARVEY, P. H. & PARTRIDGE, L. E. (eds.) *Oxford Surveys in Evolutionary Biology*. Oxford: Oxford University Press.

POMIANKOWSKI, A., IWASA, Y., & NEE, S. 1991. The evolution of costly mate preferences I. Fisher and biased mutation. *Evolution*, 45, 1422–1430.

POPAT, R., CRUSZ, S. A., MESSINA, M., WILLIAMS, P., WEST, S. A., & DIGGLE, S. P. 2012. Quorum-sensing and cheating in bacterial biofilms. *Proceedings of the Royal Society B*, 279, 4765–4771.

POPPER, K.R. 1962. *Conjectures and Refutations: The Growth of Scientific Knowledge*. New York: Basic Books.

POPPER, K. R. 1975. The rationality of scientific revolutions. *In*: HARRÉ, R. (ed.) *Problems of Scientific Revolution: Progress and Obstacles to Progress in the Sciences*. Oxford: Clarendon Press.

POULSEN, M. & BOOMSMA, J. J. 2005. Mutualistic fungi control crop diversity in fungus-growing ants. *Science*, 307, 741–744.

PRADEU, T. 2012. *The Limits of the Self: Immunology and Biological Identity*. New York: Oxford University Press.

PRADEU, T. 2016. Organisms or biological individuals? Combining physiological and evolutionary individuality. *Biology and Philosophy*, 31, 797–817.

PRADEU, T., JAEGER, S., & VIVIER, E. 2013. The speed of change: towards a discontinuity theory of immunity. *Nature Reviews Immunology*, 13, 764–769.

PRADEU, T. & VIVIER, E. 2016. The discontinuity theory of immunity. *Science Immunology*, 1, aag0479.

PRANGISHVILI, D. 2013. The wonderful world of archaeal viruses. *Annual Review of Microbiology*, 67, 565–585.

PRESSMAN, A., BLANCO, C. & CHEN, I. A. 2015. The RNA World as a Model System to Study the Origin of Life. *Current Biology*, 25, R953–R963.

PREUSSGER, D., GIRI, S., MUHSAL, L. K., OÑA, L. & KOST, C. 2020. Reciprocal fitness feedbacks promote the evolution of mutualistic cooperation. *Current Biology*, 30, 3580–3590.

PRICE, G.R. 1970. Selection and covariance. *Nature* 227, 520–521.

PRICE, G. R. 1972. Fisher's 'fundamental theorem' made clear. *Annals of Human Genetics*, 36, 129–140.

PRIGOGINE, I. 1978. Time, structure, and fluctuations. *Science*, 201, 777–785.

PRIGOGINE, I. & NICOLIS, G. 1971. Biological order, structure and instabilities. *Quarterly Reviews of Biophysics*, 4, 107–148.

PROSS, A. 2012. *What is Life? How Chemistry Becomes Biology*. Oxford: Oxford University Press.

PROSSER, J. I. 2013. Think before you sequence. *Nature*, 494, 41.

PROVINE, W. B. 1971. *The Origins of Theoretical Population Genetics*. Chicago, IL: University of Chicago Press.

PROVINE, W. B. 1983. The development of Wright's theory of evolution: systematics, adaptation, and drift. *In*: GREENE, M. (ed.) *Dimensions of Darwinism*. Cambridge: Cambridge University Press.

PROVINE, W. B. 1986. *Sewall Wright and Evolutionary Biology*. Chicago, IL: University of Chicago Press.

PROVINE, W. B. 1988. Progress in evolution and meaning in life. *In*: NITECKI, M. H. (ed.) *Evolutionary Progress*. Chicago, IL: University of Chicago Press.

PROVINE, W. B. 1998. England. *In*: MAYR, E. & PROVINE, W. B. (eds.) *The Evolutionary Synthesis*. Cambridge, MA: Harvard University Press.

PULL, C. D. & MCMAHON, D. P. 2020. Superorganism immunity: a major transition in immune system evolution. *Frontiers in Ecology and Evolution*, 8: 186.

PULL, C. D., METZLER, S., NADERLINGER, E., & CREMER, S. 2018. Protection against the lethal side effects of social immunity in ants. *Current Biology*, 28, R1139–R1140.

QIU, B., DAI, X., LI, P., LARSEN, R., LI, R., PRICE, A., DING, G., TEXADA, M., ZHANG, X., ZUO, D., GAO, Q., JIANG, W., WEN, T., PONTIERI, L., GUO, C., REWITZ, K., LI, Q., LIU, W., BOOMSMA, J.J., & ZHANG, G. 2022. Canalized gene expression during development mediates caste differentiation in ants. *Nature Ecology & Evolution*, DOI: 10.1038/s41559-022-01884-y.

QIU, B., LARSEN, R. S., CHANG, N.-C., WANG, J., BOOMSMA, J. J., & ZHANG, G. 2018. Towards reconstructing the ancestral brain gene-network regulating caste differentiation in ants. *Nature Ecology & Evolution*, 2, 1782–1791.

QUELLER, D. C. 1984. Kin selection and frequency dependence: a game theoretic approach. *Biological Journal of the Linnean Society*, 23, 133–143.

QUELLER, D. C. 1985. Kinship, reciprocity and synergism in the evolution of social behaviour. *Nature*, 318, 366–367.

QUELLER, D. C. 1992a. A general model for kin selection. *Evolution*, 46, 376–380.

QUELLER, D. C. 1992b. Quantitative genetics, inclusive fitness, and group selection. *American Naturalist*, 139, 540–558.

QUELLER, D. C. 1994. Male-female conflict and parent-offspring conflict. *American Naturalist*, 144, S84–S99.

QUELLER, D. C. 1995. The spaniels of St Marx and the Panglossian paradox: a critique of a rhetorical programme. *Quarterly Review of Biology*, 70, 485–489.

QUELLER, D. C. 1997. Cooperators since life began. *Quarterly Review of Biology*, 72, 184–188.

QUELLER, D. C. 2000. Relatedness and the fraternal major transitions. *Philosophical Transactions of the Royal Society B*, 355, 1647–1655.

QUELLER, D. C. 2003. Theory of genomic imprinting conflict in social insects. *BMC Evolutionary Biology*, 3: 15.

QUELLER, D. C. 2008. The social side of wild yeast. *Nature*, 456, 589–590.

QUELLER, D. C. 2011. Expanded social fitness and Hamilton's rule for kin, kith, and kind. *Proceedings National Academy of Sciences of the United States of America*, 108, 10792–10799.

QUELLER, D. C. 2014. Joint phenotypes, evolutionary conflict and the fundamental theorem of natural selection. *Philosophical Transactions of the Royal Society B*, 369, 20130423.

QUELLER, D. C. 2017. Fundamental theorems of evolution. *American Naturalist*, 189, 345–353.

QUELLER, D. C. 2019. What life is for: a commentary on Fromhage and Jennions. *Proceedings of the Royal Society B*, 286, 20191060.

QUELLER, D. C., PONTE, E., BOZZARO, S., & STRASSMANN, J. E. 2003. Single-gene green-beard effects in the social amoeba *Dictyostelium discoideum. Science,* 299, 105–106.

QUELLER, D. C. & STRASSMANN, J. E. 2002. The many selves of social insects. *Science,* 296, 311–313.

QUELLER, D. C. & STRASSMANN, J. E. 2009. Beyond society: the evolution of organismality. *Philosophical Transactions of the Royal Society B,* 364, 3143–3155.

QUELLER, D. C., STRASSMANN, J. E., & HUGH-ES, C. R. 1988. Genetic relatedness in colonies of tropical wasps with multiple queens. *Science,* 242, 1155–1157.

QUIÑONES, A. E. & PEN, I. 2017. A unified model of hymenopteran preadaptations that trigger the evolutionary transition to eusociality. *Nature Communications,* 8: 15920.

RADZVILAVICIUS, A. L., HADJIVASILIOU, Z., POMIANKOWSKI, A., & LANE, N. 2016. Selection for mitochondrial quality drives evolution of the germline. *PLoS Biology,* 14: e2000410.

RAHWAN, I., CEBRIAN, M., OBRADOVICH, N., BONGARD, J., BONNEFON, J.-F., BREAZEAL, C., CRANDALL, J. W., CHRISTAKIS, N. A., COUZIN, I. D., JACKSON, M. O., JEN-NINGS, N. R., KAMAR, E., KLOUMANN, I. M., LAROCHELLE, H., LAZER, D., MCELREATH, R., MISLOVE, A., PARKES, D. C., PENTLAND, A. S., ROBERTS, M. E., SHARIFF, A., TENEN-BAUM, J. B., & WELLMAN, M. 2019. Machine behaviour. *Nature,* 568, 477–486.

RAINEY, P. B. & KERR, B. 2011. Conflict among levels of selection as fuel for the evolution of individuality. *In:* CALCOTT, B. & STERELNY, K. (eds.) *The Major Transitions in Evolution Revisited.* Cambridge MA: MIT Press.

RAJAKUMAR, R., KOCH, S., COUTURE, M., FAVÉ, M.-J., LILLICO-OUACHOUR, A., CHEN, T., DE BLASIS, G., RAJAKUMAR, A., OUEL-LETTE, D., & ABOUHEIF, E. 2018. Social regulation of a rudimentary organ generates complex worker-caste systems in ants. *Nature,* 562, 574–577.

RANG, C. U., PENG, A. Y., & CHAO, L. 2011. Temporal dynamics of bacterial aging and rejuvenation. *Current Biology,* 21, 1813–1816.

RAO, V. & NANJUNDIAH, V. 2011. J.B.S. Haldane, Ernst Mayr and the beanbag genetics dispute. *Journal of the History of Biology,* 44, 233–281.

RATCLIFF, W. C., DENISON, R. F., BORRELLO, M., & TRAVISANO, M. 2012. Experimental evolution of multicellularity. *Proceedings of the National Academy of Sciences of the United States of America,* 109, 1595–1600.

RATCLIFF, W. C., HERRON, M. D., HOWELL, K., PENTZ, J. T., ROSENZWEIG, F., & TRAVISANO, M. 2013. Experimental evolution of an alternating uni- and multicellular life cycle in *Chlamydomonas reinhardtii. Nature Communications,* 4: 2742.

RATNIEKS, F. L. W. 1988. Reproductive harmony via mutual policing by workers in eusocial Hymenoptera. *American Naturalist,* 132, 217–236.

RATNIEKS, F. L. W., FOSTER, K. R., & WENSE-LEERS, T. 2006. Conflict resolution in insect societies. *Annual Review of Entomology,* 51, 581–608.

RATNIEKS, F. L. W., FOSTER, K. R., & WENSE-LEERS, T. 2011. Darwin's special difficulty: the evolution of "neuter insects" and current theory. *Behavioral Ecology and Sociobiology,* 65, 481–492.

RATNIEKS, F. L. W. & REEVE, H. K. 1992. Conflict in single-queen hymenopteran societies: the structure of conflict and processes that reduce conflict in advanced eusocial species. *Journal of Theoretical Biology,* 158, 33–65.

RATNIEKS, F. L. W. & VISSCHER, P. K. 1989. Worker policing in the honeybee. *Nature,* 342, 796–797.

RAYMOND, B., WEST, S. A., GRIFFIN, A. S., & BONSALL, M. B. 2012. The dynamics of cooperative bacterial virulence in the field. *Science,* 337, 85–88.

REEVE, H. K. 1991. *Polistes.* In: ROSS, K. G. & MATHEW, R. W. (eds.) *The Social Biology of Wasps.* Ithaca, NY: Cornell University Press.

REGENBERG, B., EBBESEN HANGHØJ, K., SCHERZ ANDERSEN, K., & BOOMSMA, J. J. 2016. Clonal yeast biofilms can reap competitive advantages through cell differentiation without being obligatory multicellular. *Proceedings of the Royal Society B,* 283, 20161303.

REMNANT, E. J., ASHE, A., YOUNG, P. E., BUCH-MANN, G., BEEKMAN, M., ALLSOPP, M. H., SUTER, C. M., DREWELL, R. A., & OLDROYD, B. P. 2016. Parent-of-origin effects on genome-wide DNA methylation in the Cape honey bee (*Apis mellifera capensis*) may be confounded by allele-specific methylation. *BMC Genomics*, 17: 226.

RENDUELES, O., AMHERD, M., & VELICER, G. J. 2015a. Positive frequency-dependent interference competition maintains diversity and pervades a natural population of cooperative microbes. *Current Biology*, 25, 1673–1681.

RENDUELES, O., ZEE, P. C., DINKELACKER, I., AMHERD, M., WIELGOSS, S., & VELICER, G. J. 2015b. Rapid and widespread de novo evolution of kin discrimination. *Proceedings of the National Academy of Sciences of the United States of America*, 112, 9076–9081.

REVELY, L., SUMNER, S., & EGGLETON, P. 2021. The plasticity and developmental potential of termites. *Frontiers in Ecology and Evolution*, 9: 552624.

REYNOLDS, A. S. 2018. *The Third Lens: Metaphor and the Creation of Modern Cell Biology*. Chicago, IL: University of Chicago Press.

REYNOLDS, A. 2007. The theory of the cell state and the question of cell autonomy in nineteenth and early twentieth-century biology. *Science in Context*, 20, 71–95.

REYNOLDS, A. 2008. Ernst Haeckel and the theory of the cell state: remarks on the history of a bio-political metaphor. *History of Science*, 46, 123–152.

REZNICK, D. N. 2011. *The Origin Then and Now: An Interpretive Guide to the Origin of Species*. Princeton, NJ: Princeton University Press.

REZNICK, D. & TRAVIS, J. 1996. The empirical study of adaptation in natural populations. *In*: ROSE, M. R. & LAUDER, G. V. (eds.) *Adaptation*. San Diego, CA: Academic Press.

RICHARDS, R. J. 1987. *Darwin and the Emergence of Evolutionary Theories of Mind and Behavior*. Chicago, IL: University of Chicago Press.

RICHARDSON, T. O., KAY, T., BRAUNSCHWEIG, R., JOURNEAU, O. A., RÜEGG, M., MCGREGOR, S., DE LOS RIOS, P., & KELLER, L. 2021. Ant behavioral maturation is mediated by a stochastic transition between two fundamental states. *Current Biology*, 31, 2253–2260.

RICHERSON, P.J., BALDINI, R., BELL, A. V., DEMPS, K., FROST, K., HILLIS, V., MATHEW, S., NEWTON, E. K., NAAR, N., NEWSON, L., ROSS, C., SMALDINO, P. E., WARING, T. M., & ZEFFERMAN, M. 2016. Cultural group selection plays an essential role in explaining human cooperation: a sketch of the evidence. *Behavioral and Brain Sciences*, 39: e30.

RICHERSON, P. J., BOYD, R., & HENRICH, J. 2003. The cultural evolution of human cooperation. *In*: HAMMERSTEIN, P. (ed.) *The Genetic and Cultural Evolution of Cooperation*. Cambridge, MA: MIT Press.

RICHMOND, M. L. 2006. The 1909 Darwin Celebration: reexamining evolution in the light of Mendel, mutation, and meiosis. *Isis*, 97, 447–484.

RIDLEY, M. 2004. *Evolution*, 3rd edition. Oxford: Blackwell Science Ltd.

RIEHL, C. 2013. Evolutionary routes to non-kin cooperative breeding in birds. *Proceedings of the Royal Society B*, 280, 20132245.

RILEY, M. A. & WERTZ, J. E. 2002. Bacteriocins: evolution, ecology, and application. *Annual Review of Microbiology*, 56, 117–137.

RITTSCHOF, C. C. & ROBINSON, G. E. 2016. Behavioral genetic toolkits: towards the evolutionary origins of complex phenotypes. *Current Topics in Developmental Biology*, 119, 157–204.

ROBERTS, W.C. 1944. Multiple mating of queen bees proved by progeny and flight tests. *Gleanings in Bee Culture* 72, 255–259, 303.

RODRIGUES, A. M. M. & GARDNER, A. 2021. Reproductive value and the evolution of altruism. *Trends in Ecology and Evolution*, 37, 346–358.

ROGER, A. J., MUÑOZ-GÓMEZ, S. A., & KAMIKAWA, R. 2017. The origin and diversification of mitochondria. *Current Biology*, 27, R1177–R1192.

ROGER, A. J., SUSKO, E., & LEGER, M. M. 2021. Evolution: reconstructing the timeline of eukaryogenesis. *Current Biology*, 31, R193–R196.

ROKAS, A. 2008a. The origins of multicellularity and the early history of the genetic toolkit for animal development. *Annual Review of Genetics*, 42, 235–251.

ROKAS, A. 2008b. The molecular origins of multicellular transitions. *Current Opinion in Genetics and Development*, 18, 472–478.

ROMERALO, M., SKIBA, A., GONZALEZ-VOYER, A., SCHILDE, C., LAWAL, H., KEDZIORA, S., CAVENDER, J. C., GLÖCKNER, G., URUSHIHARA, H., & SCHAAP, P. 2013. Analysis of phenotypic evolution in Dictyostelia highlights developmental plasticity as a likely consequence of colonial multicellularity. *Proceedings of the Royal Society B*, 280: 20130976.

ROMIGUIER, J., CAMERON, S. A., WOODART, S. H., FISCHMAN, B. J., & KELLER, L. 2015. Phylogenomics controlling for base compositional bias reveals a single origin of eusociality in corbiculate bees. *Molecular Biology and Evolution*, 33, 670–678.

ROOT-BERNSTEIN, M. & ROOT-BERNSTEIN, R. 2015. The ribosome as a missing link in the evolution of life. *Journal of Theoretical Biology*, 367, 130–158.

ROPER, M., ELLISON, C., TAYLOR, J. W., & GLASS, N. L. 2011. Nuclear and genome dynamics in multinucleate ascomycete fungi. *Current Biology*, 21, R786–R793.

ROSE, M. R. 1991. *Evolutionary Biology of Aging*. Oxford: Oxford University Press.

ROSENGAUS, R. B., MALAK, T., & MACKINTOSH, C. 2013. Immune-priming in ant larvae: social immunity does not undermine individual immunity. *Biology Letters*, 9: 20130563.

ROSENGAUS, R. B., TRANIELLO, J. F. A., & BULMER, M. S. 2011. Ecology, behavior and evolution of disease resistance in termites. *In*: BIGNELL, D. E., ROISIN, Y., & LO, N. (eds.) *Biology of Termites: A Modern Synthesis*. Dordrecht: Springer.

ROSS-GILLESPIE, A., GARDNER, A., WEST, S. A., & GRIFFIN, A. S. 2007. Frequency dependence and cooperation: theory and a test with bacteria. *American Naturalist*, 170, 331–342.

ROUSSET, F. 2004. *Genetic Structure and Selection in Subdivided Populations*. Princeton, NJ: Princeton University Press.

ROUSSET, F. 2015. Regression, least squares, and the general version of inclusive fitness. *Evolution*, 69, 2963–2970.

RUBENSTEIN, D. R. & ABBOT, P. (eds.) 2017. *Comparative Social Evolution*. Cambridge: Cambridge University Press.

RUECKERT, S. & LEANDER, B. S. 2008. Morphology and molecular phylogeny of *Haplozoon praxillellae* n. sp. (Dinoflagellata): a novel intestinal parasite of the maldanid polychaete *Praxillella pacifica* Berkeley. *European Journal of Protistology*, 44, 299–307.

RUIZ-GONZÁLEZ, M. X., BRYDEN, J., MORET, Y., REBER-FUNK, C., SCHMID-HEMPEL, P. & BROWN, M. J. F. 2012. Dynamic transmission, host quality, and population structure in a multi-host parasite of bumblebees. *Evolution*, 66, 3053–3066.

RUIZ-TRILLO, I. & NEDELCU, A. M. (eds.) 2015. *Evolutionary Transitions to Multicellular Life: Principles and Mechanisms*. Dordrecht: Springer.

RUMBAUGH, K. P., DIGGLE, S. P., WATTERS, C. M., ROSS-GILLESPIE, A., GRIFFIN, A. S., & WEST, S. A. 2009. Quorum sensing and the social evolution of bacterial virulence. *Current Biology*, 19, 341–345.

RUSE, M. 1988. Molecules to men: evolutionary biology and thoughts of progress. *In*: NITECKI, M. H. (ed.) *Evolutionary Progress*. Chicago, IL: Chicago University Press.

RUSE, M. 2003. *Darwin and Design: Does Evolution Have a Purpose?* Cambridge, MA: Harvard University Press.

RUSSELL, B. 1945. *A History of Western Philosophy*. New York: Simon and Schuster.

RUSSELL, E. S. 1916. *Form and Function: A Contribution to the History of Animal Morphology*. London: John Murray.

RUST, J. & WAPPLER, T. 2016. Palaeontology: the point of no return in the fossil record of eusociality. *Current Biology*, 26, R159–R161.

RYAN, M. J. 2009. The evolution of behavior, and integrating it towards a complete and correct understanding of behavioral biology. *In*: BOLHUIS, J. J. & VERHULST, S. (eds.) *Tinbergen's Legacy*. Cambridge: Cambridge University Press.

SACHS, J. L. & SIMMS, E. L. 2006. Pathways to mutualism breakdown. *Trends in Ecology & Evolution*, 21, 585–592.

SACHS, J. L., MUELLER, U. G., WILCOX, T. P., & BULL, J. J. 2004. The evolution of cooperation. *Quarterly Review of Biology*, 79, 135–160.

SAGAN, L. 1967. On the origin of mitosing cells. *Journal of Theoretical Biology*, 14, 225–274.

SANN, M., MEUSEMANN, K., NIEHUIS, O., ESCALONA, H.E., MOKROUSOV, M., OHL, M., PAULI, T., SCHMID-EGGER, C. 2021. Reanalysis of the apoid wasp phylogeny with additional taxa and sequence data confirms the placement of Ammoplanidae as sister to bees. *Systematic Entomology* 46, 558–569.

SANTELICES, B. 2004. Mosaicism and chimerism as components of intraorganismal genetic heterogeneity. *Journal of Evolutionary Biology*, 17, 1187–1188.

SANTORELLI, L. A., THOMPSON, C. R. L., VILLEGAS, E., SVETZ, J., DINH, C., PARIKH, A., SUCGANG, R., KUSPA, A., STRASSMANN, J. E., QUELLER, D. C., & SHAULSKY, G. 2008. Facultative cheater mutants reveal the genetic complexity of cooperation in social amoebae. *Nature*, 451, 1107–1110.

SANTOS, F. P., SANTOS, F. C., & PACHECO, J. M. 2018. Social norm complexity and past reputations in the evolution of cooperation. *Nature*, 555, 242–245.

SCHAAP, P., WINCKLER, T., NELSON, M., ALVAREZ-CURTO, E., ELGIE, B., HAGIWARA, H., CAVENDER, J., MILANO-CURTO, A., ROZEN, D. E., DINGERMANN, T., MUTZEL, R., & BALDAUF, S. L. 2006. Molecular phylogeny and evolution of morphology in the social amoebas. *Science*, 314, 661–663.

SCHACHT, R. & KRAMER, K. L. 2019. Are we monogamous? A review of the evolution of pair-bonding in humans and its contemporary variation cross-culturally. *Frontiers in Ecology and Evolution*, 7: 230.

SCHEURING, I. & YU, D. W. 2012. How to assemble a beneficial microbiome in three easy steps. *Ecology Letters*, 15, 1300–1307.

SCHILDE, C., LAWAL, H. M., KIN, K., SHIBANO-HAYAKAWA, I., INOUYE, K., & SCHAAP, P. 2019. A well supported multi gene phylogeny of 52 dictyostelia. *Molecular Phylogenetics and Evolution*, 134, 66–73.

SCHLISSEL, G., KRZYZANOWSKI, M. K., CAUDRON, F., BARRAL, Y., & RINE, J. 2017. Aggregation of the Whi3 protein, not loss of heterochromatin, causes sterility in old yeast cells. *Science*, 355, 1184–1187.

SCHMID-HEMPEL, P. 1998. *Parasites in Social Insects*. Princeton, NJ: Princeton University Press.

SCHMIDT, C. 2013. Molecular phylogenies of ponerine ants (Hymenoptera: Formicidae: Ponerinae). *Zootaxa*, 3647, 201–250.

SCHOEMAKER, P. J. H. 1991. The quest for optimality: a positive heuristic of science? *Behavioral and Brain Sciences*, 14, 205–215.

SCHOETERS, E. & BILLEN, J. 2000. The importance of the spermathecal duct in bumblebees. *Journal of Insect Physiology*, 46, 1303–1312.

SCHRÖDINGER, E. 1944. *What is Life? The Physical Aspect of the Living Cell*. Cambridge: Cambridge University Press.

SCHUMPETER, J. A. 1926. *The Theory of Economic Development*. Cambridge, MA: Harvard University Press.

SCHWARZ, M. P., BULL, N. J., & HOGENDOORN, K. 1998. Evolution of sociality in the allodapine bees: a review of sex allocation, ecology and evolution. *Insectes Sociaux*, 45, 349–368.

SCHWILLE, P. 2011. Bottom-up synthetic biology: engineering in a tinkerer's world. *Science*, 333, 1252–1254.

SCOTT, T. W. & WEST, S. A. 2019. Adaptation is maintained by the parliament of genes. *Nature Communications*, 10: 5163.

SCOTT-PHILLIPS, T. C., DICKINS, T. E., & WEST, S. A. 2011. Evolutionary theory and the ultimate-proximate distinction in the human behavioral sciences. *Perspectives on Psychological Science*, 6, 38–47.

SEBÉ-PEDRÓS, A., DEGNAN, B. M., & RUIZ-TRILLO, I. 2017. The origin of Metazoa: a unicellular perspective. *Nature Reviews Genetics*, 18, 498–512.

SEELEY, T. D. 1989. The honey bee colony as a superorganism. *American Scientist*, 77, 546–553.

SEGER, J. 1981. Kinship and covariance. *Journal of Theoretical Biology*, 91, 191–213.

SEGER, J. & STUBBLEFIELD, J. W. 1996. Optimization and adaptation. *In*: ROSE, M. R. & LAUDER, G. V. (eds.) *Adaptation*. San Diego, CA: Academic Press.

SEGERSTRÅLE, U. 2000. *Defenders of the Truth: The Battle for Science in the Sociobiology Debate and Beyond*, Oxford; New York, Oxford University Press.

SEGERSTRALE, U. 2013. *Nature's Oracle*. Oxford: Oxford University Press.

SELLA, G. & HIRSH, A. E. 2005. The application of statistical physics to evolutionary biology. *Proceedings of the National Academy of Sciences of the United States of America*, 102, 9541–9546.

SHAKIBA, N., FAHMY, A., JAYAKUMARAN, G., MCGIBBON, S., DAVID, L., TRCKA, D., ELBAZ,

J., PURI, M. C., NAGY, A., VAN DER KOOY, D., GOYAL, S., WRANA, J. L., & ZANDSTRA, P. W. 2019. Cell competition during reprogramming gives rise to dominant clones. *Science*, 364, 354 and eaan0925.

SHANNON, C. E. 1948. A mathematical theory of communication. *Bell Systems Technology Journal*, 27, 379–423, 623–656.

SHAULSKY, G. & KESSIN, R. H. 2007. The cold war of the social amoebae. *Current Biology*, 17, R684–R692.

SHEIKH, S., THULIN, M., CAVENDER, J. C., ESCALANTE, R., KAWAKAMI, S., LADO, C., LANDOLT, J. C., NANJUNDIAH, V., QUELLER, D. C., STRASSMANN, J. E., SPIEGEL, F. W., STEPHENSON, S. L., VADELL, E. M., & BALDAUF, S. L. 2018. A New classification of the dictyostelids. *Protist*, 169, 1–28.

SHEN, S.-F., EMLEN, S. T., KOENIG, W. D., & RUBENSTEIN, D. R. 2017. The ecology of cooperative breeding behaviour. *Ecology Letters*, 20, 708–720.

SHERMAN, P. W. 1977. Nepotism and the evolution of alarm calls. *Science*, 197, 1246–1253.

SHERMAN, P. W., JARVIS, J. U. M., & ALEXANDER, R. D. (eds.) 1991. *The Biology of the Naked Mole Rat*. Princeton, NJ: Princeton University Press.

SHERMAN, P. W., LACEY, E. A., REEVE, H. K., & KELLER, L. 1995. The eusociality continuum. *Behavioral Ecology*, 6, 102–108.

SHERWIN, W. B., CHAO, A., JOST, L., & SMOUSE, P. E. 2017. Information theory broadens the spectrum of molecular ecology and evolution. *Trends in Ecology and Evolution*, 32, 948–963.

SHI, H. & HUANG, K. C. 2019. Chromosome organization: making room in a crowd. *Current Biology*, 29, R630–R632.

SHINTANI, M., SANCHEZ, Z. K., & KIMBARA, K. 2015. Genomics of microbial plasmids: classification and identification based on replication and transfer systems and host taxonomy. *Frontiers in Microbiology*, 6: 242.

SIBLY, R. M. & CALOW, P. 1985. Classification of habitats by selection pressures: a synthesis of life-cycle and r/K theory. *In*: SIBLY, R. M. & SMITH, R. H. (eds.) *Behavioural Ecology: Ecological consequences of adaptive behaviour*. Oxford, England: Blackwell Scientific Publications.

SIBLY, R. M. & SMITH, R. H. (eds.) 1985. *Behavioural Ecology: Ecological consequences of adaptive behaviour*, Oxford, England: Blackwell Scientific Publications.

SILBERFELD, T., LEIGH, J. W., VERBRUGGEN, H., CRUAUD, C., DE REVIERS, B., & ROUSSEAU, F. 2010. A multi-locus time-calibrated phylogeny of the brown algae (Heterokonta, Ochrophyta, Phaeophyceae): investigating the evolutionary nature of the "brown algal crown radiation". *Molecular Phylogenetics and Evolution*, 56, 659–674.

SIMPSON, C., HERRERA-CUBILLA, A., & JACKSON, J. B. C. 2020. How colonial animals evolve. *Science Advances*, 6, eaaw9530.

SIMPSON, G. G. 1941. The role of the individual in evolution. *Journal of the Washington Academy of Sciences*, 31, 1–20.

SIMPSON, G. G. 1944. *Tempo and Mode in Evolution*. New York: Columbia University Press.

SIMPSON, G. G. 1949. *The Meaning of Evolution*. New Haven, CT: Yale University Press.

SIMPSON, G. G. 1953. *The Major Features of Evolution*. New York: Columbia University Press.

SINGH, M., GLOWACKI, L., & WRANGHAM, R. 2016. Self-interested agents create, maintain, and modify group-functional culture. Open peer commentary on Richerson et al. (2016). *Behavioral and Brain Sciences*, 39: e52.

SKIPPER, R. A. 2009. Revisiting the Fisher–Wright controversy. *In*: CAIN, J. & RUSE, M. (eds.) Descended from Darwin: Insights into the History of Evolutionary Studies, 1900–1970. Philadelphia, PA: American Philosophical Society.

SLACK, J. M. W. 2002. Conrad Hal Waddington: the last renaissance biologist? *Nature Reviews Genetics*, 3, 889–895.

SLEIGH, C. 2004. The ninth mortal sin: the Lamarckism of W.M. Wheeler. *In*: LUSTIG, A., RICHARDS, R. W. & RUSE, M. (eds.) *Darwinian Heresies*. Cambridge: Cambridge University Press.

SMITH, J., VAN DIJKEN, J. D., & ZEE, P. C. 2010. A generalization of Hamilton's rule for the evolution of microbial cooperation. *Science*, 328, 1700–1703.

SMITH, P. & SCHUSTER, M. 2019. Public goods and cheating in microbes. *Current Biology*, 29, R442–R447.

SMITH, S. M., KENT, D. S., BOOMSMA, J. J., & STOW, A. J. 2018. Monogamous sperm storage and permanent worker sterility in a long-lived ambrosia beetle. *Nature Ecology & Evolution*, 2, 1009–1018.

SMITH, W. P. J., BRODMANN, M., UNTERWEGER, D., DAVIT, Y., COMSTOCK, L. E., BASLER, M., & FOSTER, K. R. 2020. The evolution of tit-for-tat in bacteria via the type VI secretion system. *Nature Communications*, 11: 5395.

SMOCOVITIS, V. B. 1994. Disciplining evolutionary biology: Ernst Mayr and the founding of the Society for the Study of Evolution and *Evolution* (1939–1950). *Evolution*, 48, 1–8.

SMOCOVITIS, V. B. 1996. *Unifying Biology: The Evolutionary Synthesis and Evolutionary Biology*. Princeton, NJ: Princeton University Press.

SMOCOVITIS, V. B. 1999. The 1959 Darwin Centennial Celebration in America. *Osiris*, 14, 274–323.

SMOCOVITIS, V. B. 2016. The unifying vision: Julian Huxley, evolutionary humanism, and the evolutionary synthesis. *In*: SOMSEN, G. & KAMMINGA, H. (eds.) *Pursuing the Unity of Science: Ideology and Scientific Practice Between the Great War and the Cold War*. London: Routledge.

SMUKALLA, S., CALDARA, M., POCHET, N., BEAUVAIS, A., GUADAGNINI, S., YAN, C., VINCES, M. D., JANSEN, A., PREVOST, M. C., LATGÉ, J.-P., FINK, G. R., FOSTER, K. R. & VERSTREPEN, K. J. 2008. *FLO1* is a variable green beard gene that drives biofilm-like cooperation in budding yeast. *Cell*, 135, 726–737.

SOBER, E. 1984. *The Nature of Selection*. Cambridge, MA: MIT Press.

SOBER, E. 1991. Extremum descriptions, process laws and minimality heuristics. Open peer commentary on Schoemaker (1991). *Behavioral and Brain Sciences*, 14, 232–233.

SOBER, E. 2015. *Ockham's Razors: A User's Manual*. Cambridge: Cambridge University Press.

SOBER, E. & WILSON, D. S. 1998. *Unto Others*. Cambridge, MA: Harvard University Press.

SOJO, V., HERSCHY, B., WHICHER, A., CAMPRUBÍ, E., & LANE, N. 2016. The origin of life in alkaline hydrothermal vents. *Astrobiology*, 16, 181–197.

SOJO, V., POMIANKOWSKI, A., & LANE, N. 2014. A bioenergetic basis for membrane divergence in Archaea and Bacteria. *PLoS Biology*, 12: e1001926.

SÖLCH, D. 2016. Wheeler and Whitehead: process biology and process philosophy in the early twentieth century. *Journal of the History of Ideas*, 77, 489–507.

SONNENSCHEIN, C. & SOTO, A. M. 2020. Over a century of cancer research: inconvenient truths and promising leads. *PLoS Biology*, 18: e3000670.

SPANG, A., SAW, J. H., JORGENSEN, S. L., ZAREMBA-NIEDZWIEDZKA, K., MARTIJN, J., LIND, A. E., VAN EIJK, R., SCHLEPER, C., GUY, L., & ETTEMA, T. J. G. 2015. Complex archaea that bridge the gap between prokaryotes and eukaryotes. *Nature*, **521**, 173–179.

SPEARE, L., CECERE, A. G., GUCKES, K. R., SMITH, S., WOLLENBERG, M. S., MANDEL, M. J., MIYASHIRO, T., & SEPTER, A. N. 2018. Bacterial symbionts use a type VI secretion system to eliminate competitors in their natural host. *Proceedings of the National Academy of Sciences of the United States of America*, 115, E8528–E8537.

SPENCER, H. 1881. *The Principles of Sociology, Volume 1*. New York: Appleton and Company.

SPERBER, D. & BAUMARD, N. 2012. Moral reputation: an evolutionary and cognitive perspective. *Mind and Language*, 27, 495–518.

SPERBER, D. 1996. *Explaining Culture: A Naturalistic Approach*. Oxford: Blackwell.

SPRINGER, S. A., CRESPI, B. J., & SWANSON, W. J. 2011. Beyond the phenotypic gambit: molecular behavioural ecology and the evolution of genetic architecture. *Molecular Ecology*, 20, 2240–2257.

STAJICH, J. E., BERBEE, M. L., BLACKWELL, M. E., HIBBETT, D. S., JAMES, T. Y., SPATAFORA, J. W., & TAYLOR, J. W. 2009. The fungi. *Current Biology* 19, R840–R845.

STANIER, R. Y. & VAN NIEL, C. B. 1962. The concept of a bacterium. *Archiv für Mikrobiologie*, 42, 17–35.

STEARNS, S. C. 1976. Life-history tactics: a review of ideas. *Quarterly Review of Biology*, 51, 3–47.

STEARNS, S. C. (ed.) 1987. *The Evolution of Sex and its Consequences*. Basel: Birkhäuser.

STEARNS, S. C. 1992. *The Evolution of Life Histories*. Oxford: Oxford University Press.

STEARNS, S. C. 2002. Less would have been more. *Evolution*, 56, 2339–2345.

STEARNS, S. C. 2007. Are we stalled part way through a major evolutionary transition from individual to group? *Evolution*, 61, 2275–2280.

STEARNS, S. C. 2008. How the European Society for Evolutionary Biology and the *Journal of Evolutionary Biology* were founded. *Journal of Evolutionary Biology*, 21, 1449–1451.

STEARNS, S. C. 2011. *George Christopher Williams 1926–2010. Biographical Memoir*. Washington, DC: National Academy of Sciences.

STEARNS, S. C. & HOEKSTRA, R. F. 2005. *Evolution: An Introduction*. Oxford: Oxford University Press.

STEARNS, S. C. & MEDZHITOV, R. 2016. *Evolutionary Medicine*. Oxford: Oxford University Press.

STEBBINS, G. L. 1950. *Variation and Evolution in Plants*. New York: Columbia University Press.

STEIDINGER, B. S. & BEVER, J. D. 2016. Host discrimination in modular mutualisms: a theoretical framework for meta-populations of mutualists and exploiters. *Proceedings of the Royal Society B*, 283, 20152428.

STERN, C. D. 2019. The 'omics revolution: how an obsession with compiling lists is threatening the ancient art of experimental design. *BioEssays*, 41, 1900168.

STEWART, E. J., MADDEN, R., PAUL, G., & TADDEI, F. 2005. Aging and death in an organism that reproduces by morphologically symmetric division. *PLoS Biology*, 3: e45.

STRASSMANN, J. E. 2000. Bacterial cheaters. *Nature*, 404, 555–556.

STRASSMANN, J. E. 2001. The rarity of multiple mating by females in the social Hymenoptera. *Insectes Sociaux*, 48, 1–13.

STRASSMANN, J. E. 2014. Tribute to Tinbergen: the place of animal behavior in biology. *Ethology*, 120, 123–126.

STRASSMANN, J. E. 2016. Kin discrimination in *Dictyostelium* social amoebae. *Journal of Eukaryotic Microbiology*, 63, 378–383.

STRASSMANN, J. E., HUGHES, C. R., QUELLER, D. C., TURILLAZZI, S., CERVO, R., DAVIS, S. K., & GOODNIGHT, K. F. 1989. Genetic relatedness in primitively eusocial wasps. *Nature*, 342, 268–269.

STRASSMANN, J. E. & QUELLER, D. C. 2007. Insect societies as divided organisms: the complexities of purpose and cross-purpose. *Proceedings of the National Academy of Sciences of the United States of America*, 104, 8619–8626.

STRASSMANN, J. E. & QUELLER, D. C. 2010. The social organism: congresses, parties, and committees. *Evolution*, 64, 605–616.

STRASSMANN, J. E., ZHU, Y., & QUELLER, D. C. 2000. Altruism and social cheating in the social amoeba *Dictyostelium discoideum*. *Nature*, 408, 965–967.

STROEYMEYT, N., GRASSE, A. V., CRESPI, A., MERSCH, D. P., CREMER, S., & KELLER, L. 2018. Social network plasticity decreases disease transmission in a eusocial insect. *Science*, 362, 941–945.

STURTEVANT, A. H. 1938. Essays on evolution. II. On the effects of selection on social insects. *Quarterly Review of Biology*, 13, 74–76.

STURTEVANT, A. H. 1965. *A History of Genetics*. New York: Harper & Row.

SUDDENDORF, T. 2013. *The Gap: The Science of What Separates Us from Other Animals*. New York: Basic Books.

SUMMERS, K. & CRESPI, B. E. 2013. *Human Social Evolution: The Foundational Works of Richard D. Alexander*. Oxford: Oxford University Press.

SUN, D., CHAI, S., HUANG, X., WANG, Y., XIAO, L., XU, S., & YANG, G. 2022. Novel genomic insights into body size evolution in cetaceans and a resolution of Peto's paradox. *American Naturalist*, 199, E28–E42.

SUNG, W., TUCKER, A. E., DOAK, T. G., CHOI, E., THOMAS, W. K., & LYNCH, M. 2012. Extraordinary genome stability in the ciliate *Paramecium tetraurelia*. *Proceedings of the National Academy of Sciences of the United States of America*, 109, 19339–19344.

SVIREZHEV, Y. M. 1972. Optimum principles in population genetics. *In*: RATNER, V. A. (ed.) *Studies on Theoretical Genetics*. Novosibirsk: Academy of Sciences, USSR.

SZABÓ, P., SCHEURING, I., CZÁRÁN, T. & SZATHMÁRY, E. 2002. In silico simulations reveal that replicators with limited dispersal evolve towards higher efficiency and fidelity. *Nature*, 420, 340–343.

SZATHMÁRY, E. 2015. Toward major evolutionary transitions theory 2.0. *Proceedings of the National Academy of Sciences of the United States of America*, 112, 10104–10111.

SZATHMÁRY, E. & DEMETER, L. 1987. Group selection of early replicators and the origin of life. *Journal of Theoretical Biology*, 128, 463–486.

SZATHMÁRY, E. & MAYNARD SMITH, J. 1995. The major evolutionary transitions. *Nature*, 374, 227–232.

SZNYCER, D., XYGALATAS, D., AGEY, E., ALAMI, S., AN, X.-F., ANANYEVA, K. I., ATKINSON, Q. D., BROITMAN, B. R., CONTE, T. J., FLORES, C., FUKUSHIMA, S., HITOKOTO, H., KHARITONOV, A. N., ONYISHI, C. N., ONYISHI, I. E., ROMERO, P. P., SCHROCK, J. M., SNODGRASS, J. J., SUGIYAMA, L. S., TAKEMURA, K., TOWNSEND, C., ZHUANG, J.-Y., AKTIPIS, C. A., CRONK, L., COSMIDES, L., & TOOBY, J. 2018. Cross-cultural invariances in the architecture of shame. *Proceedings of the National Academy of Sciences of the United States of America*, 115, 9702–9707.

TACHIBANA, M., SPARMAN, M., SRITANAUDOMCHAI, H., MA, H., CLEPPER, L., WOODWARD, J., LI, Y., RAMSEY, C., KOLOTUSHKINA, O., & MITALIPOV, S. 2009. Mitochondrial gene replacement in primate offspring and embryonic stem cells. *Nature*, 461, 367–372.

TAMAKI, C., TAKATA, M., & MATSUURA, K. 2021. The lose-to-win strategy of the weak: intraspecific parasitism via egg abduction in a termite. *Biology Letters*, 17: 20210540.

TANG, Q., PANG, K., YUAN, X., & XIAO, S. 2020. A one-billion-year-old multicellular chlorophyte. *Nature Ecology and Evolution*, 4, 543–549.

TANSKANEN, A. O., DANIELSBACKA, M., & ROTKIRCH, A. 2021. Kin detection cues and sibling relationship quality in adulthood: the role of childhood co-residence duration and maternal perinatal association. *Evolution and Human Behavior*, 42, 481–490.

TAUTZ, J. 2008. *The Buzz About Bees: Biology of a Superorganism*. Berlin: Springer.

TAX, S. (ed.) 1960a. *Evolution After Darwin, Volume 1: The Evolution of Life*. Chicago, IL: University of Chicago Press.

TAX, S. (ed.) 1960b. *Evolution After Darwin, Volume 2: The Evolution of Man*. Chicago, IL: University of Chicago Press.

TAX, S. & CALLENDER, C. (eds.) 1960. *Evolution After Darwin, Volume 3*: Issues in Evolution. Chicago, IL: University of Chicago Press.

TAYLOR, P. D. 1996. Inclusive fitness arguments in genetic models of behaviour. *Journal of Mathematical Biology*, 34, 654–674.

TAYLOR, P. D. & FRANK, S. A. 1996. How to make a kin selection model. *Journal of Theoretical Biology*, 180, 27–37.

TESCHENDORFF, A. E. & ENVER, T. 2017. Single-cell entropy for accurate estimation of differentiation potency from a cell's transcriptome. *Nature Communications*, 8: 15599.

TESCHENDORFF, A. E., SOLLICH, P., & KUEHN, R. 2014. Signalling entropy: a novel network-theoretical framework for systems analysis and interpretation of functional omic data. *Methods*, 67, 282–293.

THAXTER, R. 1892. Contributions from the Cryptogamic Laboratory of Harvard University. XVIII. On the Myxobacteriaceae, a new order of Schizomycetes. *Botanical Gazette* 1892, 389–406.

THOMPSON, J. N. 2005. *The Geographic Mosaic of Coevolution*. Chicago, IL: University of Chicago Press.

THORNHILL, R. 1983. Cryptic female choice and its implications in the scorpionfly *Harpobittacus nigriceps*. *American Naturalist*, 122, 765–788.

TINBERGEN, N. 1932. Über die Orientierung des Bienenwolfes (*Philanthus triangulum* Fabr.). *Zeitschrift für Vergleichende Physiologie*, 16, 305–334.

TINBERGEN, N. 1960. Behavior, systematics, and natural selection. *In*: TAX, S. (ed.) *Evolution After Darwin, Volume 1: The Evolution of Life*. Chicago, IL: University of Chicago Press.

TINBERGEN, N. 1963. On aims and methods of ethology. *Zeitschrift für Tierpsychologie*, 20, 410–433.

TINBERGEN, N. 1965. Behavior and natural selection. *In*: MOORE, J. A. (ed.) *Ideas in Modern Biology*. Garden City, NY: Natural History Press.

TOLLIS, M., BODDY, A. M., & MALEY, C. C. 2017. Peto's paradox: how has evolution solved the problem of cancer prevention? *BMC Biology*, 15: 60.

TOOBY, J. & COSMIDES, L. 1990. The past explains the present: emotional adaptations and the structure of ancestral environments. *Ethology and Sociobiology*, 11, 375–424.

TOOBY, J. & COSMIDES, L. 2016. Human cooperation shows the distinctive signatures of adaptations to small-scale social life. Open peer commentary on Richerson et al. (2016). *Behavioral and Brain Sciences*, 39: e54.

TOOBY, J., COSMIDES, L., & BARRETT, H.C. 2003. The second law of thermodynamics is the first law of psychology: evolutionary developmental psychology and the theory of tandem, coordinated inheritances: Comment on Lickliter and Honeycutt (2003). *Psychological Bulletin*, 129, 858–865.

TOTH, A. L. & ROBINSON, G. E. 2007. Evo-devo and the evolution of social behavior. *Trends in Genetics*, 23, 334–341.

TOWNSEND, C. R., HARPER, J. L. & BEGON, M. 2005. *Ecology: From Individuals to Ecosystems*, 4th edition. Oxford: Blackwell Science.

TRAGUST, S., MITTEREGGER, B., BARONE, V., KONRAD, M., UGELVIG, L. V., & CREMER, S. 2013. Ants disinfect fungus-exposed brood by oral uptake and spread of their poison. *Current Biology*, 23, 76–82.

TRANTER, C., LEFEVRE, L., EVISON, S. E. F. & HUGHES, W. O. H. 2014. Threat detection: contextual recognition and response to parasites by ants. *Behavioral Ecology*, 26, 396–405.

TRIVERS, R. 1985. *Social Evolution*. Menlo Park, CA: Benjamin Cummings Publishing Company.

TRIVERS, R. 2004. Mutual benefits at all levels of life. *Science*, 304, 964–965.

TRIVERS, R. 2011. *The Folly of Fools: The Logic of Deceit and Self-Deception in Human Life*. New York: Basic Books.

TRIVERS, R. 2015. *Wild Life*. New Brunswick, NJ: Biosocial Research.

TRIVERS, R. L. 1971. The evolution of reciprocal altruism. *Quarterly Review of Biology*, 46, 35–57.

TRIVERS, R. L. 1972. Parental investment and sexual selection. *In*: CAMPBELL, B. (ed.) *Sexual Selection and the Descent of Man 1871–1971*. Chicago, IL: Aldine.

TRIVERS, R. L. 1974. Parent-offspring conflict. *American Zoologist*, 14, 249–264.

TRIVERS, R. L. & HARE, H. 1976. Haplodiploidy and the evolution of the social insects. *Science*, 191, 249–263.

TRIVERS, R. L. & WILLARD, D. E. 1973. Natural selection of parental ability to vary the sex-ratio of offspring. *Science*, 179, 90–92.

TURELLI, M. & BARTON, N. H. 2006. Will population bottlenecks and multilocus epistasis increase additive genetic variance? *Evolution*, 60, 1763–1776.

TURILLAZZI, S. 1991. The Stenogastrinae. *In*: ROSS, K. G. & MATTHEWS, R. W. (eds.) *The Social Biology of Wasps*. Ithaca, NY: Cornell University Press.

TURILLAZZI, S. & WEST-EBERHARD, M. J. (eds.) 1996. *Natural History and Evolution of Paper-Wasps*. New York: Oxford University Press.

TURING, A. M. 1952. The chemical basis of morphogenesis. *Philosophical Transactions of the Royal Society B*, 237, 37–72.

TURNER, P. E. & CHAO, L. 1999. Prisoner's dilemma in an RNA virus. *Nature*, 398, 441–443.

ULLER, T., ENGLISH, S., & PEN, I. 2015. When is incomplete epigenetic resetting in germ cells favoured by natural selection? *Proceedings of the Royal Society B*, 282, 20150682.

ULLER, T., MOCZEK, A. P., WATSON, R. A., BRAKEFIELD, P. M., & LALAND, K. N. 2018. Developmental bias and evolution: a regulatory network perspective. *Genetics*, 209, 949–966.

UMEN, J. G. 2014. Green algae and the origins of multicellularity in the plant kingdom. *Cold Spring Harbor Perspectives in Biology*, 6: a016170.

VAIDYA, N., MANAPAT, M. L., CHEN, I. A., XULVI-BRUNET, R., HAYDEN, E. J., & LEHMAN, N. 2012. Spontaneous network formation among cooperative RNA replicators. *Nature*, 491, 72–77.

VAN BAALEN, M. 2013. Biological information: why we need a good measure and the challenges ahead. *Interface Focus*, 3, 20130030.

VAN DER HAVE, T. M., BOOMSMA, J. J., & MENKEN, S. B. J. 1988. Sex-investment ratios and relatedness in the monogynous ant *Lasius niger* (L). *Evolution*, 42, 160–172.

VAN NOORDWIJK, A. J. & DE JONG, G. 1986. Acquisition and allocation of resources: their influence on variation in life history tactics. *American Naturalist*, 128, 137–142.

VAN VALEN, L. 1973. A new evolutionary law. *Evolutionary Theory*, 1, 1–30.

VASCONCELOS, M., HOLLIS, K., NOWBAHARI, E., & KACELNIK, A. 2012. Pro-sociality without empathy. *Biology Letters*, 8, 910–912.

VASSE, M., TORRES-BARCELÓ, C., & HOCHBERG, M. E. 2015. Phage selection for bacterial cheats leads to population decline. *Proceedings of the Royal Society B*, 282: 20152207.

VEHRENCAMP, S. L. 1983. A model for the evolution of despotic versus egalitarian societies. *Animal Behaviour*, 31, 667–682.

VELICER, G. J., KROOS, L., & LENSKI, R. E. 1998. Loss of social behaviors by *Myxococcus xanthus* during evolution in an unstructured habitat. *Proceedings of the National Academy of Sciences of the United States of America*, 95, 12376–12380.

VELICER, G. J., KROOS, L., & LENSKI, R. E. 2000. Developmental cheating in the social bacterium *Myxococcus xanthus*. *Nature*, 404, 598–601.

VELICER, G. J. & PLUCAIN, J. 2016. Evolution: bacterial territoriality as a byproduct of kin discriminatory warfare. *Current Biology*, 26, R364–R366.

VELICER, G. J. & YU, Y.-T. N. 2003. Evolution of novel cooperative swarming in the bacterium *Myxococcus xanthus*. *Nature*, 425, 75–78.

VILLESEN, P., MURAKAMI, T., SCHULTZ, T. R., & BOOMSMA, J. J. 2002. Identifying the transition between single and multiple mating of queens in fungus-growing ants. *Proceedings of the Royal Society B*, 269, 1541–1548.

VOS, M. & VELICER, G. J. 2008. Isolation by distance in the spore-forming soil bacterium *Myxococcus xanthus*. *Current Biology*, 18, 386–391.

VOS, M. & VELICER, G. J. 2009. Social conflict in centimeter- and global-scale populations of the bacterium *Myxococcus xanthus*. *Current Biology*, 19, 1763–1767.

VOTAW, H. R. & OSTROWSKI, E. A. 2017. Stalk size and altruism investment within and among populations of the social amoeba. *Journal of Evolutionary Biology*, 30, 2017–2030.

VREEBURG, S., NYGREN, K., & AANEN, D. K. 2016. Unholy marriages and eternal triangles: how competition in the mushroom life cycle can lead to genomic conflict. *Philosophical Transactions of the Royal Society B*, 371, 20150533.

WADDINGTON, C. H. 1956. *The Strategy of the Genes*. London: George Allen & Unwin Ltd.

WADDINGTON, C. H. 1958. Theories of evolution. *In*: BARNETT, S. A. (ed.) *A Century of Darwin*. London: William Heinemann, Ltd.

WADDINGTON, C. H. 1960. Evolutionary adaptation. *In*: TAX, S. (ed.) *Evolution after Darwin, Volume 1: The Evolution of Life*. Chicago, IL: University of Chicago Press.

WADDINGTON, C. H. 1966. *Principles of Development and Differentiation*. New York: MacMillan.

WADE, M. J. 1979. The evolution of social interactions by family selection. *American Naturalist*, 113, 399–417.

WADE, M. J. 1980. Kin selection: its components. *Nature*, 210, 665–667.

WADE, M. J. 1985. Soft selection, hard selection, kin selection, and group selection. *American Naturalist*, 125, 61–73.

WADE, M. J. 1992. Heritability: historical perspectives. *In*: FOX KELLER, E. & LLOYD, E. A. (eds.) *Keywords in Evolutionary Biology*. Cambridge, MA: Harvard University Press.

WADE, M. J. 2002. A gene's eye view of epistasis, selection and speciation. *Journal of Evolutionary Biology*, 15, 337–346.

WADE, M. J. 2016. *Adaptation in Metapopulations*. Chicago, IL: University of Chicago Press.

WAGNER, A. 2005. *Robustness and Evolvability in Living Systems*. Princeton, NJ: Princeton University Press.

WAGNER, A. 2011. *The Origins of Evolutionary Innovations*. Oxford: Oxford University Press.

WAGNER, A. 2017. Information theory, evolutionary innovations and evolvability. *Philosophical Transactions of the Royal Society B*, 372, 20160416.

WAGNER, D. E., WEINREB, C., COLLINS, Z. M., BRIGGS, J. A., MEGASON, S. G., & KLEIN, A. M. 2018. Single-cell mapping of gene expression landscapes and lineage in the zebrafish embryo. *Science*, 360, 981–987.

WAGNER, G. P. 2014. Homology, Genes, and Evolutionary Innovation. Princeton, NJ: Princeton University Press.

WAGNER, G. P. 2016. What is "homology thinking" and what is it for? *Journal of Experimental Zoology*, 326B, 3–8.

WALLACE, D. C. 2018. Mitochondrial genetic medicine. *Nature Genetics*, 50, 1642–1649.

WALLIN, I. E. 1927. *Symbionticism and the origin of species*. Baltimore, ML, Williams & Wilkins Company.

WALSH, J. T., GARNIER, S., & LINKSVAYER, T. A. 2020. Ant collective behavior is heritable and shaped by selection. *American Naturalist*, 196, 541–554.

WANG, P., ROBERT, L., PELLETIER, J., DANG, W. L., TADDEI, F., WRIGHT, A., & JUN, S. 2010. Robust growth of *Escherichia coli*. *Current Biology*, 20, 1099–1103.

WARD, P. S., BRADY, S. G., FISHER, B. L., & SCHULTZ, T. R. 2015. The evolution of myrmicine ants: phylogeny and biogeography of a hyperdiverse ant clade (Hymenoptera: Formicidae). *Systematic Entomology*, 40, 61–81.

WATERS, C. K. & VAN HELDEN, A. (eds.) 1992. *Julian Huxley: Biologist and Statesman of Science*. Houston, TX: Rice University Press.

WATSON, J. D. & CRICK, F. H. C. 1953. Molecular structure of nucleic acids: a structure for deoxyribose nucleic acid. *Nature*, 171, 737–738.

WATSON, P. 2010. *The German Genius*. London: Simon & Schuster.

WAYNFORTH, D. 2020. Kin-based alloparenting and infant hospital admissions in the UK Millennium cohort. *Evolution, Medicine, and Public Health*, [2020], 72–81.

WCISLO, W. T. & DANFORTH, B. N. 1997. Secondarily solitary: the evolutionary loss of social behavior. *Trends in Ecology and Evolution*, 12, 468–474.

WEIGERT, M. & KÜMMERLI, R. 2017. The physical boundaries of public goods cooperation between surface-attached bacterial cells. *Proceedings of the Royal Society B*, 284, 20170631.

WEINBERG, S. 2015. *To Explain the World: The Discovery of Modern Science*. New York: Harper Collins.

WEISMANN, A. 1889. *Essays upon Heredity and Kindred Biological Problems*. Oxford: Clarendon Press.

WEISMANN, A. 1892. *Das Keimplasma. Eine Theorie der Vererbung*. Jena: Gustav Fischer Verlag.

WEISMANN, A. 1893. The all-sufficiency of natural selection: a reply to Herbert Spencer. *The Contemporary Review*, 64, 309–338 and 596–610.

WEISMANN, A. 1904. *The Evolution Theory, Volume 1*. London: Edward Arnold.

WEISS, P. 1967. 1 + 1 ≠ 2. *In*: QUARTON, G. C., MELNECHUK, T., & SCHMITT, F. O. (eds.) *The Neurosciences*. New York: The Rockefeller University Press.

WENT, F. W. 1968. The size of man. *American Scientist*, 56, 400–413.

WELCH, J. J. 2017. What's wrong with evolutionary biology? *Biology and Philosophy*, 32, 263–279.

WELLS, H. G., HUXLEY, J. S., & WELLS, G. P. 1931. *The Science of Life*. London: Cassell and Company Ltd.

WENSELEERS, T. & RATNIEKS, F. L. W. 2006. Enforced altruism in insect societies. *Nature*, 444, 50.

WENSELEERS, T., GARDNER, A. & FOSTER, K. R. 2010. Social evolution theory: a review of methods and approaches. *In*: SZÉKELY, T., MOORE, A. J. & KOMDEUR, J. (eds.) *Social Behaviour: Genes, Ecology and Evolution*. Cambridge: Cambridge University Press.

WERNEGREEN, J. J. 2013. First impressions in a glowing host-microbe partnership. *Cell Host & Microbe*, 14, 121–123.

WERNER, G. D. A. & KIERS, E. T. 2015. Partner selection in the mycorrhizal mutualism. *New Phytologist*, 205, 1437–1442.

WERNER, G. D. A., STRASSMANN, J. E., IVENS, A. B. F., ENGELMOER, D. J. P., VERBRUGGEN, E., QUELLER, D. C., NOË, R., JOHNSON, N. C., HAMMERSTEIN, P., & KIERS, E. T. 2014. Evolution of microbial markets. *Proceedings of the National Academy of Sciences of the United States of America*, 111, 1237–1244.

WERREN, J. H. 2011. Selfish genetic elements, genetic conflict, and evolutionary innovation.

Proceedings of the National Academy of Sciences of the United States of America, 108, 10863–10870.

WERREN, J. H., ZHANG, W., & GUO, L. R. 1995. Evolution and phylogeny of *Wolbachia*: reproductive parasites of arthropods. *Proceedings of the Royal Society B*, 261, 55–63.

WEST, S. A. 2009. *Sex Allocation*. Princeton, NJ: Princeton University Press.

WEST, S. A. & BUCKLING, A. 2003. Cooperation, virulence and siderophore production in bacterial parasites. *Proceedings of the Royal Society B*, 270, 37–44.

WEST, S. A., DIGGLE, S. P., BUCKLING, A., GARDNER, A., & GRIFFIN, A. S. 2007c. The social lives of microbes. *Annual Review of Ecology, Evolution and Systematics*, 38, 53–77.

WEST, S. A., EL MOUDEN, C., & GARDNER, A. 2011. Sixteen common misconceptions about the evolution of cooperation in humans. *Evolution and Human Behavior*, 32, 231–262.

WEST, S. A., FISHER, R. M., GARDNER, A., & KIERS, E. T. 2015. Major evolutionary transitions in individuality. *Proceedings of the National Academy of Sciences of the United States of America*, 112, 10112–10119.

WEST, S. A. & GARDNER, A. 2010. Altruism, spite, and green beards. *Science*, 327, 1341–1344.

WEST, S. A. & GARDNER, A. 2013. Adaptation and inclusive fitness. *Current Biology*, 23, R577–R584.

WEST, S. A., GRIFFIN, A. S., & GARDNER, A. 2007a. Social semantics: altruism, cooperation, mutualism, strong reciprocity and group selection. *Journal of Evolutionary Biology*, 20, 415–432.

WEST, S. A., GRIFFIN, A. S., & GARDNER, A. 2007b. Evolutionary explanations for cooperation. *Current Biology*, 17, R661–R672.

WEST, S. A., GRIFFIN, A. S., & GARDNER, A. 2008. Social semantics: how useful has group selection been? *Journal of Evolutionary Biology*, 21, 374–385.

WEST, S. A., GRIFFIN, A. S., GARDNER, A., & DIGGLE, S. P. 2006. Social evolution theory for microorganisms. *Nature Reviews Microbiology*, 4, 597–607.

WEST, S. A., KIERS, E. T., PEN, I., & DENISON, R. F. 2002a. Sanctions and mutualism stability: when should less beneficial mutualists be tolerated? *Journal of Evolutionary Biology*, 15, 830–837.

WEST, S. A., KIERS, E. T., SIMMS, E. L., & DENISON, R. F. 2002c. Sanctions and mutualism stability: why do rhizobia fix nitrogen? *Proceedings of the Royal Society B*, 269, 685–694.

WEST, S. A., MURRAY, M. G., MACHADO, C. A., GRIFFIN, A. S., & HERRE, E. A. 2001. Testing Hamilton's rule with competition between relatives. *Nature*, 409, 510–513.

WEST, S. A., PEN, I., & GRIFFIN, A. S. 2002b. Cooperation and competition between relatives. *Science*, 296, 72–75.

WEST, T., SOJO, V., POMIANKOWSKI, A., & LANE, N. 2017. The origin of heredity in protocells. *Philosophical Transactions of the Royal Society B*, 372, 20160419.

WEST-EBERHARD, M. J. 1978. Polygyny and the evolution of social behavior in wasps. *Journal of the Kansas Entomological Society*, 51, 832–856.

WEST-EBERHARD, M. J. 1987. Flexible strategy and social evolution. *In*: ITO, Y., BROWN, J. L., & KIKKAWA, J. (eds.) *Animal Societies: Theories and Facts*. Tokyo: Japan Scientific Societies Press.

WEST-EBERHARD, M. J. 2003. *Developmental Plasticity and Evolution*. New York: Oxford University Press.

WESTNEAT, D. F. 1987. Extra-pair fertilizations in a predominantly monogamous bird: genetic evidence. *Animal Behaviour*, 35, 877–886.

WESTNEAT, D. F. 1990. The ecology and evolution of extra-pair copulations in birds. *Current Ornithology* 7, 331–369.

WHEELER, W. M. 1902. 'Natural history', 'oecology' or 'ethology'? *Science*, 15, 971–976.

WHEELER, W. M. 1904. The obligations of the student of animal behavior. *The Auk*, 21, 251–255.

WHEELER, W. M. 1905. Ethology and the mutation theory. *Science*, 21, 535–540.

WHEELER, W. M. 1906. The Kelep excused. *Science*, 23, 348–350.

WHEELER, W. M. 1910. *Ants: Their Structure, Development and Behavior*. New York: Columbia University Press.

WHEELER, W. M. 1911. The ant-colony as an organism. *Journal of Morphology*, 22, 307–325.

WHEELER, W. M. 1920a. The organization of research. *Science*, 53, 53–67.

WHEELER, W. M. 1920b. The Termitodoxa, or biology and society. *The Scientific Monthly*, 10, 113–124.

WHEELER, W. M. 1920–1921. On instincts. *The Journal of Abnormal Psychology*, 15, 295–318.

WHEELER, W. M. 1923a. The dry-rot of our academic biology. *Science*, 57, 61–71.

WHEELER, W. M. 1923b. Social life among the insects. New York, Harcourt, Brace & Co.

WHEELER, W. M. 1926. Emergent evolution and the social. *Science*, 64, 433–440.

WHEELER, W. M. 1928a. Emergent evolution and the development of societies. New York, W. W. Norton & Co.

WHEELER, W. M. 1928b. *The Social Insects*. New York: Harcourt, Brace & Co.

WHEELER, W. M. 1929. Present tendencies in biological theory. *The Scientific Monthly*, 28, 97–109.

WHEELER, W. M. 1934. Biology and society: animal societies. *The Scientific Monthly*, 39, 289–301.

WHITELEY, M., DIGGLE, S. P., & GREENBERG, E. P. 2017. Progress in and promise of bacterial quorum sensing research. *Nature*, 551, 313–320.

WHITEN, A., HORNER, V. & DE WAAL, F. B. M. 2005. Conformity to cultural norms of tool use in chimpanzees. *Nature*, 437, 737–740.

WHITMAN, C. O. 1893. The inadequacy of the cell-theory of development. *Journal of Morphology*, 8, 639–658.

WHON, T. W., KIM, M.-S., ROH, S. W., SHIN, N.-R., LEE, H.-W., & BAE, J.-W. 2012. Metagenomic characterization of airborne viral DNA diversity in the near-surface atmosphere. *Journal of Virology*, 86, 8221–8231.

WICKEN, J. S. 1986. Entropy and evolution: ground rules for discourse. *Systematic Zoology*, 35, 22–36.

WIELGOSS, S., FIEGNA, F., RENDUELES, O., YU, Y.-T. N., & VELICER, G. J. 2018. Kin discrimination and outer membrane exchange in *Myxococcus xanthus*: a comparative analysis among natural isolates. *Molecular Ecology*, 27, 3146–3158.

WIELGOSS, S., WOLFENSBERGER, R., SUN, L., FIEGNA, F., & VELICER, G. J. 2019. Social genes are selection hotspots in kin groups of a soil microbe. *Science*, 363, 1342–1345.

WILD, G. & WEST, S. A. 2007. A sex allocation theory for vertebrates: combining local resource competition and condition-dependent allocation. *American Naturalist*, 170, E112–E128.

WILKINS, A. S. & HOLLIDAY, R. 2009. The evolution of meiosis from mitosis. *Genetics*, 181, 3–12.

WILLIAMS, G. C. 1957. Pleiotropy, natural selection, and the evolution of senescence. *Evolution*, 11, 398–411.

WILLIAMS, G. C. 1966a. *Adaptation and Natural Selection*. Princeton, NJ: Princeton University Press.

WILLIAMS, G. C. 1966b. Natural selection, the cost of reproduction, and a refinement of Lack's principle. *American Naturalist*, 100, 687–690.

WILLIAMS, G. C. (ed.) 1971. *Group Selection*. Chicago, IL: Aldine Atherton Inc.

WILLIAMS, G. C. 1975. *Sex and Evolution*. Princeton, NJ: Princeton University Press.

WILLIAMS, G. C. 1985. A defense of reductionism in evolutionary biology. *In*: DAWKINS, R. & RIDLEY, M. (eds.) Oxford Surveys in Evolutionary Biology, Volume 2. Oxford: Oxford University Press.

WILLIAMS, G. C. 1992. *Natural Selection: Domains, Levels, and Challenges*. New York: Oxford University Press.

WILLIAMS, G. C. 1996. *Adaptation and Natural Selection*, 2nd edition. Princeton, NJ: Princeton University Press.

WILLIAMS, G. C. & WILLIAMS, D. C. 1957. Natural selection of individually harmful social adaptations among sibs with special reference to social insects. *Evolution*, 11, 32–39.

WILLIAMS, N. 2010. Aristotle's lagoon. *Current Biology*, 20, R84–R85.

WILLIAMS, T. A., COX, C. J., FOSTER, P. G., SZÖLLŐSI, G. J., & EMBLEY, T. M. 2020. Phylogenomics provides robust support for a two-domains tree of life. *Nature Ecology & Evolution*, 4, 138–147.

WILLIAMS, T. A., FOSTER, P. G., COX, C. J., & EMBLEY, T. M. 2013. An archaeal origin of eukaryotes supports only two primary domains of life. *Nature*, 504, 231–236.

WILSON, D. S. 1975. A theory of group selection. *Proceedings National Academy of Sciences of the United States of America*, 72, 143–146.

WILSON, D. S. 1990. Weak altruism, strong group selection. *Oikos*, 59, 135–140.

WILSON, D. S. & SOBER, E. 1989. Reviving the superorganism. *Journal of Theoretical Biology*, 136, 337–356.

WILSON, E. O. 1967. The superorganism concept and beyond. *L'effet de groupe chez des animaux*, 1967, 27–39.

WILSON, E. O. 1971. *The Insect Societies*. Cambridge, MA: Harvard University Press.

WILSON, E. O. 1975. *Sociobiology: The New Synthesis*. Cambridge, MA: Belknap Press of Harvard University Press.

WILSON, E. O. 1978. *On Human Nature*. Cambridge, MA: Harvard University Press.

WILSON, E. O. 1990. Success and Dominance in Ecosystems: The Case of the Social Insects. *Excellence in Ecology*, Volume 2. Oldendorf/Luhe: Ecology Institute.

WILSON, E. O. 1992. *The Diversity of Life*, Cambridge, MA, Belknap Press of Harvard University.

WILSON, E. O. & BOSSERT, W. H. 1971. *A Primer of Population Biology*. Oxford: Oxford University Press.

WOESE, C. R. & FOX, G. E. 1977. Phylogenetic structure of the prokaryotic domain: the primary kingdoms. *Proceedings of the National Academy of Sciences of the United States of America*, 74, 5088–5090.

WOLF, J. B., BRODIE III, E. D., CHEVERUD, J. M., MOORE, A. J., & WADE, M. J. 1998. Evolutionary consequences of indirect genetic effects. *Trends in Ecology & Evolution*, 13, 64–69.

WOLF, J. B., HOWIE, J. A., PARKINSON, K., GRUENHEIT, N., MELO, D., ROZEN, D., & THOMPSON, C. R. L. 2015. Fitness trade-offs result in the illusion of social success. *Current Biology*, 25, 1086–1090.

WOLF, Y. I. & KOONIN, E. V. 2007. On the origin of the translation system and the genetic code in the RNA world by means of natural selection, exaptation, and subfunctionalization. *Biology Direct*, 2: 14.

WOLLENBERG, M. S. & RUBY, E. G. 2009. Population structure of *Vibrio fischeri* within the light organs of *Euprymna scolopes* squid from two Oahu (Hawaii) populations. *Applied Environmental Microbiology*, 75, 193–202.

WOODGER, J. H. 1929. *Biological Principles: A Critical Study*. London: Routledge & Kegan Paul Ltd.

WOODGER, J. H. 1945. On biological transformations. *In*: LE GROS CLARK, W. E., MEDAWAR, P. B. & Thompson, D. (eds.) *Essays on Growth and Form*. Oxford: Clarendon Press.

WORSLEY, S. F., INNOCENT, T. M., HEINE, D., MURRELL, J. C., YU, D. W., WILKINSON, B., HUTCHINGS, M. I., BOOMSMA, J. J. & HOLMES, N. A. 2019. Symbiotic partnerships and their chemical interactions in the leafcutter ants (Hymenoptera: Formicidae). *Myrmecological News* 27, 59–74.

WORSLEY, S. F., INNOCENT, T. M., HOLMES, N. A., AL-BASSAM, M. M., SCHIØTT, M., WILKINSON, B., MURRELL, J. C., BOOMSMA, J. J., YU, D. W., & HUTCHINGS, M. I. 2021. Competition-based screening helps to secure the evolutionary stability of a defensive microbiome. *BMC Biology*, 19, 205.

WOYCIECHOWSKI, M. & KUSZEWSKA, K. 2012. Swarming generates rebel workers in honeybees. *Current Biology*, 22, 707–711.

WRANGHAM, R. 2019. *The Goodness Paradox*. London: Profile Books.

WRANGHAM, R. 2021. Targeted conspirational killing, human self-domestication and the evolution of groupishness. *Evolutionary Human Sciences*, 3, e26.

WRAY, G. A., HOEKSTRA, H. E., FUTUYMA, D. J., LENSKI, R. E., MACKAY, T. F. C., SCHLUTER, D., & STRASSMANN, J. E. 2014. Does evolutionary theory need a rethink? COUNTERPOINT: no, all is well. *Nature*, 514, 161–164.

WREGE, P. H. & EMLEN, S. T. 1987. Biochemical determination of parental uncertainty in white-fronted bee-eaters. *Behavioral Ecology and Sociobiology*, 20, 153–160.

WRIGHT, S. 1931. Evolution in Mendelian populations. *Genetics*, 16, 97–159.

WRIGHT, S. 1932. The roles of mutation, inbreeding, crossbreeding and selection in evolution. *Proceedings of the Sixth International Congress on Genetics*, 1, 356–366.

WRIGHT, S. 1949. Adaptation and selection. *In*: JEPSEN, G. L., MAYR, E., & SIMPSON, G. G.

(eds.) *Genetics, Paleontology, and Evolution*. Princeton, NJ: Princeton University Press.

WRIGHT, S. 1953. Gene and organism. *American Naturalist*, 87, 5–18.

WYATT, G. A. K., KIERS, E. T., GARDNER, A., & WEST, S. A. 2014. A biological market analysis of the plant-mycorrhizal symbiosis. *Evolution*, 68, 2603–2618.

WYATT, G. A. K., WEST, S. A., & GARDNER, A. 2013. Can natural selection favour altruism between species? *Journal of Evolutionary Biology*, 26, 1854–1865.

WYNNE, C.D.L. 2004. The perils of Anthropomorphism. *Nature*, 428, 606.

WYNNE-EDWARDS, V. C. 1962. *Animal Dispersion in Relation to Social Behaviour*. Edinburgh: Oliver & Boyd.

YAGOUND, B., DOGANTZIS, K. A., ZAYED, A., LIM, J., BROEKHUYSE, P., REMNANT, E. J., BEEKMAN, M., ALLSOPP, M. H., AAMIDOR, S. E., DIM, O., BUCHMANN, G., & OLDROYD, B. P. 2020. A single gene causes thelytokous parthenogenesis, the defining feature of the cape honeybee *Apis mellifera capensis*. *Current Biology*, 30, 2248–2259.

YAGUCHI, H., KOBAYASHI, I., MAEKAWA, K., & NALEPA, C. A. 2021. Extra-pair paternity in the wood-feeding cockroach *Cryptocercus punctulatus* Scudder: social but not genetic monogamy. *Molecular Ecology*, 30, 6743–6758.

YOON, H. S., HACKETT, J. D. & BHATTACHARYA, D. 2006. A genomic and phylogenetic perspective on endosymbiosis and algal origin. *Journal of Applied Phycology*, 18, 475–481.

YU, X., ZHANG, X., ZHAO, P., PENG, X., CHEN, H., BLECKMANN, A., BAZHENOVA, A., SHI, C., DRESSELHAUS, T., & SUN, M. 2021. Fertilized egg cells secrete endopeptidases to avoid polytubey. *Nature*, 592, 433–437.

ZAHAVI, A. 1975. Mate selection—a selection for a handicap. *Journal of Theoretical Biology* 53, 205–214.

ZAHAVI, A. 1977. The cost of honesty (further remarks on the handicap principle). *Journal of Theoretical Biology*, 67, 603–605.

ZHANG, T., JANDÉR, K. C., HUANG, J.-F., WANG, B., ZHAO, J.-B., MIAO, B.-G., PENG, Y.-Q., & HERRE, E. A. 2021. The evolution of parasitism from mutualism in wasps pollinating the fig, *Ficus microcarpa*, in Yunnan Province, China. *Proceedings of the National Academy of Sciences of the United States of America*, 118, 32: e2021148118.

ZHOU, C., LI, R. & KENNEDY, B. K. 2014. Life history: mother-specific proteins that promote aging. *Current Biology* 24, R1162–R1164.

ZINK, A. G. 2015. Kin selection and the evolution of mutualisms between species. *Ethology*, 121, 823–830.

ZUCCHI, R., SAKAGAMI, S. F., NOLL, F. B., MECHI, M. R., MATEUS, S., BAIO, M. V., & SHIMA, S. N. 1995. *Agelaia vicina*, a swarm-founding polistine with the largest colony size among wasps and bees (Hymenoptera: Vespidae). *Journal of the New York Entomological Society*, 103, 129–137.

ZUCKERMAN, S. 1992. Comments and recollections. *In*: WATERS, C. K. & VAN HELDEN, A. (eds.) *Julian Huxley: Biologist and Statesman of Science*. Houston, TX: Rice University Press.

ZUREK, W. H. 2009. Quantum Darwinism. *Nature Physics*, 5, 181–188.

Index

Note: *f* indicates an illustration; *t* a table